A Modern Introduction to Fuzzy Mathematics

A Modern Introduction to Fuzzy Mathematics

Apostolos Syropoulos
Theophanes Grammenos

This edition first published 2020

Registered Office
John Wiley & Sons, Inc., 111 River Street, Hoboken, NJ 07030, USA

Editorial Office
111 River Street, Hoboken, NJ 07030, USA

For details of our global editorial offices, customer services, and more information about Wiley products visit us at www.wiley.com.

Wiley also publishes its books in a variety of electronic formats and by print-on-demand. Some content that appears in standard print versions of this book may not be available in other formats.

Library of Congress Cataloging-in-Publication Data

Names: Syropoulos, Apostolos, author.
Title: A modern introduction to fuzzy mathematics / Apostolos Syropoulos, Theophanes Grammenos.
Description: First edition. | New York : Wiley, 2020. | Includes bibliographical references and index.
Identifiers: LCCN 2020004300 (print) | LCCN 2020004301 (ebook) | ISBN 9781119445289 (cloth) | ISBN 9781119445302 (adobe pdf) | ISBN 9781119445296 (epub)
Subjects: LCSH: Fuzzy mathematics.
Classification: LCC QA248.5 .S97 2020 (print) | LCC QA248.5 (ebook) | DDC 511.3/13—dc23
LC record available at https://lccn.loc.gov/2020004300
LC ebook record available at https://lccn.loc.gov/2020004301

Cover Design: Wiley
Cover Image: Courtesy of Apostolos Syropoulos

Set in 9.5/12.5pt STIXTwoText by SPi Global, Chennai, India

10 9 8 7 6 5 4 3 2 1

To my son Demetrios-Georgios,
to Koula and Linda.
Apostolos Syropoulos

To my wife Andromache for her
endurance, tolerance, and care.
Theophanes Grammenos

Contents

Preface

In mathematics, we investigate the properties of abstract objects (e.g. numbers, geometric shapes, spaces) and the possible relationships between these objects. One basic characteristic of mathematics is that all these properties and relationships are *absolute*. Thus, properties and relationships are either true or false. Nothing else is meaningful. In other words, in mathematics there is no room for *vagueness*, for *randomness*, and for *extremely small quantities*. By introducing one of these qualities into mathematics, one can create *alternative* mathematics [1]. But how do we introduce vagueness into mathematics? One very simple way to achieve this is to allow notions like "small," "large," and "few." However, another way is to modify the most basic object of mathematics, that is, to modify sets. In this respect, fuzzy mathematics is a form of alternative mathematics since it is based on a generalization of set membership. Simply put, in fuzzy mathematics, an element may belong to a degree to a set, while in ordinary mathematics, it either belongs or does not belong to a set. This simple idea has been applied to most fields of mathematics and so we can talk about fuzzy mathematics.

Even today, many researchers and thinkers consider fuzzy mathematics as a tool that can be used instead of probability theory to reason about or to work with a specific system. This text is based on the idea that vagueness is a basic notion and thus tries to present fuzzy mathematics as a form of alternative mathematics and not as an alternative to probability theory. In addition, this text is an introduction to fuzzy mathematics. This simply means that we have tried to cover as many fields of fuzzy mathematics as possible. Thus, this text is a compendium but not a handbook of fuzzy mathematics.

Chapter 1 of this book explains what vagueness means from a philosophical point of view. Also, it demonstrates the connection between vagueness and fuzzy mathematics.

In Chapter 2, we introduce the notion of fuzzy set as well as the basic set operations. In addition, we introduce a number of variants or extensions of fuzzy sets. The chapter concludes with a section marked with a star. There are a few such sections in the book and these are optional readings as they deal with quite advanced ideas.

Fuzzy number are special kinds of fuzzy sets that, in a way, have been introduced to generalize the notion of a number. In Chapter 3, we present various forms of fuzzy numbers and the basic arithmetic operations between them. We introduce linguistic variables, that is, terms such as "small," "heavy," "tall." Also, we present fuzzy equations (e.g. simple equations like $ax + b = c$, where known and unknown quantities are fuzzy numbers) and how one can solve them. The chapter concludes with applications of fuzzy numbers. In general, most chapters that follow have a final section that presents applications of the main material presented in the specific chapter.

Fuzzy relations are a very important subject that is presented in Chapter 4. We discuss fuzzy relations, the Cartesian product of fuzzy sets and related notions, and fuzzy orders. Since graphs can be described by relations, we also discuss fuzzy graphs. Also, since precategories are also described by graphs, we discuss fuzzy categories.

Chapter 5 is devoted to possibility theory, that is, generally speaking, the fuzzy "version" of a probability theory. Also, we compare probability and possibility theories in order to see their differences.

Chapter 6 discusses fuzzy statistics. In this chapter, we discuss fuzzy random variables (in a way as something that corresponds to vague randomness …) and all related notions such as fuzzy regression and fuzzy point estimation.

In Chapter 7, we discuss many-valued and fuzzy logics. We do not just present truth values and the basic logical operations, but we present complete logical systems. In addition, we discuss approximate reasoning and try to see what is a logic of vagueness.

Although computability theory is a very basic part of logic, we discuss in a separate chapter fuzzy computation (Chapter 8). In particular, we discuss fuzzy automata, fuzzy Turing machines, and other fuzzy models of computation.

In Chapter 9, we give a taste of fuzzy abstract algebra theory. We present fuzzy groups, fuzzy rings, fuzzy vector spaces, fuzzy normed spaces, and fuzzy Lie algebras.

Chapter 10 introduces the basic notions and ideas of fuzzy metric spaces and fuzzy topology. In addition, we briefly discuss fuzzy Banach spaces and fuzzy Hilbert spaces.

Fuzzy geometry is introduced in Chapter 11. We discuss the notion of fuzzy points and the distance between them, fuzzy lines, fuzzy circles, and fuzzy polygons.

Chapter 12 introduces the reader to fuzzy calculus. In particular, we discuss fuzzy functions, integrals and derivatives of fuzzy functions, and fuzzy limits of sequences and function. Furthermore, fuzzy (ordinary and partial) differential equations are presented.

The book includes two appendices: the first briefly presents fuzzy approximation and the second gives a taste of fuzzy chaos and fuzzy fractals.

Each chapter starts with a section that describes the nonfuzzy concepts whose fuzzy counterparts are presented in the rest of the chapter. We felt this was necessary since we present many and quite diverse topics and we cannot expect everyone to be familiar with all these notions and ideas. Also, most chapters have some exercises at the end. Readers are invited to work on them if they want to deepen their understanding of the ideas presented in the corresponding chapter. However, the chapter on fuzzy computation does not include exercises since the subject is not mature enough.

Apostolos Syropoulos
Xanthi, Greece, November 2019

Theophanes Grammenos
Volos, Greece, November 2019

Reference

1 Van Bendegen, J.P. (2005). Can there be an alternative mathematics, really? In: *Activity and Sign: Grounding Mathematics Education* (ed. M.H. Hoffmann, J. Lenhard, and F. Seeger), 349–359. Boston, MA: Springer US. [cited on page(s) xiii]

Acknowledgments

We would like to thank Kathleen Pagliaro who believed in this project and helped us in every possible way to realize it. Also, we thank Christina E. Linda, our project editor, for her help and assistance. In addition, we thank Andromache Spanou for carefully reading drafts of the text and suggesting improvements and corrections. Also, we thank Athanasios Margaris and Basil K. Papadopoulos for reading parts of the book and making comments and suggestions.

1

Introduction

Vagueness is a fundamental property of this world. Vague objects are real objects and exist in the real world. Fuzzy mathematics is mathematics of vagueness. The core of fuzzy mathematics is the idea that objects have a property to some degree.

1.1 What Is Vagueness?

When we say that something is vague, we mean that its properties and capacities are not sharply determined. In different words, a vague concept is one that is characterized by fuzzy boundaries (i.e. there are cases where it is not clear if an object has or does not have a specific property or capacity). Jiri Benovsky [27] put forth an objection to this idea by claiming that everybody who thinks that there are ordinary objects must accept that they are vague, whereas everybody must accept the existence of sharp boundaries to ordinary objects. This does not lead to a contradiction since the two claims do not concern the same "everybody".

The Sorites Paradox (σόφισμα τοῦ σωρείτη), which was introduced by Eubulides of Miletus (Εὐβουλίδης ὁ Μιλήσιος),[1] is a typical example of an argument that demonstrates what fuzzy boundaries are. The term "σωρείτες" (sorites) derives from the Greek word σωρός (soros), which means "heap." The paradox is about the number of grains of wheat that makes a heap. All agree that a single grain of wheat does not comprise a heap. The same applies for two grains of wheat as they do not comprise a heap, etc. However, there is a point where the number of grains becomes large enough to be called a heap, but there is no general agreement as to where this occurs. Although there is no precise definition of vagueness, still most people would agree that adjectives such as tall, old, short, and young, express vague concepts since, for instance, a person who is 6 years old is definitely young but can

1 Ancient Greek philosopher, teacher of Demosthenes and a student of Euclid, famous for his paradoxes. He lived in the fourth century BC.

A Modern Introduction to Fuzzy Mathematics, First Edition.
Apostolos Syropoulos and Theophanes Grammenos.

we say the same for a person who is 30 years old? Moreover, there are objects that one can classify as vague. For example, a cloud is vague since its boundaries are not sharp. Also, a dog is a vague object since it loses hair all the time and so it is difficult to say what belongs to it.

To a number of people, these arguments look like sophisms. Others consider vagueness as a *linguistic* phenomenon, that is, something that exists only in the realm of natural languages and gives us greater expressive power. And there are others that think that vagueness is a property of the world. In summary, there are three views regarding the nature of vagueness[2] : the *ontic* view, the *semantic* view, and the *epistemic* view. According to the ontic view the world itself is vague and, consequently, language is vague so to describe the world. The semantic view asserts that vagueness exists only in our language and our thoughts. In a way, this view is similar to the mental constructions of intuitionism, that is, things that exist in our minds but not in the real world. On the other hand, the epistemic view asserts that vagueness exists because we do not know where the boundaries exist for a "vague" concept. So we wrongly assume they are vague. In this book, we assume that onticism about vagueness is the right view. In different words, we believe that there are vague objects and that vagueness is a property of the real world. It seems that semanticism is shared by many people, engineers in particular who use fuzzy mathematics, while if epistemicism is true, then there is simply no need for fuzzy mathematics, and this book is useless.

Let us consider countries and lakes. These geographical objects do not have sharply defined boundaries since natural phenomena (e.g. drought or heavy rainfalls) may alter the volume of water contained in a lake. Thus, one can think these are vague objects. Nevertheless, vagueness can emerge from other unexpected observations. In 1967, Benoît Mandelbrot [208] argued that the measured length of the coastline of Great Britain (or any island for that matter) depends on the scale of measurement. Thus, Great Britain is a vague object since its boundaries are not sharp. Nevertheless, one may argue that here there is no genuine geographical vagueness, instead this is just a problem of representation. A response to this argument was put forth by Michael Morreau [224]. Obviously, if the existence of vague objects is a matter of representation, then there are obviously no vague objects including animals. Consider Koula the dog. Koula has hair that she will lose tonight, so it is a questionable part of her. Because Koula has many such questionable parts (e.g. nails, whisker), she is a vague dog. Assume that Koula is not a vague dog. Instead, assume that there are many precise mammals that must be dogs because they differ from each other around the edges of the hair. Obviously, all these animals are dogs that differ slightly when compared to Koula. All of

2 See [247] for a general introduction to vagueness and the ideas briefly presented in this paragraph.

these candidates are dogs, and they have very small differences between them. If vagueness is a matter of representation, then, wherever I own a dog, I own at least a thousand dogs. Clearly, this is not the case.

Gareth Evans [119] presented an argument that *proves* that there are no vague objects. Evans used the modality operator ∇ to express indeterminacy. Thus, $\nabla\phi$ is read as *it is indeterminate whether* ϕ. The dual of ∇ is the Δ operator and $\Delta\phi$ is read as *it is determinate that* ϕ. Evans started his argument with the following premise:

$$\nabla(a = b). \tag{1.1}$$

This means that it is true that it is indeterminate whether a and b are identical. Next, he transformed this expression to an application of some sort of λ-abstraction:

$$\lambda x.\nabla(x = a)b. \tag{1.2}$$

Of course, it is a fact that it is not indeterminate whether a is identical to a:

$$\neg\nabla(a = a). \tag{1.3}$$

Using this "trick" to derive formula (1.2), one gets

$$\neg\lambda x.\nabla(x = a)a. \tag{1.4}$$

Finally, he used the *identity of indiscernibles* principle to derive from (1.2) and (1.4):

$$\neg(a = b)$$

meaning that a and b are not identical. So we started by assuming that it is indeterminate whether a and b are identical and concluded that they are not identical. In different words, indeterminate identities become nonidentities, which makes no sense, therefore, the assumption makes no sense. The identity of indiscernibles principle (see [125] for a thorough discussion of this principle) states that if, for every property F, object x has F if and only if object y has F, then x is identical to y. This principle was initially formulated by Wilhelm Gottfried Leibniz.

A first response to this argument is that the logic employed to deliver this proof is not really adequate. Francis Jeffry Pelletier [240] points out that when one says that an object is vague, then this means that there is a predicate that neither applies nor does not apply to it. Thus when you have a meaningful predicate Fa, it makes no sense to make it indeterminate by just prefixing it with the ∇ operator. Although this logic is not appropriate for vagueness, still this does not refute Evan's argument.

Edward Jonathan Lowe [196] put forth an argument that is a response to Evans' "proof":

> Suppose (to keep matters simple) that in an ionization chamber a free electron a is captured by a certain atom to form a negative ion which, a

> short time later, reverts to a neutral state by releasing an electron b. As I understand it, according to currently accepted quantum-mechanical principles there may simply be no objective fact of the matter as to whether or not a is identical with b.

The idea behind this example is that "identity statements represented by '$a = b$' are 'ontically' indeterminate in the quantum mechanical context" [126] (for a thorough discussion of the problem of identity in physics see [127]). Lowe's argument prompted a series of responses, nevertheless, we are not going to describe them here and the interested reader can read a summary of these responses in [83]. In a way, these responses culminated to a revised à la Evans proof based on Lowe's initial argument:

(i) At t_1, $\nabla(a$ has been emitted$)$.
(ii) So at t_1, $\lambda x.\nabla(x$ has been emitted$)a$.
(iii) But at t_1, $\neg\nabla(b$ has been emitted$)$.
(iv) So at t_1, $\lambda x.\neg\nabla(x$ has been emitted$)b$.
(v) Therefore, $a \neq b$.

It is possible to provide a reinterpretation of quantum mechanics that does not use probabilities but possibilities instead [277]. This reinterpretation assumes that vagueness is a fundamental property of the world. Although the ideas involved are very simple, still they require a good background in quantum mechanics and in ideas that we are going to present in this book.

1.2 Vagueness, Ambiguity, Uncertainty, etc.

We explained that a vague concept is one that is characterized by fuzzy boundaries, still this is not a precise definition as the Sorites Paradox has demonstrated. Of course, one could say that it is an oxymoron to expect a precise definition which by its very nature is not precise. Not so surprisingly, at least to these authors, is the definition that is provided by Otávio Bueno and Mark Colyvan [51]:

Definition 1.2.1 A predicate is vague just in case it can be employed to generate a sorites argument.

To be fair, even the authors admit that this is not a new idea, but they were the first to systematically defend this idea. This definition of vague predicates is very useful in order to distinguish vagueness from ambiguity and uncertainty.

The term *ambiguity* refers to something that has more than one possible meaning, which may cause confusion. Of course, we encounter ambiguity only in language, but if vagueness is omnipresent, then it is definitely present in language. And this is the reason why it is confused with ambiguity. Examples of ambiguous sentences include *Sarah gave a bath to her dog wearing a pink T-shirt* and *Mary ate the biscuits on the couch*. The first sentence is ambiguous because it is not clear who was wearing the pink T-shirt: Sarah or the dog? The second sentence is ambiguous because it is not clear if the biscuits were on the couch or if Mary brought them and ate them on the couch. Of course, one can come up with many more examples of ambiguous sentences, but none of them can be employed to generate a sorites paradox.

Imprecision is a notion that is closer to vagueness than ambiguity. Typically, we employ this term when talking about things whose boundaries are not precise. Of course, not precise does not mean fuzzy. It would be sufficient to say that we encounter imprecision when our tools are not precise enough to perform a given task (e.g. measuring the length of a tiny object).

Generality is yet another word that is close to vagueness. General terms are words like "chair." There are many and different kinds of chairs, nevertheless, all the chairs that one may imagine are still chairs. Thus, in generality we focus on a few common characteristics and use them to make up classes of objects.

It is not widely accepted that chance plays an important role in nature. There are people who believe that our universe is completely *deterministic*, but there are others who assume that the universe is *nondeterministic*. In such a universe, randomness has a central role to play. In a way, randomness is a guarantee that we have no way to say what will happen next. Naive probability theory [i.e. not the one formulated by Andrei Nikolaevich Kolmogorov [179] (Андрей Николаевич Колмогоров)] is an attempt to make predictions about future events. Naturally, these predictions depend on the assumptions we make. But then the real problem is to what extent these assumptions are meaningful. Certainly, we will not discuss what meaningful actually means...

One more term that is remotely related to vagueness is *uncertainty*. This notion is related to a situation where the consequences, the extent of circumstances, the conditions, or the events are unpredictable. In quantum mechanics, the uncertainty principle is, roughly, about our inability to accurately measure at the same time the location and the momentum of a subatomic particle (e.g. an electron). Figure 1.1 shows how an electron looks like at any given moment. Strictly speaking, an electron is a wave and not a "solid" particle, but for the sake of the argument, we can assume that it is a sphere that vibrates extremely fast. So an electron would be a vague object if it would be in all these positions at the same time, but the uncertainty springs from the fact that it moves so fast that we cannot spot its place. In reality, an electron as a wave is in all places at the same time. Moreover, one can say that an electron is actually a cloud, so an electron is actually a

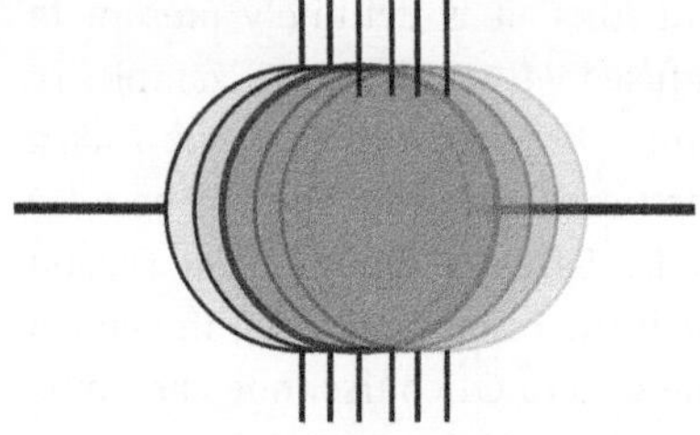

Figure 1.1 A drawing depicting an electron.

vague object. Interestingly, there is a deep connection between uncertainty and vagueness through possibility theory.

When Lotfi Aliasker Zadeh, the founder of fuzzy mathematics, introduced fuzzy sets [305], he justified his work by using examples of vague concepts like "the class of beautiful women," and "the class of tall men." Later on, people working on fuzzy mathematics were divided into two groups: those who believe that fuzzy mathematics are mathematics of vagueness, and those who believe that fuzziness and vagueness are two different things. Members of the second group argue that there is some sort of misunderstanding between the two communities:

> One of the reasons for the misunderstanding between fuzzy sets and the philosophy of vagueness may lie in the fact that Zadeh was trained in engineering mathematics, not in the area of philosophy.
>
> Dubois (2012) [108]

We find this argument completely silly as there is no engineering mathematics, but just mathematics. Also, one cannot compare (presumably applied) mathematics with philosophy as one cannot compare apples with flowers. On the other hand, we are in favor of an idea that is closer to what the first group is advocating.

1.3 Vagueness and Fuzzy Mathematics

Conferences and workshops are usually very interesting events. Typically, one has the chance to meet people and discuss new ideas. Some years ago, the first author was invited to talk about fuzzy computing in a workshop. After the talk, he had a chat with someone who really liked the idea of vagueness in computing. She noted that even our hardware is vague since we assume that it operates within a specific range, but in fact this is just an approximation that makes our life easier. Of course, hardware is not vague for this reason, but it gives an idea of why vagueness actually matters.

If we accept the ontic view of vagueness, then we must explore its use in science. This simply means to recognize that objects are vague and therefore their properties are not exact. For example, one of the first "applications" of vagueness was the definition of fuzzy algorithms and fuzzy conceptual computing devices (see [275] for details). In addition, vagueness should be used in order to provide alternative (and possibly better?) interpretations of quantum phenomena and all sciences that are affected by them (e.g. chemistry and biology at the molecular level). But even if we accept the semantic view, then vagueness should be relevant to law.

In our own opinion, vagueness is a fundamental property of this world. On the other hand, all sciences use the language of mathematics to express laws, make predictions, etc. Thus, if one wants to introduce vagueness in an "exact" science, there is a need for a *new* language. The question, of course, is: What kind of language should this be? Many people agree that *sets* can be used to describe the universe of all mathematical objects. Others believe that *categories* can play this role. And there are some who suggest that *types* should play this role. By extending the properties of the most fundamental building block, one can create new mathematics. Zadeh [305] extended the notion of set membership and so he defined his fuzzy sets. More specifically, an element belongs to a fuzzy set to some degree that is usually a number between 0 and 1. The next question is how this extension introduces vagueness into mathematics?

Consider a cloud, which is a vague object. At a given moment, we can consider that the cloud is a strange solid object in three-dimensional space. Obviously, there are points that are definitely inside the cloud and points that are definitely outside the cloud. However, there are points for which we cannot say with certainty whether they are inside or outside the cloud. For these points, one can employ *linguistic modifiers*, that is, words such as "more", or "less", to express their membership degree or one can use a number that clearly expresses to what extent a point belongs to the cloud. In summary, fuzzy sets are a quantitative way to describe borderline cases. At this point, let us say that one can approach borderline cases by using two nonvague solids: one that is contained in the vague object and the other that contains the vague object. These two nonvague objects are called the *lower* and the *upper* approximation, and they form a *rough* set. These sets have been introduced by Zdzisław Pawlak [237].

Saunders Mac Lane, who cofounded category theory with Samuel Eilenberg, noted that "problems, generalizations, abstraction and just plain curiosity are some of the forces driving the development of the Mathematical network" [201, pp. 438–439]. Then, he goes on as follows [201, pp. 439–440]:

> Not all outside influences are really fruitful. For example, one engineer came up with the notion of a *fuzzy set*—a set X where a statement $x \in X$ of membership may be neither true nor false but lies somewhere in between, say between 0 and 1. It was hoped that this ingenious notion would lead to all sorts of fruitful applications, to fuzzy automata, fuzzy decision theory

> and elsewhere. However, as yet most of the intended applications turn out to be just extensive exercises, not actually applicable; there has been a spate of such exercises. After all, if all Mathematics can be built up from sets, then a whole lot of variant (or, should we say, deviant) Mathematics can be built by fuzzifying these sets.

Of course, the fact that there are books like this one means that Mac Lane's prediction was not verified by reality. And yes, there are fuzzy automata and fuzzy decision theory is an established interdisciplinary topic. However, Mac Lane was not alone, as many researchers and scholars have expressed negative views about fuzzy sets. One obviously should learn to listen to the opposite view and try to understand the relevant arguments. But there are cases where this is simply impossible. A few years ago, the first author submitted a paper to a prestigious journal devoted to the international advancement of the theory and application of fuzzy sets. The paper described a model of computation built on the idea that vagueness is a fundamental property of our world. To his surprise, the paper was rejected mainly because three (!) reviewers insisted that it is a "fact that there is an equivalence between fuzzy set theory and probability theory." The author was sure that the editor did not like him for some reason, so he asked the reviewers to reject the paper and they found the most stupid argument to do so. Of course, no one can be sure about anything, but the fact remains that a prestigious journal on fuzzy sets confuses fuzzy set theory with probability theory. And this is not the only confusion as we have already seen.

Contra to these nonsensical ideas, ontic vagueness and fuzzy mathematics are very serious and interesting ideas that took the scientific community by storm and now most mathematical structures have been *fuzzyfied*. In certain cases, some have gone too far. For example, since vagueness and probabilities are about different things, it makes sense to talk about fuzzy probabilities, however, it does not make sense to talk about fuzzy rough sets or rough fuzzy sets. These are completely nonsensical ideas as we are talking about two different mathematical methods to describe vagueness. Similarly, a car can have either a diesel engine or a gasoline engine or a compressed natural gas engine, but not a "hybrid" engine that burns either gasoline or petrol. So our motto could be expressed as *generalization is useful, but too much generalization may lead to nonsense.*

Exercises

1.1 Find an argument similar to the Sorites Paradox.

1.2 Name five vague properties and five vague objects.

1.3 Find two ambiguous sentences.

1.4 Name one more idea that led to the development of a new branch of mathematics.

1.5 Explain why vagueness and probability theory are different and unrelated things.

1.6 Multisets form another generalization of sets where elements may occur more than one time. Does it make sense to talk about fuzzy multisets?

2

Fuzzy Sets and Their Operations

Based on the assumption that all mathematics can be built up from sets, one has to first define fuzzy sets and their variations. However, one needs to be familiar with certain notions and ideas in order to fully understand the notions that will be introduced. Thus, we start by exploring truth values and their algebras and continue with t-norms and t-conorms.

2.1 Algebras of Truth Values

Aristotle (Ἀριστοτέλης) was the first thinker who made a systematic study of logic in his Organon (Ὄργανον), a collection of six works on logic. However, the transformation of logic into something that looks like an algebra was started by the German philosopher and mathematician Gottfried Wilhelm Leibniz, who is primarily known as one of the two people who devised the differential calculus. Later on, George Boole in his *The Mathematical Analysis of Logic* [35] introduced in a way the truth values **1** (true) and **0** (false) and algebraic operations between them. This work was important because it showed that truth values can be represented by numbers and the various logical operations as "ordinary" number operations. To be accurate, Boole represented *true* with "$x = 1$" and *false* with "$x = 0$," where x is any *variable*. Having identified truth values with numbers, we are able to define conjunction, disjunction, and negation as arithmetic operations. In this way, one builds algebraic structures that describe truth values and the operations between them. However, numbers are *ordered* and this is used to define the logical operations. In general, these structures are studied in *lattice theory*[1] and in the rest of this section, we are going to present basic ideas and results.

1 The "standard" reference on lattice theory is [30], but we have opted to use the definitions and follow the presentation of [290].

A Modern Introduction to Fuzzy Mathematics, First Edition.
Apostolos Syropoulos and Theophanes Grammenos.

2.1.1 Posets

Definition 2.1.1 A *partially ordered set* or just *poset* is a set P equipped with a binary relation[2] $\leq$ (i.e. $\leq$ is a subset of $P \times P$) that for all $a, b, c \in P$ has the following properties:

Reflexivity $a \leq a$;
Transitivity if $a \leq b$ and $b \leq c$, then $a \leq c$;
Antisymmetry if $a \leq b$ and $b \leq a$, then $a = b$.

One can say that each element of a poset is a logical proposition and the operator "$\leq$" is the "$\Rightarrow$" operator. Thus, when $x \leq y$, this actually means that x entails y.

Example 2.1.1 The set $P = \{1, 2, \dots, \}$ is a poset when $a \leq b$ has the usual meaning. Also, the set of all subsets of some set X and the operation $\subseteq$ form a poset. Thus, $A \leq B$ if and only if $A \subseteq B$.

We have seen how the truth values can be expressed as numbers and how the entails operator is represented. Let us continue with the *and* operator.

Definition 2.1.2 Assume that P is a poset, $X \subseteq P$, and $y \in P$. Then, y is a *greatest lower bound* (glb or inf or meet) for X if and only if:

- when $x \in X$, then $y \leq x$, that is, y is a *lower bound* of X, and
- when z is any other lower bound for X, then $z \leq y$.

Typically, one writes $y = \bigwedge X$. The expression $\bigwedge\{a_1, a_2, \dots, a_n\}$ can also be written as $a_1 \wedge a_2 \wedge \cdots \wedge a_n$.

Proposition 2.1.1 *Assume that P is a poset and $X \subseteq P$. Then, X can have at most one glb.*

Proof: Assume that y and y' are two glbs of X. Since both are glbs, it holds that $y \leq y'$ and $y' \leq y$ and by antisymmetry $y = y'$. □

The least upper bound corresponds to *or*.

Definition 2.1.3 Assume that P is a poset, $X \subseteq P$, and $y \in P$. Then, y is a *least upper bound* (lub or sup or join) for X if and only if:

- when $x \in X$, then $x \leq y$, that is, y is an *upper bound* of X, and
- when z is any other upper bound for X, then $y \leq z$.

---

2 The notion of a relation and the various properties a relation may have are briefly presented in Section 4.1.

Typically, one writes $y = \bigvee X$. The expression $\bigvee\{a_1, a_2, \dots, a_n\}$ can also be written as $a_1 \vee a_2 \vee \cdots \vee a_n$. It is easy to prove the following proposition:

Proposition 2.1.2 *Assume that P is a poset and $X \subseteq P$. Then, X can have at most one lub.*

Example 2.1.2 Consider the set $\{0, 1\}$. This set is trivially a poset. The glb of this set is $\min(0, 1) = 0$, while its lub is $\max(0, 1) = 1$.

2.1.2 Lattices

Definition 2.1.4 A poset P is a *lattice* if and only if every finite subset of P has both a glb and a lub.

In what follows, the symbol **1** will denote the *top* (greatest) element and **0** the bottom (least) element of a poset. The operators $\wedge$ and $\vee$ have the following properties:

Commutativity	$x \wedge y = y \wedge x$	$x \vee y = y \vee x$
Associativity	$(x \wedge y) \wedge z = x \wedge (y \wedge z)$	$(x \vee y) \vee z = x \vee (y \vee z)$
Unit laws	$x \wedge \mathbf{1} = x$	$x \vee \mathbf{0} = x$
Idempotence	$x \wedge x = x$	$x \vee x = x$
Absorption	$x \wedge (x \vee y) = x$	$x \vee (x \wedge y) = x$

Definition 2.1.5 A lattice P is *distributive* if and only if for all $x, y, z \in P$ it holds that

$$x \wedge (y \vee z) = (x \wedge y) \vee (x \wedge z).$$

Example 2.1.3 Consider a set A and the set that consists of all subsets of A. The latter is called the *powerset* of A and it is denoted by $\mathscr{P}(A)$. Given $A_1, A_2 \in \mathscr{P}(A)$, then $A_1 \wedge A_2 = A \cap B$ and $A_1 \vee A_2 = A_1 \cup A_2$. It is not difficult to show that the powerset is a lattice.

Definition 2.1.6 A poset P with all lubs and all glbs is called a *complete lattice.*

Definition 2.1.7 A *Heyting algebra* is a lattice with **0** and **1** that has to each pair of elements a and b an *exponential* element b^a. This element is usually written as $a \Rightarrow b$ and has the following property:

$$c \leq (a \Rightarrow b) \quad \text{if and only if} \quad c \wedge a \leq b.$$

That is, $a \Rightarrow b$ is a lub for all those elements c with $c \wedge a \leq b$.

2.1.3 Frames

Definition 2.1.8 A poset P is a *frame* if and only if

(i) every subset has a lub;
(ii) every finite subset has a glb; and
(iii) the operator $\wedge$ distributes over $\vee$:

$$x \wedge \bigvee Y = \bigvee \{x \wedge y | y \in Y\}.$$

Note that the structure just defined is called by some authors *locale* or *complete Heyting algebra* by some authors.

Example 2.1.4 The set {false, true}, where false $\leq$ true, is a frame (why?).

2.2 Zadeh's Fuzzy Sets

In 1965, Zadeh published a paper entitled "Fuzzy Sets" [305] where he introduced his fuzzy sets. His motivation for introducing them is best explained in the introduction of his paper:

> More often than not, the classes of objects encountered in the real physical world do not have precisely defined criteria of membership. For example, the class of animals clearly includes dogs, horses, birds, etc., as its members, and clearly excludes such objects as rocks, fluids, plants, etc. However, such objects as starfish, bacteria, etc., have an ambiguous status with respect to the class of animals.

From this statement, one can say that he actually supported the semantic view of vagueness. In classical set theory, one can define a set using one of the following methods:

List method. A set is defined by naming all its members. This method can be used only for finite sets. For example, any set A, whose members are the elements $a_1, a_2, \ldots, a_n$, where n is small enough, is usually written as

$$A = \{a_1, a_2, \ldots, a_n\}.$$

Rule method. A set is defined by specifying a property that is satisfied by all its members. A common notation expressing this method is

$$A = \{x | P(x)\},$$

where the symbol | denotes the phrase "such that," and $P(x)$ designates a proposition of the form "x has the property P."

Characteristic function. Assume that X is a set, which is called a *universe*, and $A \subseteq X$. Then, the *characteristic function* $\chi_A : X \to \{0, 1\}$ of A is defined as follows:

$$\chi_A(a) = \begin{cases} 1, & \text{if } a \in A, \\ 0, & \text{if } a \notin A. \end{cases}$$

For example, the characteristic function of the set of positive even numbers less or equal than 10 can be defined as follows:

$$\chi_{\mathscr{M}}(x) = \begin{cases} \dfrac{1+(-1)^x}{2}, & \text{when } 0 < x \leq 10, \\ 0, & \text{otherwise.} \end{cases}$$

Zadeh had opted to introduce his fuzzy sets by employing an extension of the characteristic function:

Definition 2.2.1 Let X be a *universe* (i.e. an arbitrary set). A *fuzzy subset* A of X, is characterized by a function $A : X \to [0, 1]$, which is called the *membership function*. For every $x \in X$, the value $A(x)$ is called a *degree to which element x belongs to the fuzzy subset A.*

We say that a fuzzy subset is *characterized* by a function or not and that it *is* just a function because Zadeh used this term. However, we know that $\mathscr{P}(X)$, the powerset of X, is isomorphic to 2^X, the set of all functions from X to $\{0, 1\}$, and similarly, $\mathscr{F}(X)$, the set of all fuzzy subsets of X, is isomorphic to the set of all fuzzy characteristic functions, $[0, 1]^X$. Therefore, there is no need to distinguish between "is" and "characterized." If there is at least one element y of a fuzzy subset A such that $A(y) = 1$, then A is called *normal*. Otherwise, it is called *subnormal*. The *height* of a fuzzy subset A is the maximum membership degree, that is,

$$h(A) = \max_{x \in X}\{A(x)\},$$

provided that X is finite, or

$$h(A) = \bigvee_{x \in X}\{A(x)\},$$

when X is not finite. For reasons of simplicity, the term "fuzzy set" is preferred over the term "fuzzy subset."

When the universe X is a set with few elements, it is customary to write down a fuzzy set A as follows:

$$A = a_1/x_1 + a_2/x_2 + \cdots + a_n/x_n,$$

where $x_i \in X$, a_i is the degree to which x_i belongs to A, and a_i/x_i means that x_i belongs to A with a degree that is equal to a_i. Obviously, the symbol "+" does not denote addition, but it is some sort of *metasymbol*. Alternatively, one can write down a fuzzy set as follows:

$$A = \sum_{i=1}^{n} A(x_i)/x_i.$$

In case X is not finite (e.g. it is an interval of real numbers), a fuzzy set can be written down as

$$A = \int_X a_i/x_i.$$

Note that the symbols $\sum$ and $\int$ do not denote summation or integration. In the first case, x ranges over a set of discrete values, while in the second case, it ranges over a continuum.

Example 2.2.1 Consider the following five squares:

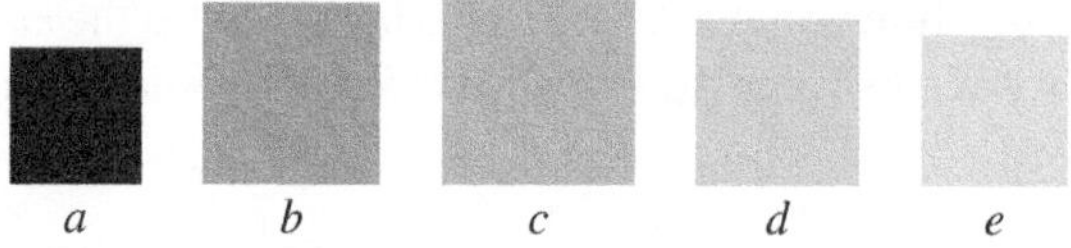

Assume that the set $\{a, b, c, d, e\}$ forms a universe. Then, we can form a fuzzy subset of black squares of this set. Let us call this set B. Then, the membership values of this set are $B(a) = 1$, $B(b) = 0.8$, $B(c) = 0.6$, $B(d) = 0.4$, and $B(e) = 0.3$.

In Example 2.2.1, we assigned the membership degrees by consulting a gray color code table, but, in general, there is no method by which one can assign membership degrees to elements of a fuzzy subset. Instead, one should employ some rule of thumb to compute the membership degrees. Since the membership degree can be any number that belongs to $[0, 1]$, it follows that numbers like $\pi/6$, $e/8$, $\sqrt{2}/4$, and so on, can be used as membership degrees. However, these numbers do not have a finite decimal representation, and some people are not very comfortable with this idea. Instead, it is quite common to assume that membership degrees should be elements of the set $\mathbb{Q} \cap [0, 1]$, where $\mathbb{Q}$ is the set of rational numbers. On the other hand, irrational numbers are used very frequently in physics in order to describe the natural world.

The empty fuzzy subset of some universe X is a set $\emptyset$ such that $\emptyset(x) = 0$, for all $x \in X$. A fuzzy subset A of X such that $A(x) = 1$ for all $x \in X$ is called a *crisp* or *sharp* set (i.e. an ordinary set).

Definition 2.2.2 Assume that $A : X \to [0, 1]$ and $B : X \to [0, 1]$ are two fuzzy sets of X. Then,

- their *union* is

$$(A \cup B)(x) = \max\{A(x), B(x)\};$$

- their *intersection* is

$$(A \cap B)(x) = \min\{A(x), B(x)\};$$

- the *complement*[3] of A is the fuzzy set

$$A^{\complement}(x) = 1 - A(x), \quad \text{for all } x \in X;$$

- their algebraic product is

$$(AB)(x) = A(x) \cdot B(x);$$

- A is a *subset* of B, denoted by $A \subseteq B$, if and only if

$$A(x) \leq B(x), \quad \text{for all } x \in X;$$

- the *scalar* cardinality[4] of A is

$$\text{card}(A) = \sum_{x \in X} A(x) \text{ and}$$

- the fuzzy *powerset* of X (i.e. the set of all ordinary fuzzy subsets of X) is denoted by $\mathscr{F}(X)$.

Example 2.2.2 Suppose that X is the set of all pupils of some class. Then, we can construct the fuzzy subset of tall pupils and the fuzzy subset of obese pupils. Let us call these sets T and O, respectively. Then, $T \cup O$ is the fuzzy set of pupils that are either tall or obese, while $T \cap O$ is the fuzzy subset of pupils that are tall and obese. In fact, there is a relation between set operations and logical connectives (see Section 7.1).

The operations between fuzzy sets have a number of properties that are similar to the properties of crisp sets. For example, it is trivial to prove that union and intersection are commutative operations:

$$A \cup B = B \cup A \quad \text{and} \quad A \cap B = B \cap A.$$

Also, it is easy to show that the De Morgan's laws, named after Augustus De Morgan,

$$(A \cup B)^{\complement} = A^{\complement} \cap B^{\complement} \quad \text{and} \quad (A \cap B)^{\complement} = A^{\complement} \cup B^{\complement}$$

---

3 Recall that if A is a subset of a universe X, then set of all those elements of X that do not belong to A is called the *complement* of A. This is denoted by $A^{\complement}$.

4 The cardinality of a set A is denoted by card(A) and it is equal to the number of elements of A.

are valid. Indeed, we want to prove that $(A \cup B)^{\mathrm{C}}(x) = A^{\mathrm{C}}(x) \cap B^{\mathrm{C}}(x)$ for all $x \in X$. We note that the membership degree of $(A \cup B)^{\mathrm{C}}(x)$ is

$$\begin{aligned} 1 - \max(A(x), B(x)) &= 1 + \min(-A(x), -B(x)) \\ &= \min(1 - A(X), 1 - B(x)), \end{aligned}$$

which is the membership degree of $A^{\mathrm{C}} \cap B^{\mathrm{C}}$. Similarly, we can prove the validity of the second rule. However, the *law of the excluded middle* and the *law of contradiction*

$$\begin{aligned} &A \cup A^{\mathrm{C}} = \chi_X \quad \text{(law of the excluded middle)}, \\ &A \cap A^{\mathrm{C}} = \varnothing \quad \text{(law of contradiction)}, \end{aligned}$$

are not valid. Clearly, if $A(x) \neq 1$, then $\max\{A(x), 1 - A(x)\} \neq 1$ but $\chi_X(x) = 1$. Similarly, if $A(x) \neq 0$, then $\min\{A(x), 1 - A(x)\} \neq 0$, but $\varnothing(x) = 0$. These laws are associated with two fundamental laws of logic. Since sets and their operations are models of logic, it follows that these laws are not valid in a fuzzy way of thinking. However, this is not true unless you assume that contradictions are meaningful.

An important property of fuzzy sets is convexity:

Definition 2.2.3 A fuzzy set $A : X \to [0,1]$ is *convex* if and only if

$$A[\lambda x_1 + (1 - \lambda)x_2] \geq \min\{A(x_1), A(x_2)\},$$

for all $x_1, x_2 \in X$ and all $\lambda \in [0, 1]$. If

$$A[\lambda x_1 + (1 - \lambda)x_2] > \min\{A(x_1), A(x_2)\},$$

then A is *strongly convex*.

Proposition 2.2.1 *If $A : X \to [0,1]$ and $B : X \to [0,1]$ are convex fuzzy sets, so is their intersection.*

The proof follows from the convexity of A and B and the "fact" that

$$C[\lambda x_1 + (1 - \lambda)x_2] = \min\{A[\lambda x_1 + (1 - \lambda)x_2], B[\lambda x_1 + (1 - \lambda)x_2]\},$$

where $C = A \cap B$.

Over the years, it became clear that the scalar cardinality of fuzzy sets is problematic. Ronald Robert Yager [303] discussed some of these problems and proposed alternative formulations of the cardinality of a fuzzy set. The first problem that he mentioned is that the scalar cardinality is not always a natural number. Another problem is demonstrated by the following fuzzy set that is assumed to be the fuzzy subset of intelligent people of the set X of people:

$$A = 0.1/x_1 + 0.1/x_2 + 0.1/x_3 + \cdots + 0.1/x_{10} + 0/x_{11} + 0/x_{12} + 0/x_{13} + \cdots .$$

If we ask how many intelligent people are in X using the scalar cardinality, then, surprisingly, the answer is 1. Yager presented a fuzzy cardinality and an alternative "crisp" cardinality.

Suppose that $X = \{x_1, x_2, \dots, x_n\}$ is a crisp set whose cardinality is n. Also, assume that

$$A = a_1/x_1 + a_2/x_2 + \cdots + a_n/x_n$$

is a fuzzy subset of X. In addition, assume that $\pi(j)$ is a permutation of the a_i such that $a_{\pi(j)}$ is the jth largest of the a_i. In what follows, we will write b_j instead of $a_{\pi(j)}$. Thus, b_j is the jth largest of the membership degrees of A. We define the fuzzy set B_A as follows:

$$B_A(0) = 1,$$
$$B_A(j) = b_j,\ j = 1, \dots, n,$$
$$B_A(j) = 0,\ j > n.$$

Then, B_A can be written as follows:

$$B_A = \frac{1}{0} + \frac{b_1}{1} + \frac{b_2}{2} + \cdots + \frac{b_n}{n} + \frac{0}{n+1} + \frac{0}{n+2} + \cdots .$$

Let $d_j = 1 - b_{j+1}$ be elements of a fuzzy set D_A. Then, we define this fuzzy set as follows:

$$D_A = \frac{1-b_1}{0} + \frac{1-b_2}{1} + \frac{1-b_3}{2} + \cdots + \frac{1-b_n}{n-1} + \frac{1}{n} + \frac{1}{n+1} + \cdots .$$

For the fuzzy set $E_A = B_A \cup D_A$, it holds that $E_A(j)$ is the degree to which j is exactly the number of elements of A. The fuzzy set E_A is the "fuzzy" cardinality of A.

2.3 α-Cuts of Fuzzy Sets

Two very important concepts of fuzzy set theory, which have been introduced by Zadeh [305, 311], are the concepts of an *α-cut* (pronounced *alpha-cut*), which is also known as α-level, and of a *strong α-cut*. An α-cut is a crisp set that consists of those elements that have membership degree greater than or equal to a specific value. More specifically:

Definition 2.3.1 Suppose that $A : X \to [0, 1]$ is a fuzzy subset of X. Then, for any $\alpha \in [0, 1]$, the α-cut ${}^{\alpha}A$ and the strong α-cut ${}^{\alpha+}A$ are the crisp sets

$${}^{\alpha}A = \{x | (x \in X) \wedge (A(x) \geq \alpha)\}$$

and

$${}^{\alpha+}A = \{x | (x \in X) \wedge (A(x) > \alpha)\},$$

respectively.

From the mathematical point of view, the α-cuts of a fuzzy set and its membership function are equivalent characterizations. By Definition 2.3.1, if we know the membership function, then we can determine the α-cuts. Conversely, if we know all the α-cuts ${}^{a}A$, then we can get each value $A(x)$ as the largest value α for which $x \in {}^{a}A$. However, Vladik Kreinovich [182] showed that only α-cuts are required to ensure algorithmic fuzzy data processing. Thus, α-cuts are more important than membership functions from an algorithmic point of view.

An interesting property of α-cuts is that for any fuzzy set $A : X \to [0, 1]$ if $a_1 < a_2$, then

$${}^{a_1}A \supseteq {}^{a_2}A \quad \text{and} \quad {}^{a_1+}A \supseteq {}^{a_2+}A.$$

In addition, the following properties hold:

Theorem 2.3.1 *Assume that A and B are two fuzzy subsets of X. Then, for all* $a, b \in [0, 1]$

(i) ${}^{a+}A \subseteq {}^{a}A$;
(ii) ${}^{a}(A \cap B) = {}^{a}A \cap {}^{a}B$ *and* ${}^{a}(A \cup B) = {}^{a}A \cup {}^{a}B$;
(iii) ${}^{a+}(A \cap B) = {}^{a+}A \cap {}^{a+}B$ *and* ${}^{a+}(A \cup B) = {}^{a+}A \cup {}^{a+}B$; *and*
(iv) ${}^{a}(A^{\complement}) = ({}^{(1-a)+}A)^{\complement}$.

Proof: The proof that follows is based on the proof given in [178].

(i) ${}^{a}A$ contains all elements of ${}^{a+}A$ and possibly more elements that have the property $A(x) = a$. Thus, ${}^{a+}A \subseteq {}^{a}A$.

(ii) Let us prove the first equality. For all $x \in {}^{a}(A \cap B)$ we have that $(A \cap B)(x) \geq a$ and this means that $\min[A(x), B(x)] \geq a$. This implies that $A(x) \geq a$ and $B(x) \geq a$. Thus, $x \in {}^{a}A \cap {}^{a}B$ and, consequently, ${}^{a}(A \cap B) \subseteq {}^{a}A \cap {}^{a}B$. On the other hand, for all $x \in {}^{a}A \cap {}^{a}B$, this implies that $x \in {}^{a}A$ and $x \in {}^{a}B$ and this means that $A(x) \geq a$ and $B(x) \geq a$. Therefore, $\min[A(X), B(x)] \geq a$ and this means that $(A \cap B)(x) \geq a$. From this, we deduce that $x \in {}^{a}(A \cap B)$ and, consequently, ${}^{a}A \cap {}^{a}B \subseteq {}^{a}(A \cap B)$. So ${}^{a}(A \cap B) = {}^{a}A \cap {}^{a}B$.
The second equality can be proved similarly. For all $x \in {}^{a}(A \cup B)$, we have that $\max[A(x), B(x)] \geq a$. This implies that $A(x) \geq a$ or $B(x) \geq a$. Thus, $x \in {}^{a}A \cup {}^{a}B$ and, consequently, ${}^{a}(A \cup B) \subseteq {}^{a}A \cup {}^{a}B$. On the other hand, for all $x \in {}^{a}A \cup {}^{a}B$, we have that $x \in {}^{a}A$ or $x \in {}^{a}B$ and this means that $A(x) \geq a$ or $B(x) \geq a$. Therefore, $\max[A(X), B(x)] \geq a$ and this means that $(A \cup B)(x) \geq a$. From this, we deduce that $x \in {}^{a}(A \cup B)$ and, consequently, ${}^{a}A \cup {}^{a}B \subseteq {}^{a}(A \cup B)$. So ${}^{a}(A \cup B) = {}^{a}A \cup {}^{a}B$.

(iii) By replacing $\geq$ with $>$ in the previous proof, we obtain the proof of these properties.

(iv) For all $x \in {}^{a}(A^{C})$, we have that $1 - A(x) = A^{C}(x) \geq a$; that is, $A(x) \leq 1 - a$. This implies that $x \notin {}^{(1-a)+}A$ but $x \in ({}^{(1-a)+}A)^{C}$, which means that ${}^{a}(A^{C}) \subseteq ({}^{(1-a)+}A)^{C}$. Similarly, for all $x \in ({}^{(1-a)+}A)^{C}$, we have that $x \notin {}^{(1-a)+}A$. Therefore, $A(x) \leq 1 - a$ and $1 - A(x) \geq a$. That is, $A^{C}(x) \geq a$, which implies that $x \in {}^{a}(A^{C})$. Therefore, $({}^{(1-a)+}A)^{C} \subseteq {}^{a}(A^{C})$ and so $({}^{(1-a)+}A)^{C} = {}^{a}(A^{C})$. □

Another important result, which we present without proof (see Ref. [178] for the proof) is the following:

Theorem 2.3.2 *For every fuzzy set $A : X \to [0, 1]$,*

$$A = \bigcup_{\alpha \in [0,1]} {}^{\alpha}A.$$

A special kind of α-cut is the *support* of a fuzzy set A. In particular, the support of A, denoted by supp(A), is the set of all elements of X that have nonzero membership degrees, that is, the support of A is its strong α-cut ${}^{0+}A$. Another special kind of α-cut is the *core* of a fuzzy set A. More specifically, the core of A, denoted by core(A) is the α-cut ${}^{1}A$, that is, the crisp set that contains elements with degree of membership equal to 1. Further, a fuzzy set A can be reconstructed from its α-cut sets as follows:

$$A(x) = \bigvee \{\alpha | x \in {}^{\alpha}A\}.$$

Also, a fuzzy set A is *bounded* if and only if all the sets ${}^{\alpha}A$ are bounded for all $\alpha > 0$.

We can define a cardinality of a fuzzy set A using its α-cuts. In particular, Yager [303] presented the following definition:

$$\text{card}_{\text{f}}(A) = \frac{0}{\text{card}({}^{0}A)} + \cdots + \frac{\alpha}{\text{card}({}^{\alpha}A)} + \cdots + \frac{1}{\text{card}({}^{1}A)}.$$

For example, if the fuzzy set A is defined as follows:

$$A = \frac{1}{x_1} + \frac{1}{x_2} + \frac{0.8}{x_3} + \frac{0.6}{x_4} + \frac{0.6}{x_5} + \frac{0.3}{x_6} + \frac{0.2}{x_7} + \frac{0}{x_8},$$

then

$$\text{card}_{\text{f}}(A) = \frac{0}{8} + \frac{0.2}{7} + \frac{0.3}{6} + \frac{0.6}{5} + \frac{0.8}{3} + \frac{1}{2}.$$

2.4 Interval-valued and Type 2 Fuzzy Sets

From a certain point of view, it is quite unnatural to assign specific numbers as membership degrees to elements of some fuzzy subset. Just as meteorologists usually predict that the temperature will be within a specific range, it is more "natural" to use intervals as membership degrees. Thus, instead of using the set $[0, 1]$, if we opt to use the set

$$I([0,1]) = \{(a,b) \mid (a, b \in [0,1]) \wedge (a \leq b)\},$$

where (a, b) is an open interval (i.e. the set $\{x \mid a < x < b\}$), we can define an extension of fuzzy sets. *Interval-valued* fuzzy sets have been defined by Roland Sambuc [253]:

Definition 2.4.1 An interval-valued fuzzy set A is a function $A : X \to I([0, 1])$, where X is some universe, and $A(x) = [A_*(x), A^*(x)]$.

If $(a, b), (c, d) \in I([0, 1])$, then $(a, b) \leq (c, d)$ if $a \leq c$ and $b \leq d$.

Definition 2.4.2 Assume that $A, B : X \to I([0, 1])$ are two interval-valued fuzzy sets. Then,

- their *union* is

$$(A \cup B)(x) = \left[\max(A_*(x), B_*(x)), \max(A^*(x), B^*(x))\right];$$

- their *intersection* is

$$(A \cap B)(x) = \left[\min(A_*(x), B_*(x)), \min(A^*(x), B^*(x))\right];$$

- the *complement* of A is the interval-valued fuzzy set

$$A^{\complement}(x) = \left[1 - A^*(x), 1 - A_*(x)\right].$$

There are various approaches to the cardinality of interval-valued fuzzy sets. The interested reader should consult the literature for more information (e.g. see Ref. [97]).

Remark 2.4.1 In general, one can say that interval-valued fuzzy sets can model uncertainty (i.e. lack of information). Thus, each element of the universe is mapped to a closed subinterval of $[0, 1]$, since we have incomplete knowledge of this element.

Type 2 fuzzy sets are an extension of ordinary fuzzy sets and were introduced by Zadeh [311]. In fact, Zadeh introduced type n fuzzy sets as follows:

Definition 2.4.3 A fuzzy set is of type n, $n = 2, 3, \ldots$, if its membership function ranges over fuzzy sets of type n–l. The membership function of a fuzzy set of type 1 ranges over the interval $[0, 1]$.

Since only type 2 fuzzy sets are used in certain tasks, in the rest of this section, we will discuss only type 2 fuzzy sets. Our brief presentation is based on [216].

Definition 2.4.4 A type 2 fuzzy set $\tilde{A}$ is characterized by a type 2 membership function $A(x, u)$, where $x \in X$ and $u \in J_x \subseteq [0, 1]$, that is,

$$\tilde{A} = \{((x, u), A(x, u)) | \forall x \in X, \forall u \in J_x \subseteq [0, 1]\},$$

where $0 \leq A(x, u) \leq 1$. Alternatively, one can use the following notation to denote the same thing:

$$A = \int_{\forall x \in X} \int_{\forall u \in J_x \subseteq [0,1]} \frac{A(x, u)}{(x, u)}.$$

Recall that the symbol $\int$ does not denote integration. It merely denotes that x and u range over a continuum. Also, when variables range over a set of discrete values, we use the symbol $\sum$ instead.

By specifying the lower and upper limit in the $\int$ or $\sum$ operators, one specifies that a variable ranges over a closed interval or a denumerable set whose endpoints are these two limits, respectively. Also, the expression $A(x, u)/(x, u)$ is not some sort of division but a generic way to denote the pair "return value"/"argument."

Example 2.4.1 We borrow an example of a type 2 fuzzy set from [281]. First of all one should note that

$$\sum_{\forall x \in X} \sum_{\forall u \in J_x \subseteq [0,1]} \frac{A(x, u)}{(x, u)} = \sum_{\forall x \in X} \left[\sum_{\forall u \in J_x \subseteq [0,1]} \frac{A(x, u)}{u} \right] \Big/ x.$$

This equality is rather important as the type 2 fuzzy set is defined using the right-hand side of this equation.

Assume that $X = \{0, 1, 2, \dots, 10\}$. Then, a type 2 fuzzy set $\tilde{A}$ in X is the following:

$$\begin{aligned}
\tilde{A} = {} & [1/0 + 0.9/0.1 + 0.8/0.2] \ /1 \\
& + [1/0 + 0.8/0.1 + 0.6/0.2 + 0.4/0.3 + 0.2/0.4] \ /2 \\
& + [0.3/0.2 + 0.6/0.3 + 0.8/0.4 + 1/0.5 + 0.8/0.6 + 0.6/0.7 + 0.3/0.8]/3 \\
& + [0.2/0.6 + 0.4/0.7 + 0.6/0.8 + 0.8/0.9 + 1/1] \ /4 \\
& + [0.2/0.8 + 0.6/0.9 + 1/1] \ /5 \\
& + [0.2/0.6 + 0.4/0.7 + 0.6/0.8 + 0.8/0.9 + 1/1] \ /6 \\
& + [0.3/0.2 + 0.6/0.3 + 0.8/0.4 + 1/0.5 + 0.8/0.6 + 0.6/0.7 + 0.3/0.8] \ /7 \\
& + [1/0 + 0.8/0.1 + 0.6/0.2 + 0.4/0.3 + 0.2/0.4] \ /8 \\
& + [1/0 + 0.9/0.1 + 0.8/0.2] \ /9.
\end{aligned}$$

Table 2.1 A table representation of the type 2 fuzzy set of Example 2.4.1.

1.0	0	0	0	0	1	1	1	0	0	0	0
0.9	0	0	0	0	0.8	0.6	0.8	0	0	0	0
0.8	0	0	0	0.3	0.6	0.2	0.6	0.3	0	0	0
0.7	0	0	0	0.6	0.4	0	0.4	0.6	0	0	0
0.6	0	0	0	0.8	0.2	0	0.2	0.8	0	0	0
0.5	0	0	0	1	0	0	0	1	0	0	0
0.4	0	0	0.2	0.8	0	0	0	0.8	0.2	0	0
0.3	0	0	0.4	0.6	0	0	0	0.6	0.4	0	0
0.2	0	0.8	0.6	0.3	0	0	0	0.3	0.6	0.8	0
0.1	0	0.9	0.8	0	0	0	0	0	0.8	0.9	0
0.0	1	1	1	0	0	0	0	0	1	1	1
	0	1	2	3	4	5	6	7	8	9	10

Note that 0 and 10 are not members of $\tilde{A}$, that is, they belong to $\tilde{A}$ with degree $1/0$ (Table 2.1).

2.5 Triangular Norms and Conorms

It is not difficult to prove that the *unit interval*, that is, the closed interval $[0, 1]$, equipped with min as the gld operator and max as the lub operator, is a frame. However, it is quite possible to define and use other functions as gld and lub operators. In fact, any function that has some specific properties can be used as a gld or a lub operator. These functions are known in the literature as *triangular norms*, or just *t*-norms, and *triangular conorms*, or just *t*-conorms, respectively [152, 296].

Definition 2.5.1 A *t*-norm is a binary operation $*\colon [0,1] \times [0,1] \to [0,1]$ that satisfies at least the following conditions for all $a, b, c \in [0, 1]$:

Boundary condition $a * 1 = a$ and $a * 0 = 0$.
Monotonicity $b \leq c$ implies $a * b \leq a * c$.
Commutativity $a * b = b * a$.
Associativity $a * (b * c) = (a * b) * c$.

A *t*-norm is *Archimedean* if and only if

(i) $*$ is continuous, and
(ii) $a * a < a$, for all $a \in (0, 1)$.

As it was noted previously, there are many functions that have these properties and the following are examples of some t-norms that are used frequently in practice:

standard intersection or Gödel t-norm $a * b = \min(a, b)$.
product t-norm $a * b = a \cdot b$.
bounded difference or Łukasiewicz t-norm $a * b = \max(0, a + b - 1)$.

drastic intersection t-norm $a * b = \begin{cases} a, & \text{when } b = 1, \\ b, & \text{when } a = 1, \\ 0, & \text{otherwise.} \end{cases}$

The dual of t-norm is a t-conorm:

Definition 2.5.2 A t-conorm is a binary operation $\bigstar : [0,1] \times [0,1] \to [0,1]$ that satisfies at least the following conditions for all $a, b, c \in [0,1]$:

Boundary condition $a \bigstar 0 = a$ and $a \bigstar 1 = 1$.
Monotonicity $b \leq c$ implies $a \bigstar b \leq a \bigstar c$.
Commutativity $a \bigstar b = b \bigstar a$.
Associativity $a \bigstar (b \bigstar c) = (a \bigstar b) \bigstar c$.

A t-conorm is *Archimedean* if and only if

(i) $\bigstar$ is continuous, and
(ii) $a \bigstar a > a$, for all $a \in (0,1)$.

The following are examples of some t-conorms that are frequently used in practice:

Standard union $a \bigstar b = \max(a, b)$.
Algebraic sum $a \bigstar b = a + b - ab$.
Bounded sum $a \bigstar b = \min(1, a + b)$.

Drastic union $a \bigstar b = \begin{cases} a, & \text{when } b = 0, \\ b, & \text{when } a = 0, \\ 1, & \text{otherwise.} \end{cases}$

A t-norm (t-conorm, respectively) is called *strong* if it is continuous and strictly decreasing (increasing, respectively). A strong t-norm $*$ and a strong t-conorm $\bigstar$ satisfy the following conditions

$$x * x < x \quad \text{and} \quad x \bigstar x > x \qquad \text{for all } x \in (0,1),$$

respectively.

It is even possible to define extensions of the standard complementation operator:

Definition 2.5.3 A fuzzy complement function is a continuous function $\eta : [0,1] \to [0,1]$ that satisfies at least the following conditions for all $a, b, c \in [0,1]$:

Boundary condition $\eta(0) = 1$ and $\eta(1) = 0$.
Monotonicity $a \le b$ implies $\eta(a) \ge \eta(b)$.
Involution $\eta(\eta(a)) = a$ for all $a \in [0,1]$.

In fact, a *t*-conorm ★ is the dual of a *t*-norm $*$ with respect to some complement operation η. The members of a *dual triple* $(*, \bigstar, \eta)$ must satisfy the De Morgan laws, that take the following form:

$$\eta(A(x) * B(x)) = \eta(A(x)) \bigstar \eta(B(x)),$$
$$\eta(A(x) \bigstar B(x)) = \eta(A(x)) * \eta(B(x)),$$

for all $x \in X$.

Example 2.5.1 The following is an example of a dual triple borrowed from [296]:

$$a * b = \max\left[\frac{a + b - 1 + \lambda ab}{1 + \lambda}, 0\right],$$
$$a \bigstar b = \min(a + b + \lambda ab, 1),$$
$$\eta(a) = \frac{1 - a}{1 + \lambda a},$$

where $\lambda > -1$. As an exercise, choose a value for λ and check if the De Morgan's laws hold.

2.6 *L*-fuzzy Sets

L-fuzzy sets form a natural extension of fuzzy sets that was introduced by Joseph Amadee Goguen [142], and this is why they are also known as *Goguen sets*. Note that Goguen was, during his doctoral studies at the University of California, Berkeley, a student of Zadeh. Goguen defined *L*-fuzzy sets as follows:

Definition 2.6.1 Assume that L is a frame and X is a universe. Then, a function $A : X \to L$ is an L-fuzzy set. Naturally, for every $x \in X$, the value $A(x) \in L$ is the degree to which element x belongs to the L-fuzzy subset A.

Operations between *L*-fuzzy sets are straightforward extensions of the corresponding operations of ordinary fuzzy sets.

Definition 2.6.2 Assume that L is a frame and that $A : X \to L$ and $B : X \to L$ are two L-fuzzy subsets of X. Then,

- their *union* is

$$(A \cup B)(x) = A(x) \vee B(x);$$

- their *intersection* is

$$(A \cap B)(x) = A(x) \wedge B(x);$$

- the *complement* of A is the L-fuzzy subset $A^{\mathsf{C}}(x)$ for which

$$A^{\mathsf{C}}(x) \vee A(x) = \mathbf{1} \text{ and } A^{\mathsf{C}}(x) \wedge A(x) = \mathbf{0}, \ \forall x \in X,$$

where $\mathbf{1}$ and $\mathbf{0}$ are the top and the bottom elements of L.

2.7 "Intuitionistic" Fuzzy Sets and Their Extensions

Krassimir T. Atanassov's [14, 15] "intuitionistic" fuzzy sets are another extension of fuzzy sets. Recall that for an ordinary fuzzy set $A : X \to [0, 1]$, its complement is the fuzzy set $1 - A$. Atanassov advocated the idea that one cannot compute the complement of a fuzzy set, but instead one has to define it just like we define a fuzzy set.

Definition 2.7.1 An "intuitionistic" fuzzy set A is characterized by the expression

$$\{\langle x, \mu_A(x), \nu_A(x)\rangle | x \in X\},$$

where X is a universe, $\mu_A : X \to [0, 1]$ is a function called the *membership* function and $\nu_A : X \to [0, 1]$ is another function called the *nonmembership* function. Moreover, for all $x \in X$ it must hold that $0 \leq \mu_A(x) + \nu_A(x) \leq 1$.

Remark 2.7.1 For any fuzzy set $B : X \to [0, 1]$, there is an "intuitionistic" fuzzy set

$$\{\langle x, B(x), 1 - B(x)\rangle | x \in X\}.$$

The term "intuitionistic" appears in quotes because it is a misnomer. It has been used in "intuitionistic" fuzzy set theory, because in intuitionistic logic (see Definition 7.7.1) there is no complementarity between a proposition and its negation.

Definition 2.7.2 Assume that $\{\langle x, \mu_A(x), \nu_A(x)\rangle | x \in X\}$ and $\{\langle x, \mu_B(x), \nu_B(x)\rangle | x \in X\}$ are two "intuitionistic" fuzzy sets A and B. Then,

- $A \subset B$ if and only if for all $x \in X$

$$\mu_A(x) \leq \mu_B(x) \quad \text{and} \quad \nu_A(x) \geq \nu_B(x);$$

- $A = B$ if and only if

$$A \subset B \quad \text{and} \quad B \subset A;$$

- their *union* is

$$(A \cup B) = \{\langle x, \max[\mu_A(x), \mu_B(x)], \min[\nu_A(x), \nu_B(x)]\rangle | x \in X\};$$

- their *intersection* is

$$(A \cap B) = \{\langle x, \min[\mu_A(x), \mu_B(x)], \max[\nu_A(x), \nu_B(x)]\rangle | x \in X\};$$

- the *complement* of A is the "intuitionistic" fuzzy set $A^{\complement}$

$$A^{\complement} = \{\langle x, \mu_A(x), \mu_A(x)\rangle | x \in X\};$$

- their *sum* is

$$(A + B) = \{\langle x, \mu_A(x) + \mu_B(x) - \mu_A(x) \cdot \mu_B(x), \nu_A(x) \cdot \nu_B(x)\rangle | x \in X\};$$

- their *product* is

$$(A \cdot B) = \{\langle x, \mu_A(x) \cdot \mu_B(x), \nu_A(x) + \nu_B(x) - \nu_A(x) \cdot \nu_B(x)\rangle | x \in X\};$$

- the *necessity* of A is the "intuitionistic" fuzzy set $\Box A$

$$\Box A = \{\langle x, \mu_A(x), 1 - \mu_A(x)\rangle | x \in X\};$$

- the *possibility* of A is the "intuitionistic" fuzzy set $\Diamond A$

$$\Diamond A = \{\langle x, 1 - \nu_A(x), \mu_A(x)\rangle | x \in X\}.$$

It is easy to define "intuitionistic" L-fuzzy sets:

Definition 2.7.3 Assume that L is a frame. Then, an "intuitionistic" L-fuzzy set A is a triplet (X, μ_A, ν_A), where X is a universe, the function $\mu_A : X \to L$ is the *membership* function and the function $\nu_A : X \to L$ is the *nonmembership* function. Moreover, for all $x \in X$, it must hold that $\mu_A(x) \leq N(\nu_A(x))$, where $N : L \to L$ is an involutive order reversing operation.

Of course, one can define "intuitionistic" interval-valued fuzzy sets and other generalizations, but we will not present all these generalizations and the interested reader should consult [15].

"Intuitionistic" fuzzy sets have been met with great skepticism, mainly because part of the scientific community favors the idea that they are a special case of L-fuzzy sets. For example, one can say that the set $A = \{\langle x, \mu_A(x), \nu_A(x)\rangle | x \in X\}$ is actually a nice way to write the membership function of any L-fuzzy set

$A' : X \to [0,1]^2$. In particular, the idea is that $\mu_A(x)$ and $\nu_A(x)$ are the first and the second coordinates of A' (see Ref. [293] for more details). Mathematically, this is a sound argument, but this means almost nothing! "Intuitionistic" fuzzy sets have been developed to promote a specific idea – the membership and nonmembership functions cannot be complementary, in general.

In 1998, Florentin Smarandache introduced another generalization of fuzzy sets by extending "intuitionistic" fuzzy sets. He opted to call these new structures *neutrosophic* sets [262, 292]. The term *neutrosophy*, which was coined by Smarandache, derives from the French *neutre* that, in turn, derives from the Latin *neuter*, which means neutral, and from the Greek *sophia*, which means wisdom. Thus, the term "neutrosophic" means knowledge of neutral thought. Neutrosophic sets are a generalization of fuzzy sets where the various membership degrees are drawn from the *nonstandard unit interval*. The term *nonstandard* refers to nonstandard analysis [144]. This is a theory that studies the set of *hyperreals*, ${}^*\mathbb{R}$, which is an extension of the set of real numbers, $\mathbb{R}$, that contains infinitely large and small numbers. A nonzero number ϵ is defined to be *infinitely small* if

$$|\epsilon| < \frac{1}{n} \text{ for all } n = 1, 2, 3, \ldots .$$

The reciprocal $\omega = \frac{1}{\epsilon}$ will be *infinitely large*. A number b is called *appreciable* if it is limited but not infinitely small, that is, $r < |b| < s$ for some positive real numbers r and s. Moreover, the number $b + \epsilon$ is also appreciable. Thus, the numbers $1 + \epsilon$ and $0 - \epsilon$ are appreciable. The nonstandard unit interval is the interval $(0 - \epsilon, 1 + \epsilon)$. Let us denote this interval with the symbol J^*.

Definition 2.7.4 Assume that X is a universe. Then, a *neutrosophic* set A in X is characterized by three functions: a membership function $T_A : X \to J^*$, an indeterminate-membership function $I_A : X \to J^*$, and a nonmembership function $F_A : X \to J^*$. The value $T_A(x)$ is the degree to which x belongs to A, $F_A(x)$ is the degree to which x does not belong to A, and $I_A(x)$ is the degree to which we are sure about the status of the element x.

It has become common practice to use $[0, 1]$ instead of J^*, nevertheless we use J^* for the rest of this short presentation.

Suppose that S_1 and S_2 are two sets that are subsets either of $\mathbb{R}$ or of ${}^*\mathbb{R}$. Then, we define the following operations:

$$\begin{aligned}
S_1 \oplus S_2 &= \{x | x = s_1 + s_2, \text{ where } s_1 \in S_1 \text{ and } s_2 \in S_2\}, \\
\{1+\epsilon\} \oplus S_2 &= \{x | x = (1+\epsilon) + s_2, \text{ where } s_2 \in S_2\}, \\
S_1 \ominus S_2 &= \{x | x = s_1 - s_2, \text{ where } s_1 \in S_1 \text{ and } s_2 \in S_2\}, \\
\{1+\epsilon\} \ominus S_2 &= \{x | x = (1+\epsilon) - s_2, \text{ where } s_2 \in S_2\}, \\
S_1 \odot S_2 &= \{x | x = s_1 \cdot s_2, \text{ where } s_1 \in S_1 \text{ and } s_2 \in S_2\}.
\end{aligned}$$

Definition 2.7.5 Assume that $A : X \to J^*$ and $B : X \to J^*$ are two neutrosophic sets. Then,

- $A^{\complement}$ is the complement of A if and only if

$$\begin{aligned} \operatorname{cod} T_{A^{\complement}} &= \{1+\epsilon\} \ominus \operatorname{cod} T_A, \\ \operatorname{cod} F_{A^{\complement}} &= \{1+\epsilon\} \ominus \operatorname{cod} F_A, \\ \operatorname{cod} I_{A^{\complement}} &= \{1+\epsilon\} \ominus \operatorname{cod} I_A, \end{aligned}$$

  where cod f is the codomain or range of the function f;
- $A \subseteq B$ if and only if

$$\begin{aligned} &\bigwedge \operatorname{cod} T_A \leq \bigwedge \operatorname{cod} T_B \quad \text{and} \quad \bigvee \operatorname{cod} T_A \leq \bigvee \operatorname{cod} T_B, \\ &\bigwedge \operatorname{cod} I_A \geq \bigwedge \operatorname{cod} I_B \quad \text{and} \quad \bigvee \operatorname{cod} T_A \geq \bigvee \operatorname{cod} T_B, \\ &\bigwedge \operatorname{cod} F_A \geq \bigwedge \operatorname{cod} F_B \quad \text{and} \quad \bigvee \operatorname{cod} F_A \geq \bigvee \operatorname{cod} F_B; \end{aligned}$$

- $A \cup B$ is their union if and only if

$$\begin{aligned} \operatorname{cod} T_{A \cup B} &= \operatorname{cod} T_A \oplus \operatorname{cod} T_B \ominus \operatorname{cod} T_A \odot \operatorname{cod} T_B, \\ \operatorname{cod} I_{A \cup B} &= \operatorname{cod} I_A \oplus \operatorname{cod} I_B \ominus \operatorname{cod} I_A \odot \operatorname{cod} I_B, \\ \operatorname{cod} F_{A \cup B} &= \operatorname{cod} F_A \oplus \operatorname{cod} F_B \ominus \operatorname{cod} F_A \odot \operatorname{cod} F_B; \end{aligned}$$

- $A \cap B$ is their intersection if and only if

$$\begin{aligned} \operatorname{cod} T_{A \cap B} &= \operatorname{cod} T_A \odot \operatorname{cod} T_B, \\ \operatorname{cod} I_{A \cap B} &= \operatorname{cod} I_A \odot \operatorname{cod} I_B, \\ \operatorname{cod} F_{A \cap B} &= \operatorname{cod} F_A \odot \operatorname{cod} F_B. \end{aligned}$$

Picture fuzzy sets is another extension of "intuitionistic" fuzzy sets that was introduced by Bùi Công Cường [82].

Definition 2.7.6 A *picture fuzzy set* A on a universe X is a set of quadruples

$$A = \{(x, \mu_A(x), \eta_A(x), \nu_A(x)) | x \in X\},$$

where $\mu_A(x) \in [0,1]$ is the degree of *positive* membership of x in A, $\eta_A(x) \in [0,1]$ is the degree of *neutral* membership of x in A, and $\nu_A(x) \in [0,1]$ is the degree of *negative* membership of x in A, and where μ_A, η_A, and ν_A satisfy the following condition:

$$(\forall x \in X)(\mu_A(x) + \eta_A(x) + \nu_A(x) \leq 1).$$

The number $1 - (\mu_A(x) + \eta_A(x) + \nu_A(x))$ is the degree of *refusal* membership of x in A. PFS(X) denotes the set of all the picture fuzzy sets drawn from the universe X.

It is almost straightforward to define the various set operations.

Definition 2.7.7 Assume that A and B are two picture fuzzy sets on the same universe X. Then, we define the various set operations as follows:

- $A \subseteq B$ if and only if for all $x \in X$, $\mu_A(x) \leq \mu_B(x), \eta_A(x) \leq \eta_B(x)$, and $\nu_A(x) \geq \nu_B(x)$;
- $A = B$ if and only if $A \subseteq B$ and $B \subseteq A$;
- $A \cup B = \{(x, \max(\mu_A(x), \mu_B(x)), \min(\eta_A(x), \ \eta_B(x)), \min(\mu_A(x), \nu_B(x))|x \in X\}$;
- $A \cap B = \{(x, \min(\mu_A(x), \mu_B(x)), \min(\eta_A(x), \ \eta_B(x)), \max(\mu_A(x), \nu_B(x))|x \in X\}$; and
- $A^{\mathrm{C}} = \{(\nu_A(x), \eta_A(x), \mu_A(x))|x \in X\}$.

The Cartesian product of two picture fuzzy sets that have different universes are defined as follows:

Definition 2.7.8 Assume that X and Y are two universes, and A and B two picture fuzzy sets

$$A = \{(x, \mu_A(x), \eta_A(x), \nu_A(x))|x \in X\} \text{ and } B = \{(y, \mu_B(y), \eta_B(y), \nu_B(y))|y \in Y\}.$$

Then, the Cartesian products of these picture fuzzy sets are defined as follows:

$$A \times_1 B = \{((x, y), \mu_A(x) \cdot \mu_B(y), \eta_A(x) \cdot \eta_B(y), \mu_A(x) \cdot \nu_B(y))|x \in X, y \in Y\},$$
$$A \times_2 B = \{((x, y), \mu_A(x) \wedge \mu_B(y), \eta_A(x) \wedge \eta_B(y), \mu_A(x) \cdot \vee_B(y))|x \in X, y \in Y\}.$$

Dmitri A. Molodtsov [219] introduced *soft sets* that have as a special case fuzzy sets.

Definition 2.7.9 A pair (F, E) is a soft set (over U) if and only if F is a mapping that maps elements of E to the powerset of U.

Example 2.7.1 Assume that someone is interested in buying a house and suppose that U is the set of houses under consideration. Also, suppose that E is the set of parameters, for example E can be the following set:

$$E = \{\text{expensive, beautiful, wooden, cheap, in green surroundings, modern, in good repair, in bad repair}\}.$$

Using this set, we can define a soft set that "partitions" the houses under consideration into expensive houses, cheap houses, etc.

Example 2.7.2 Suppose that $A : U \to [0, 1]$ is a fuzzy subset and that $F(\alpha) = {}^{\alpha}A$. Then,

$$A(x) = \bigvee_{\substack{\alpha \in [0,1] \\ x \in F(\alpha)}} \alpha.$$

Not so unexpectedly, it is possible to define picture fuzzy soft sets:

Definition 2.7.10 Assume that PFS(U) is the set of all picture fuzzy sets of U. Also, assume that E is the set of parameters and $A \subseteq E$. A pair (F, A) is a *picture fuzzy soft set* over U, where $F : A \to \text{PFS}(U)$ is a mapping.

2.8 The Extension Principle

The *extension principle* defines a general method for extending crisp mathematical concepts into objects of fuzzy mathematics. In simple words, one can use this principle to transform a real-valued function into a function that has fuzzy sets as arguments. The extension principle was implicitly introduced by Zadeh [305]. A few years later, Zadeh [311] explicitly introduced it. Later on, Yager [302] provided an alternative characterization of it. We first present Zadeh's [311] formulation.

Definition 2.8.1 Assume that $f : X \to Y$ is a function and A is a fuzzy subset of X defined as follows:

$$A = a_1/x_1 + a_2/x_2 + \cdots + a_n/x_n.$$

Then, the extension principle asserts that

$$f(A) = f(a_1/x_1 + a_2/x_2 + \cdots + a_n/x_n) \equiv a_1/f(x_1) + a_2/f(x_2) + \cdots + a_n/f(x_n).$$

Zadeh [311] presented the following example to demonstrate the use of the extension principle.

Example 2.8.1 Assume that $X = \{1, 2, \ldots, 10\}$ and $Y = \{1, 2, \ldots, 10, \ldots, 20, \ldots, 100\}$ and $f : X \to Y$ is a function that squares its argument. Also, assume that the following is a fuzzy subset of X:

$$\text{small} = 1/1 + 1/2 + 0.8/3 + 0.6/4 + 0.4/5.$$

Then,

$$f(\text{small}) = \text{small}^2 = 1/1 + 1/4 + 0.8/9 + 0.6/16 + 04.25.$$

Yager [302] reformulated the extension principle as follows:

Definition 2.8.2 Suppose that X and Y are two crisp sets. Also, assume that $f : X \to Y$ is a function such that for each $x \in X$, $f(x) = y$, where $y \in Y$. Then, the extension principle allows us to use f on fuzzy subsets of X. In particular, if A is a fuzzy subset of X, then

$$f(A) = B,$$

where B is a fuzzy subset of Y defined by

$$B(y) = \bigvee_{\substack{x \in X \\ f(x)=y}} A(x),$$

while when there is no $x \in X$ such that $f(x) = y$, then $B(y) = 0$.

A modern formulation of the extension principle uses Zadeh's image and preimage operators. Note that L^X is a *function space*, that is, the set of all functions from X to L.

Definition 2.8.3 Given a lattice L and a function $f : X \to Y$, the image $f^{\rightarrow} : L^X \to L^Y$ and the preimage $f^{\leftarrow} : L^Y \to L^X$ operators for $A \in L^X$ and $B \in L^Y$ are defined by

$$f^{\rightarrow}(A)(y) = \begin{cases} \bigvee\{A(x) | x \in f^{-1}(y)\}, & \text{if } f^{-1}(y) \neq \emptyset, \\ 0, & \text{otherwise.} \end{cases}$$

and

$$f^{\leftarrow}(B) = B \circ f,$$

respectively. In addition, it is a fact that these two operators form an adjunction $f^{\rightarrow} \dashv f^{\leftarrow}$, so for all $A \in L^X$ and all $B \in L^Y$ it holds that $A \subseteq f^{\leftarrow}(f^{\rightarrow}(A))$ and $f^{\rightarrow}(f^{\leftarrow}(B)) \subseteq B$.

The following theorem reveals a relationship between α-cuts and the extension principle.

Theorem 2.8.1 *Assume that $f : X \to Y$ is an ordinary function. Then, for any $A \in [0,1]^X$ and all $a \in [0,1]$ the following hold:*

(i) ${}^{a+}f(A) = f({}^{a+}A)$;
(ii) ${}^{a}f(A) \supseteq f({}^{a}A)$.

Proof: In order to prove the equality, note that for all $y \in Y$,

$$\begin{aligned} y \in {}^{a+}f(A) &\iff [f(A)](y) > a \\ &\iff \bigvee_{\substack{x \in X \\ f(x)=y}} A(X) > a \\ &\iff (\text{there is } x_0 \in X \text{ such that } y = f(x_0) \text{ and } A(x_0) > a) \\ &\iff (\text{there is } x_0 \in X \text{ such that } y = f(x_0) \text{ and } x_0 \in {}^{a+}A) \\ &\iff y \in f({}^{a+}A). \end{aligned}$$

This proves that ${}^{a+}f(A) = f({}^{a+}A)$. Similarly, if $y \in f({}^{a}A)$, then there exists $x_0 \in {}^{a}A$ such that $y = f(x_0)$. Therefore,

$$[f(A)](y) = \bigvee_{\substack{x \in X \\ f(x)=y}} A(x) \geq A(x_0) \geq a,$$

and so $y \in {}^{a}f(A)$. This implies that $f({}^{a}A) \subseteq {}^{a}[f(A)]$. □

2.9* Boolean-Valued Sets

Rough sets are a different approach to the mathematical description of vagueness. According to this approach, we describe a vague object with an upper and a lower approximation. Think of a cloud and a rectangular cuboid that surrounds it and one that is surrounded by the cloud. The former rectangular cuboid is an upper approximation of the cloud and the latter is a lower approximation of it. However, it is not necessary to go that far to provide an alternative mathematical description of vagueness. In fact, it is possible to use an alternative generalization of the membership function (e.g. by demanding that the membership set is not a frame but something else), which will lead to a rather different model of vagueness. Obviously, one should not expect to get a set theory that is compatible with fuzzy set theory, although both the new model and fuzzy set theory describe the same thing. But why do we need such an alternative model?

As Ken Akiba [4] points out, philosophers are not really happy with fuzzy set theory, to put it mildly. The main problem of fuzzy sets is that they are *degree-functional*. In simple words, this means that if A, B, and C are fuzzy sets such that $C = A \cap B$, then x belongs to C to a degree that depends to the degrees that x belongs to A and B. In particular, suppose that T is a fuzzy set of tall persons of some group of people and F is a fuzzy set of fat people of the same group of people. Also, assume that $T(\text{Adam}) = 0.5$, and $F(\text{Adam}) = 0.5$. Then, $T^{\mathsf{C}}(\text{Adam}) = 1 - T(\text{Adam})$, $(T \cup T^{\mathsf{C}})(\text{Adam}) = 0.5$, and $(T \cup F)(\text{Adam}) = 0.5$. However, there is a problem with $(T \cup T^{\mathsf{C}})(\text{Adam}) = 0.5$ mainly because everybody is either tall or not tall, thus, $(T \cup T^{\mathsf{C}})(\text{Adam})$ should be equal to 1. Analogously, $(T \cap T^{\mathsf{C}})(\text{Adam}) = 0.5$ in fuzzy set theory, when in reality nobody has the property of being tall and not being tall at the same time. Thus, $(T \cup T^{\mathsf{C}})(\text{Adam}) = 0$!

Another problem of fuzzy set theory can be exemplified as follows: Assume that D is the set of all points on earth and E is a fuzzy subset of D. Then, the following is true:

$$E(\text{the peak of Mt. Everest}) = 1.$$

However, as we go down from the peak, we will find a point p such that $E(p) < 1$ and the question is: where does this point lie? Moreover, can we make sense of such a sharp borderline? Unfortunately, if there is such a borderline, then it seems that everything is crisp…

A solution to the first problem is to demand that for any element x

$$(A \cap A^{\mathsf{C}})(x) = 0 \quad \text{and} \quad (A \cup A^{\mathsf{C}})(x) = 1.$$

Naturally, the next question is how can we get these desiderata? Without going into all relevant details, suffice it to say that membership degrees should be drawn from a *Boolean* algebra instead of a frame.[5] A Boolean algebra is a distributive lattice that has two distinguished elements

$$0 = \bigvee \emptyset \quad \text{and} \quad 1 = \bigwedge \emptyset,$$

which are the *bottom* and *top* elements, respectively. In addition, every element x has a complement $\overline{x}$ that must satisfy the following conditions:

$$x \wedge \overline{x} = 0 \quad \text{and} \quad x \vee \overline{x} = 1.$$

In fact, any Heyting algebra that satisfies the following conditions:

$$\overline{\overline{x}} = x \quad \text{and} \quad x \vee \overline{x} = 1$$

is a Boolean algebra. For example, the set $\{0, v, \overline{v}, 1\}$, where $0 \le v \le 1$, $0 \le \overline{v} \le 1$, $v \not\le \overline{v}$, and $\overline{v} \not\le v$, is a Boolean algebra.

A Boolean-valued set theory is, one where each sentence written in the language of the theory has a value in a general Boolean algebra. Since sentences like "$x \in B$" and "$x = y$" are valid sentences in a Boolean-valued set theory, this implies that not only set membership but also set identity is vague (note that in the language of set theory an individual variable ranges over sets). In other words, a Boolean-valued model of vagueness is more general than fuzzy set theory since both set membership and element identity are vague. Although fuzzy set theory has its limitations, still we know that membership degrees are drawn from the closed interval $[0, 1]$. However, there is no specific lattice structure from which membership "degrees" should be drawn in a Boolean-valued set theory. Of course, this is not a real problem, because we can define a Boolean algebra that is general enough for our purposes. But as Akiba has pointed out, we should insist on using 1 for propositions that are logically or metaphysically necessary and 0 for propositions that are logically or metaphysically impossible. Although this book is about fuzzy set theory based on frames, we believe that the development of Boolean-valued

5 The "standard" reference on Boolean-valued models of set theory is John Lane Bell's book [25]. Later on, Jin-Wen Zhang [319] investigated the connection between fuzzy set theory and Boolean-valued set theory. Surprisingly, the work done by Joseph G. Brown [39] is not mentioned in Zhang's and Akiba's work. Essentially, Brown proposed a Boolean-valued version of fuzzy set theory. In Brown's proposal $A^{\mathsf{C}}(x) = [A(x)]^{\mathsf{C}}$, $(A \cup B)(x) = A(x) \vee B(x)$, and $(A \cap B)(x) = A(x) \wedge B(x)$.

fuzzy set theory may lead to a better mathematical understanding of vagueness. In fact, it seems that some researchers are already working in this direction (e.g. see Ref. [107]).

2.10* Axiomatic Fuzzy Set Theory

Axiomatic set theory describes the properties of sets without getting into the trouble to define what a set is or what it means for an element to be a member of a set. The theory consists of a number of axioms that should be used to define new sets and to argue about the properties of any set (see Ref. [158] for an informal introduction to axiomatic set theory, or Ref. [166] for a thorough presentation of set theory). Axiomatic set theory is known as Zermelo–Fraenkel axiomatic set theory ZF, after Ernst Friedrich Ferdinand Zermelo and Abraham Halevi (Adolf) Fraenkel whose work resulted in the formulation of the axioms of axiomatic set theory. The axioms of the theory are expressed as formulas. Atomic formulas express that either an entity belongs to another entity (i.e. $x \in y$) or that two entities are equal (i.e. $x = y$). Atomic formulas can be brought together with connectives or quantifiers to form formulas. The connectives are $\phi \wedge \psi$ (conjunction), $\phi \vee \psi$ (disjunction), $\neg\phi$ (negation), $\phi \Rightarrow \psi$ (implication), and $\phi \iff \psi$ (equivalence). The quantifiers are $\forall x\, \phi$ (universal quantifier) and $\exists x\, \phi$ (existential quantifier). Consider the Axiom of Extensionality which states that *If X and Y have the same elements, then* $X = Y$. Then, this axiom can be expressed by the following formula:

$$\forall u\ (u \in X \iff u \in Y) \Rightarrow X = Y.$$

Edward William Chapin, Jr. [68, 69] has proposed an axiomatic fuzzy set theory by replacing the atomic formula $x \in y$ with the membership relation in the language that is used to describe this new axiomatic theory. In particular, the membership relation is a ternary relation $\varepsilon(x, y, z)$ that says that "x is an element of y with degree of membership at least z." In general, $\varepsilon(x, y, z)$ should be true when $0 \le z \le y(x)$, since y denotes a fuzzy set and its membership function at the same time. This means that ε may take the degree of membership of x in y, but also all "smaller" values. The new axiomatic theory is called Za-set theory and the Axiom of Extensionality can be reformulated as follows:

$$\forall u\ \forall w\ (\varepsilon(u, X, w) \iff \varepsilon(u, Y, w)) \Rightarrow X = Y.$$

Aleš Pultr [243] pointed out that it is quite natural to talk about fuzzy equality. For example, he pointed out that two objects might be to some degree a and b, respectively. Then, this means that when we compare the color of these two objects, it makes no sense to assert that either $\text{green}_a = \text{green}_b$ is true or $\text{green}_a \neq \text{green}_b$ is true. It would be more "fuzzy" to say that $\text{green}_a =_r \text{green}_b$, that is, that green$_a$ is equal to green$_b$ with a degree at least equal to r. Thus, one should introduce a new equality relation eq(x, y, z) into the language to express equality up to some degree. Note that we propose the introduction of a new relation and not the replacement of "=" with a new one, simply because there are cases where the use of "=" is absolutely justified and meaningful.

Exercises

2.1 Show that the set $P = \{1, 2, \dots\}$ and the relation $a \leq b$, when it means that a divides b, form a poset.

2.2 Prove Proposition 2.1.2 on page 13.

2.3 Show that any finite distributive lattice is a frame.

2.4 Show that the powerset of a set A is a frame.

2.5 Define equality between two fuzzy sets.

2.6 Determine if the following are true for any fuzzy sets A, B, and C:
a. $A \cap (B \cup C) = (A \cap B) \cup (A \cap C)$;
b. $A \cup (A \cap B) = A$;
c. $A^{\mathrm{C}} \cap (A^{\mathrm{C}} \cup B^{\mathrm{C}}) = A^{\mathrm{C}}$.

2.7 Assume that A is a fuzzy subset of X. Then, prove that when $a_1 < a_2$, ${}^{a_1+}A \supseteq {}^{a_2+}A$.

2.8 Suppose that $A, B \in \mathscr{F}(X)$. Then, show that for all $a \in [0, 1]$
a. $A \subseteq B$ if and only if ${}^{a}A \subseteq {}^{a}B$;
b. $A = B$ if and only if ${}^{a}A = {}^{a}B$.

2.9 Show that $([0, 1], \min, \max)$ is a frame.

2.10 For any "intuitionistic" fuzzy set A prove the following:
a. $\Box A = [\Diamond A^{\mathrm{C}}]^{\mathrm{C}}$;
b. $\Diamond A = [\Box A^{\mathrm{C}}]^{\mathrm{C}}$;
c. $\Box\Box A = \Box A$;
d. $\Box\Diamond A = \Diamond A$;
e. $\Diamond\Box A = \Box A$;
f. $\Diamond\Diamond A = \Diamond A$.

2.11 For any two "intuitionistic" fuzzy sets A and B prove the following:

a. $\Box(A \cup B) = \Box A \cup \Box B$;

b. $\Diamond(A \cup B) = \Diamond A \cup \Diamond B$.

2.12 Show that, for picture fuzzy sets, the operations ∩ and ∪ are commutative, associative, and distributive.

2.13 Define the intersection of two fuzzy sets $A, B : X \to [0, 1]$ using the formula in Example 2.7.2 in page 32.

2.14 Based on Definition 2.8.1 of the extension principle, write down its formulation when the support of A is a continuum.

2.15 Show that $A \subseteq f^{\leftarrow}(f^{\rightarrow}(A))$ and $f^{\rightarrow}(f^{\leftarrow}(B)) \subseteq B$.

3

Fuzzy Numbers and Their Arithmetic

Fuzzy numbers are a mathematical formulation of vague statements about real numbers. For example, a statement expressing that a number is approximately 2 is modeled by a fuzzy number. Fuzzy numbers are of great importance in fuzzy systems and toward the end of this chapter, we are going to discuss how we can construct fuzzy numbers from real-world data, and how these numbers can be used.

3.1 Fuzzy Numbers

Fuzzy (real) numbers are a special kind of fuzzy sets that have been introduced by Zadeh [311].

Definition 3.1.1 A fuzzy subset A is called a fuzzy (real) number when the universe on which A is defined is the set of all real numbers $\mathbb{R}$ and it satisfies the following conditions:

(i) all the α-cuts of A are not empty for $0 \leq \alpha \leq 1$;
(ii) all the α-cuts of A are closed intervals of $\mathbb{R}$;
(iii) the support of A, that is, the set

$$^{0+}A = \{x | x \in \mathbb{R} \text{ and } A(x) > 0\},$$

is bounded.

Clearly, every fuzzy number is a convex fuzzy set since all α-cuts are closed intervals. Also, the core of a fuzzy number A should be a singleton, and we will denote its only element by $\overline{x}$. Fuzzy numbers can be either positive or negative:

Definition 3.1.2 A fuzzy number A is called positive (negative), if $A(x) = 0$, for all $x < 0$ ($x > 0$).

A Modern Introduction to Fuzzy Mathematics, First Edition.
Apostolos Syropoulos and Theophanes Grammenos.

Remark 3.1.1 Each $x \in \mathbb{R}$ can be considered as a fuzzy number $\overline{x}$ defined by

$$\overline{x}(t) = \begin{cases} 1, & \text{if } t = x, \\ 0, & \text{if } t \neq x. \end{cases}$$

A special case of fuzzy real numbers are the *discrete* fuzzy numbers. A fuzzy subset of $\mathbb{R}$ is a discrete fuzzy number if its support is finite. If A is a discrete fuzzy number and $\text{supp}(A) = \{x_1, x_2, \ldots, x_n\}$, then there is an $1 \leq i \leq n$ such that $A(x_i) = 1$. In the rest of this section, we present the various types of nondiscrete fuzzy numbers, and we mostly follow the categorization and the definitions presented in [159].

If the universe on which a fuzzy set is defined is the set $\mathbb{C}$ of complex numbers, then fuzzy sets are called *fuzzy complex numbers* [40]. The α-cuts of a fuzzy complex number Z are

$${}^{\alpha}Z = \{z \mid Z(z) > \alpha\},$$

where $0 \leq \alpha < 1$ and, when $\alpha = 1$, then we separately specify

$${}^{1}Z = \{z \mid Z(z) = 1\}.$$

Definition 3.1.3 Z is a fuzzy complex number if and only if

(i) $Z(z)$ is continuous;
(ii) ${}^{\alpha}Z, 0 \leq \alpha < 1$, is open, bounded, connected, and simply connected; and
(iii) ${}^{1}Z$ is nonempty, compact, arcwise connected, and simply connected.

Although the theory of fuzzy complex numbers is quite interesting, we are not going to present its details in this book.

3.1.1 Triangular Fuzzy Numbers

A *triangular* fuzzy number is a fuzzy set whose graph is a triangle. Usually, one specifies such a fuzzy number using the notation $\text{tfn}(\overline{x}, e_l, e_r)$, where $\overline{x}$ is the only element of the core of the fuzzy number and corresponds to the x-coordinate of the point that lies at the intersection of the altitude and the base, e_l is the length of the line segment from $\overline{x}$ to the vertex that lies to the left of it, and e_r is the length of the line segment from $\overline{x}$ to the vertex that lies to the right of it. Figure 3.1 depicts a triangular fuzzy number and its "coordinates." A triple $(\overline{x}, e_l, e_r)$ defines a fuzzy number that is characterized by the following membership function:

$$A(x) = \begin{cases} 0, & \text{if } x \leq \overline{x} - e_l, \\ 1 + (x - \overline{x})/e_l, & \text{if } \overline{x} - e_l < x < \overline{x}, \\ 1 - (x - \overline{x})/e_r, & \text{if } \overline{x} \leq x < \overline{x} + e_r, \\ 0, & \text{if } x \geq \overline{x} + e_r. \end{cases}$$

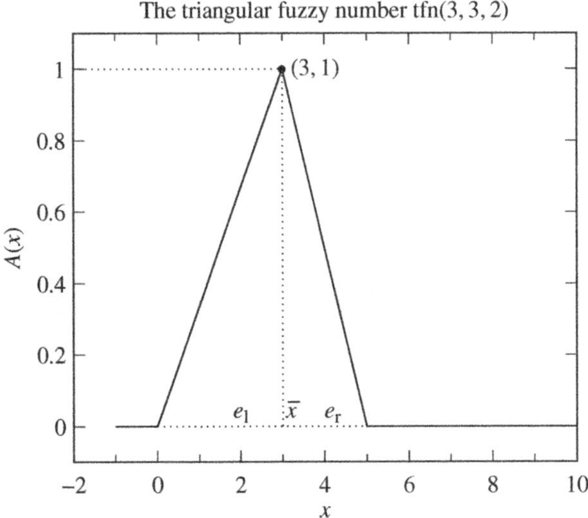

Figure 3.1 A triangular fuzzy number.

Alternatively, this membership function can be expressed as follows:

$$A(x) = \min[\max[0, 1 - (\overline{x} - x)/e_l], \max[0, 1 - (x - \overline{x})/e_r]], \quad \forall x \in \mathbb{R}. \tag{3.1}$$

In fact, this alternative formulation can be easily used to draw any triangular fuzzy number.

3.1.2 Trapezoidal Fuzzy Numbers

Trapezoidal fuzzy numbers are, of course, fuzzy sets but since their core contains more than one element, they cannot be classified as fuzzy numbers. However, it is absolutely reasonable to view these fuzzy sets as *fuzzy intervals*. On the other hand, it is a fact that the term *trapezoidal fuzzy number* is persistently used in the literature (e.g. see Ref. [20] and references therein). A trapezoidal fuzzy set is completely characterized by four real numbers $t_1 \le t_2 \le t_3 \le t_4$. We will use the notation $\text{trfn}(t_1, t_2, t_3, t_4)$ to specify the fuzzy set that is characterized by the following membership function:

$$T(x) = \begin{cases} 0, & \text{if } x < a, \\ (x - a)/(b - a), & \text{if } a \le x < b, \\ 1, & \text{if } b \le x < c, \\ (d - x)/(d - c), & \text{if } c \le x < d, \\ 0, & \text{if } x \ge d. \end{cases}$$

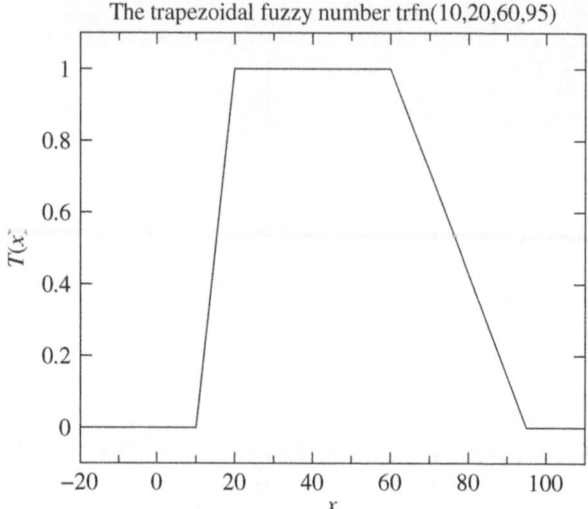

Figure 3.2 A trapezoidal fuzzy number.

Figure 3.2 depicts the trapezoidal fuzzy set $(10, 20, 60, 95)$. Note that the four numbers of the quadruple correspond to the x-coordinates of the four vertices of the resulting "trapezoid" in this specific order.

3.1.3 Gaussian Fuzzy Numbers

Gaussian fuzzy numbers are characterized by membership functions that are some special kind of a Gaussian function. The general form of these functions is

$$f(x) = a\,e^{-\frac{(x-b)^2}{2c^2}}.$$

Usually, one specifies a Gaussian fuzzy number using the notation $\mathrm{gfn}(\overline{x}, \sigma_\mathrm{l}, \sigma_\mathrm{r})$, where $\overline{x}$ is the only element of the core of the fuzzy number, and σ_l and σ_r are the left-hand and right-hand spreads that correspond to the standard deviation of the Gaussian distribution, see Figure 3.3. The membership function that characterizes any Gaussian fuzzy number has the following general form:

$$G(x) = \begin{cases} \exp[-(x-\overline{x})^2/(2\sigma_\mathrm{l}^2)], & \text{if } x < \overline{x}, \\ \exp[-(x-\overline{x})^2/(2\sigma_\mathrm{r}^2)], & \text{if } x \geq \overline{x}. \end{cases}$$

A *quasi-Gaussian* fuzzy number consists of a Gaussian fuzzy number whose membership degree is set to zero for $x < \overline{x} - 3\sigma_\mathrm{l}$ and for $x > \overline{x} + 3\sigma_\mathrm{r}$, respectively. A quasi-Gaussian fuzzy number will be written as $\mathrm{gfn}^*(\overline{x}, \sigma_\mathrm{l}, \sigma_\mathrm{r})$ and, in general, the

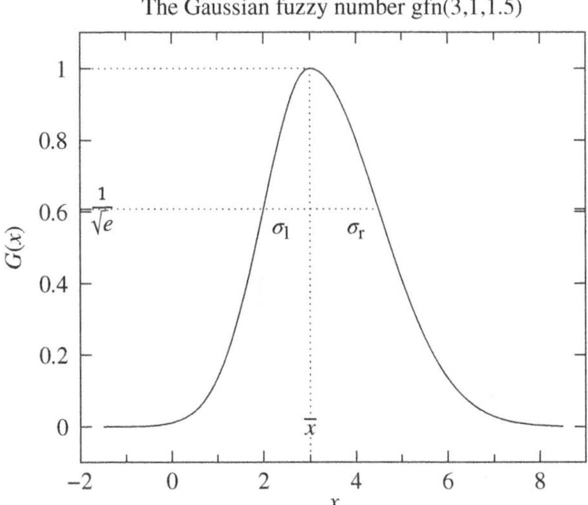

Figure 3.3 A Gaussian fuzzy number.

following function characterizes any quasi-Gaussian fuzzy number:

$$G_q(x) = \begin{cases} 0, & \text{if } x \le \overline{x} - 3\sigma_l, \\ \exp[-(x-\overline{x})^2/(2\sigma_l^2)], & \text{if } \overline{x} - 3\sigma_l < x < \overline{x}, \\ \exp[-(x-\overline{x})^2/(2\sigma_r^2)], & \text{if } \overline{x} \le x < \overline{x} + 3\sigma_r, \\ 0, & \text{if } x \ge \overline{x} + 3\sigma_r. \end{cases}$$

Note that the fuzzy number has nonzero membership values within a specific range.

3.1.4 Quadratic Fuzzy Numbers

A *quadratic* fuzzy number is yet another general form of a fuzzy number. We specify such a fuzzy number using the notation qfn($\overline{x}$, β_l, β_r). The membership function of any quadratic fuzzy number is parameterized by these three numbers:

$$Q(x) = \begin{cases} 0, & \text{if } x \le \overline{x} - \beta_l, \\ 1 - (x-\overline{x})^2/\beta_l^2 & \text{if } x - \beta_l < x < \overline{x}, \\ 1 - (x-\overline{x})^2/\beta_r^2 & \text{if } \overline{x} \le x < \overline{x} + \beta_r, \\ 0, & \text{if } x \ge \overline{x} - \beta_r. \end{cases}$$

Figure 3.4 shows exactly to what the three parameters correspond.

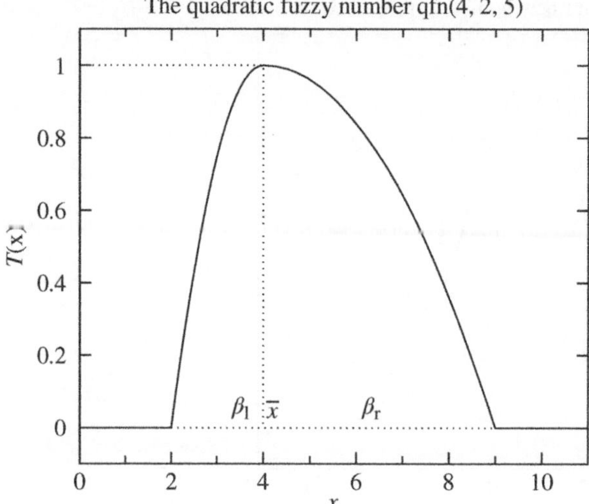

Figure 3.4 A quadratic fuzzy number.

3.1.5 Exponential Fuzzy Numbers

Exponential fuzzy numbers are yet another type of fuzzy numbers. Their membership function has the general form[1]:

$$E(x) = \begin{cases} 0, & \text{if } x < \overline{x} - a\tau_l, \\ \exp[(x - \overline{x})/\tau_l], & \text{if } \overline{x} - a\tau_l \le x < \overline{x}, \\ \exp[(\overline{x} - x)/\tau_r], & \text{if } \overline{x} \le x < \overline{x} + a\tau_r, \\ 0, & \text{if } \overline{x} + a\tau_r \le x. \end{cases}$$

Here τ_l and τ_r are the left and right spread of $\overline{x}$, respectively, and a represents a tolerance value. We will specify an exponential fuzzy number using the notation $\text{efn}(\overline{x}, \tau_l, \tau_r, a)$. Figure 3.5 depicts an exponential fuzzy number.

3.1.6 *L–R* Fuzzy Numbers

Didier Dubois and Henri Prade [109] have introduced a special form of fuzzy numbers that are dubbed *L–R* fuzzy numbers. The name derives from the fact that the graph of the membership function consists of two parts: the left and the right curve that meet at point $(\overline{x}, 1)$. Figure 3.6 shows a typical example of an *L–R* fuzzy

1 We have used the definition presented in [159] to draw an exponential fuzzy number, but the result was totally wrong. However, the definition presented in [24] yielded a real exponential fuzzy number. Thus, the definition of exponential fuzzy numbers is borrowed from [24].

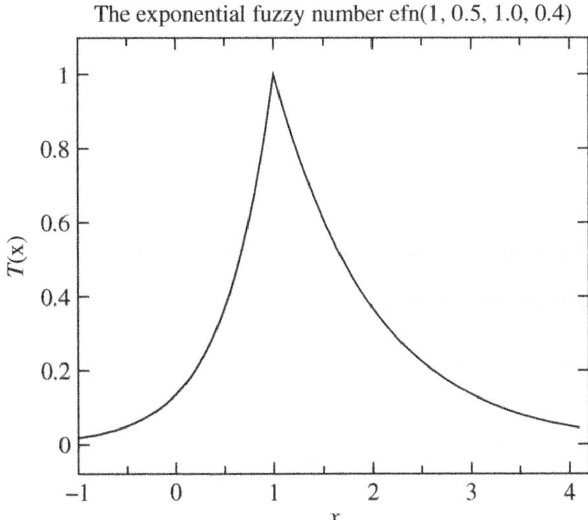

Figure 3.5 An exponential fuzzy number.

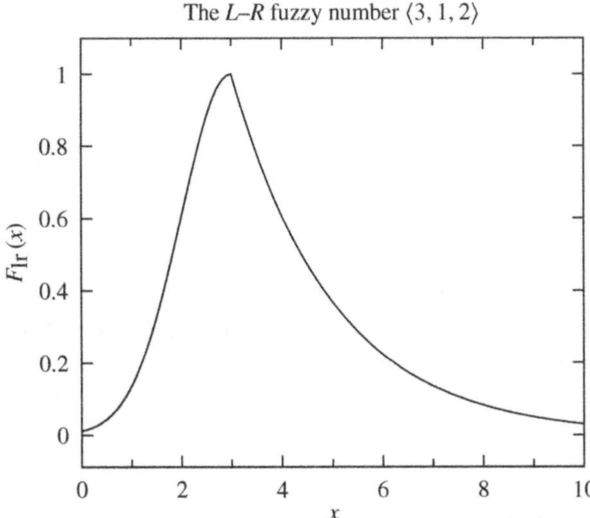

Figure 3.6 An L–R fuzzy number.

number. In general, the membership function of an L–R fuzzy number has the following form:

$$A(x) = \begin{cases} L\left[\frac{\bar{x}-x}{\alpha}\right], & \text{if } x \leq \bar{x} \text{ and } \alpha > 0, \\ R\left[\frac{x-\bar{x}}{\beta}\right], & \text{if } x \geq \bar{x} \text{ and } \beta > 0. \end{cases}$$

Clearly, not all functions can be used in place of the L and R functions. In particular, these functions must have the following properties for all $0 \leq x \in \mathbb{R}$:

(i) $L(x) \in [0, 1]$ and $R(x) \in [0, 1]$ for all x;
(ii) $L(0) = R(0) = 1$;
(iii) $L(-x) = L(x)$ and $R(-x) = R(x)$; and
(iv) L and R are decreasing in $[0, +\infty)$.

For example, in the case of the fuzzy number depicted in Figure 3.6, we have used:

$$L(x) = \exp\left[-\frac{x^2}{2}\right] \quad \text{and} \quad R(x) = e^{-x}.$$

3.1.7 Generalized Fuzzy Numbers

A generalized fuzzy number[2] $\tilde{A} = \text{gfn}(a, b, c, d; w)$ is a fuzzy subset of $\mathbb{R}$ that satisfies the following conditions:

(i) $\tilde{A}$ is a continuous mapping from $\mathbb{R}$ to the closed interval $[0, 1]$;
(ii) $\tilde{A}(x) = 0$ when $-\infty < x \leq a$;
(iii) $\tilde{A}$ is strictly increasing in $[a, b]$;
(iv) $\tilde{A} = w$ when $b \leq x \leq c$;
(v) $\tilde{A}$ is strictly decreasing in $[c, d]$;
(vi) $\tilde{A}(x) = 0$ when $d \leq x < \infty$.

Here w is supposed to be the degree of confidence of some expert's opinion. If $w = 1$, then the generalized fuzzy number $\tilde{A}$ is called a normal trapezoidal fuzzy number. If $a = b$ and $c = d$, then $\tilde{A}$ is called a crisp interval. If $b = c$, then $\tilde{A}$ is called a generalized triangular fuzzy number. If $a = b = c = d$ and $w = 1$, then $\tilde{A}$ is called a real number.

3.2 Arithmetic of Fuzzy Numbers

Given two fuzzy numbers A and B, does it make sense to compute their sum, their difference, their product, and their quotient? The answer to this question is that

2 We use the description presented in [71], since the original paper that introduced generalized fuzzy numbers was not available to us.

all four *arithmetical* operations have been extended so to make the operations $A + B$, $A - B$, $A \cdot B$, and $A \div B$ meaningful when A and B are fuzzy numbers. In particular, there are two methods to compute the operation $A \bigstar B$: one method is defined using operations on intervals, while the other method is using the extension principle.

3.2.1 Interval Arithmetic

Before we can proceed to the presentation of the methods of doing fuzzy arithmetic, we first have to learn some of the basics of interval arithmetic. In general, if $I = [a, b]$ and $J = [c, d]$ are two closed intervals and $\bigstar$ denotes any of the four arithmetic operations, then

$$I \bigstar J = \{x \bigstar y | x \in I \text{ and } y \in J\}. \tag{3.2}$$

For example, if $I = [1, 4]$ and $J = [2, 3]$ and $\bigstar$ denotes addition, then

$$[1, 4] + [2, 3] = \{x \bigstar y | x \in [1, 4] \text{ and } y \in [2, 3]\},$$

which equals [3, 7], because the end points are elements of the intervals and from this it follows that $3 = 1 + 2$ and $7 = 4 + 3$.

Using Eq. (3.2), the four arithmetic operations on closed intervals are defined as follows:

$$\begin{aligned}
[a, b] + [d, e] &= [a + d, b + e], \\
[a, b] - [d, e] &= [a - e, b - d], \\
[a, b] \cdot [d, e] &= [\min(ad, ae, bd, be), \max(ad, ae, bd, be)],
\end{aligned}$$

and, provided that $0 \notin [d, e]$,

$$\begin{aligned}
[a, b]/[d, e] &= [a, b] \cdot [1/e, 1/d] \\
&= [\min(a/d, a/e, b/d, b/e), \max(a/d, a/e, b/d, b/e)].
\end{aligned}$$

Any real number r can be considered as a degenerated interval $[r, r]$. Thus, if $\mathbf{1} = [1, 1]$, $\mathbf{0} = [0, 0]$, and $I = [a, b]$, then $I + \mathbf{0} = I$ because $[a, b] + [0, 0] = [a + 0, b + 0] = [a, b]$, and $I \cdot \mathbf{1} = I$ because $[a, b] \cdot [1, 1] = [a \cdot 1, b \cdot 1] = [a, b]$.

3.2.2 Interval Arithmetic and α-Cuts

All α-cuts of a fuzzy number are closed and bounded intervals. Assume that A and B are two fuzzy numbers and $*$ is one of the four arithmetic operations. Then, the fuzzy set $A * B$ is defined using its α-cut ${}^{\alpha}(A * B)$ as

$${}^{\alpha}(A * B) = {}^{\alpha}A * {}^{\alpha}B$$

for all $\alpha \in (0, 1]$. In case we want to divide two fuzzy numbers, it is necessary to ensure that $0 \notin {}^{\alpha}A$ for all $\alpha \in (0, 1]$. In general, it would be useful to be able to use all α-cuts. For instance, if ${}^{\alpha}A = [a_1(\alpha), a_2(\alpha)]$ and ${}^{\alpha}B = [b_1(\alpha), b_2(\alpha)]$, then

$$ {}^{\alpha}(A + B) = [a_1(\alpha) + b_1(\alpha), a_2(\alpha) + b_2(\alpha)]. $$

Of course, here we have shown how to compute specific α-cuts and not how to compute new fuzzy numbers. However, we know from Theorem 2.3.2 that the union of all α-cuts of some fuzzy set makes up the set itself. Thus, knowing all α-cuts of the sum of two fuzzy numbers means that we can easily compute the new fuzzy number.

Example 3.2.1 When one deals with triangular fuzzy numbers, it is particularly easy to find the functions that are used to compute the α-cuts. In particular, one has to compute the slope of two line segments using the formula $m = (y_1 - y_2)/(x_1 - x_2)$ and then use it to compute the equation of each line segment from the general equation $y - y_1 = m(x - x_1)$. Next, we solve this equation for x, since the α-cuts correspond to the values of y, and the result is the function we are looking for. Let us apply this idea to compute the α-cuts of the fuzzy number that is depicted in Figure 3.1, which we will call A. The slope of the left line segment is $m_l = (0 - 1)/(0 - 3) = 1/3$ and the slope of the right line segment is $m_r = (1 - 0)/(3 - 5) = -1/2$. Next, we write the equations and solve for x.

$$ y - 0 = \frac{1}{3}(x - 0) \iff x = 3y \quad \text{and} \quad y - 1 = -\frac{1}{2}(x - 3) \iff x = 5 - 2y. $$

From these we get the α-cuts: ${}^{\alpha}A = [3\alpha, 5 - 2\alpha]$. As an exercise, compute the operation $A^2 + 2A$ and then draw the resulting fuzzy number.

3.2.3 Fuzzy Arithmetic and the Extension Principle

The four arithmetic operations between fuzzy numbers can also be defined using the extension principle. The idea is that operations on real numbers are extended into operations on fuzzy real numbers. Assume that A and B are two fuzzy numbers. Then, we can define the four arithmetic operations between A and B for all $z \in \mathbb{R}$ as follows:

$$ (A + B)(z) = \bigvee_{z=x+y} \min[A(x), B(y)], $$

$$ (A - B)(z) = \bigvee_{z=x-y} \min[A(x), B(y)], $$

$$ (A \cdot B)(z) = \bigvee_{z=x \cdot y} \min[A(x), B(y)], $$

$$ (A/B)(z) = \bigvee_{z=x/y} \min[A(x), B(y)]. $$

Example 3.2.2 Assume that $A = 1/2 + 0.5/3$ and $B = 0.5/3 + 1/4$ are two *discrete* fuzzy numbers and that we want to compute $A + B$.[3] Then, z takes the following values:

z = 5. This value follows from $x + y = 2 + 3$. We have that $\min[A(2), B(3)] = \min[1, 0.5] = 0.5$. Thus, $(A + B)(5) = \bigvee_{5=2+3}[0.5] = 0.5$.
z = 6. This value follows from $x + y = 3 + 3$ or $x + y = 2 + 4$. We have that $\min[A(3), B(3)] = \min[0.5, 0.5] = 0.5$ and $\min[A(2), B(4)] = \min[1, 1] = 1$. Thus $(A + B)(6) = \bigvee_{\substack{6=3+3\\6=2+4}} [0.5, 1] = 1$.
z = 7. This value follows from $x + y = 3 + 4$. We have that $\min[A(3), B(4)] = \min[0.5, 1] = 0.5$. Thus, $(A + B)(7) = \bigvee_{7=3+4}[0.5] = 0.5$.

For all other values of z, the membership degrees are zero. Therefore, the sum of A and B is the fuzzy number $0.5/5 + 1/6 + 0.5/7$. As an exercise, compute $A - B$ and $A \cdot B$.

The technique described for discrete fuzzy numbers can be extended to nondiscrete fuzzy numbers, but it is more difficult to proceed.

3.2.4 Fuzzy Arithmetic of Triangular Fuzzy Numbers

For certain kinds of fuzzy numbers, there are special methods that can be used to compute any of the four arithmetic operations easily. This is true for triangular fuzzy numbers. Given two triangular fuzzy numbers $A = \mathrm{tfn}(\overline{x_1}, e^1_l, e^1_r)$ and $B = \mathrm{tfn}(\overline{x_2}, e^2_l, e^2_r)$, then the four arithmetic operations are defined as follows [70]:

Addition. $A + B = \mathrm{tfn}(\overline{x_1} + \overline{x_2}, e^1_l + e^2_l, e^1_r + e^2_r)$;
Subtraction. $A - B = \mathrm{tfn}(\overline{x_1} - \overline{x_2}, e^1_l - e^2_r, e^1_r - e^2_l)$;
Multiplication. $A \cdot B = \mathrm{tfn}(\overline{x_1} \cdot \overline{x_2}, \min(e^1_l e^2_l, e^1_l e^2_r, e^1_r e^2_l, e^1_r e^2_r), \max(e^1_l e^2_l, e^1_l e^2_r, e^1_r e^2_l, e^1_r e^2_r))$;
Division. $A/B = \mathrm{tfn}(\overline{x_1}/\overline{x_2}, \min(e^1_l/e^2_l, e^1_l/e^2_r, e^1_r/e^2_l, e^1_r/e^2_r), \max(e^1_l/e^2_l, e^1_l/e^2_r, e^1_r/e^2_l, e^1_r/e^2_r))$.

3.2.5 Fuzzy Arithmetic of Generalized Fuzzy Numbers

Assume that $\tilde{A}_1 = \mathrm{gfn}(a_1, b_1, c_1, d_1; w_1)$ and $\tilde{A}_2 = \mathrm{gfn}(a_2, b_2, c_2, d_2; w_2)$ are two generalized fuzzy numbers. Then, their addition is defined as follows:

$$\begin{aligned}\tilde{A}_1 + \tilde{A}_2 &= \mathrm{gfn}(a_1, b_1, c_1, d_1; w_1) + \mathrm{gfn}(a_2, b_2, c_2, d_2; w_2)\\ &= \mathrm{gfn}(a_1 + a_2, b_1 + b_2, c_1 + c_2, d_1 + d_2; \min(w_1, w_2)).\end{aligned}$$

3 For reasons of brevity, we just omit to specify the infinite numbers that have membership degree equal to zero.

Also, $\tilde{A}_1 - \tilde{A}_2$ is defined as follows:

$$\begin{aligned}\tilde{A}_1 - \tilde{A}_2 &= \text{gfn}(a_1, b_1, c_1, d_1; w_1) - \text{gfn}(a_2, b_2, c_2, d_2; w_2)\\ &= \text{gfn}(a_1 - d_2, b_1 - c_2, c_1 - b_2, d_1 - a_2; \min(w_1, w_2)).\end{aligned}$$

The multiplication $\tilde{A}_1 \times \tilde{A}_2$ is equal to $\text{gfn}(a, b, c, d, \min(w_1, w_2))$, where

$$\begin{aligned}a &= \min(a_1 \cdot a_2, a_1 \cdot d_2, d_1 \cdot a_2, d_1 \cdot d_2),\\ b &= \min(b_1 \cdot b_2, b_1 \cdot c_2, c_1 \cdot b_2, c_1 \cdot c_2),\\ c &= \max(b_1 \cdot b_2, b_1 \cdot c_2, c_1 \cdot b_2, c_1 \cdot c_2),\\ d &= \max(a_1 \cdot a_2, a_1 \cdot d_2, d_1 \cdot a_2, d_1 \cdot d_2).\end{aligned}$$

The inverse of the fuzzy number $\tilde{A}_2 = \text{gfn}(a_2, b_2, c_2, d_2; w_2)$ is

$$\frac{1}{\tilde{A}_2} = \text{gfn}\left(\frac{1}{d_2}, \frac{1}{c_2}, \frac{1}{b_2}, \frac{1}{a_2}; w_2\right),$$

where a_2, b_2, c_2, and d_2 are all nonzero positive numbers or nonzero negative numbers. If $a_1, b_1, c_1, d_1, a_2, b_2, c_2$, and d_2 are all nonzero positive real numbers, then

$$\begin{aligned}\tilde{A}_1 \div \tilde{A}_2 &= \text{gfn}(a_1, b_1, c_1, d_1; w_1) \div \text{gfn}(a_2, b_2, c_2, d_2; w_2)\\ &= \left(\frac{a_1}{d_2}, \frac{b_1}{c_2}, \frac{c_1}{b_2}, \frac{d_1}{a_2}, \min(w_1, w_2)\right).\end{aligned}$$

It was demonstrated [85] that these operations are problematic (e.g. addition does not yield the exact value). Thus, a more general description of generalized fuzzy numbers was proposed. In particular, the number $\tilde{A}_1 = \text{gfn}(a_1, b_1, c_1, d_1; w_1)$ is written as follows:

$$\tilde{A}_1(x) = \begin{cases} w_1(x - a_1)/(b_1 - a_1), & \text{when } a_1 \leq x \leq b_1,\\ w_1, & \text{when } b_1 \leq x \leq c_1,\\ w_1(x - d_1)/(c_1 - d_1), & \text{when } c_1 \leq x \leq d_1,\\ 0, & \text{otherwise,}\end{cases}$$

where $w_1 \in [0, 1]$ is the degree of confidence with respect to a decision-maker's opinion. The various arithmetic operations are defined as follows:

Addition. $A_1 + A_2 = \text{gfn}(a_3, b_3, c_3, d_3; w)$, where

$$\begin{aligned}w &= \min(w_1, w_2),\\ a_3 &= a_1 + a_2,\\ b_3 &= a_2 + c_1 + w(b_2 - a_2)/w_2,\\ c_3 &= d_2 + c_1 - w(d_2 - c_2)/w_2,\\ d_3 &= d_1 + d_2.\end{aligned}$$

Subtraction. $A_1 - A_2 = \text{gfn}(a_3, b_3, c_3, d_3; w)$, where

$$\begin{aligned}
w &= \min(w_1, w_2),\\
a_3 &= a_1 - d_2,\\
b_3 &= b_1 - d_2 + w(d_2 - c_2)/w_2,\\
c_3 &= c_1 + a_2 - w(b_2 - a_2)/w_2,\\
d_3 &= d_1 - a_2.
\end{aligned}$$

Multiplication. $A_1 \cdot A_2 = \text{gfn}(a_3, b_3, c_3, d_3; w)$, where

$$\begin{aligned}
w &= \min(w_1, w_2),\\
a_3 &= a_1 \cdot a_2,\\
b_3 &= w \cdot (b_1 \cdot b_2 - b_1 \cdot a_2)/w_2 + b_1 \cdot a_2,\\
c_3 &= w \cdot (c_1 \cdot c_2 - c_1 \cdot b_4)/w_2 + c_1 \cdot d_2,\\
d_3 &= d_1 \cdot d_2.
\end{aligned}$$

Division. $A_1 \div A_2 = \text{gfn}(a_3, b_3, c_3, d_3; w)$, where

$$\begin{aligned}
w &= \min(w_1, w_2),\\
a_3 &= a_1/d_2,\\
b_3 &= w \cdot (b_1/c_2 - b_1/d_2)/w_2 + b_1/d_2,\\
c_3 &= w \cdot (c_1/b_2 - c_1/a_2)/w_2 + c_1/a_2,\\
d_3 &= d_1/a_2,
\end{aligned}$$

and $a_1, b_1, c_1, d_1, a_2, b_2, c_2$, and d_2 are positive real numbers.

3.2.6 Comparing Fuzzy Numbers

Unfortunately, we cannot directly compare two fuzzy numbers A and B. However, we can compare them indirectly by using the operations MIN and MAX, that are obtained from the known min and max operations by using the extension principle as follows:

$$\text{MIN}(A, B)(z) = \bigvee_{z=\min(x,y)} \min[A(x), B(y)],$$

$$\text{MAX}(A, B)(z) = \bigvee_{z=\max(x,y)} \min[A(x), B(y)],$$

for $x, y, z \in \mathbb{R}$. One can use these definitions to compute the minimum and maximum of any two fuzzy numbers, but nevertheless the computations can be easy only in certain cases. Thus, we need a better mechanism to compute these two operations. Chih-Hui Chiu and Wen-June Wang [74] proved two theorems that make the computation of these two operations easier. The first theorem can be used to compute MIN.

Theorem 3.2.1 *Assume that A and B are fuzzy numbers with continuous membership functions such that $A \cap B \neq \emptyset$ and x_m is a real number such that $(A \cap B)(x_m) \geq (A \cap B)(x)$ for all $x \in \mathbb{R}$, $A(x_m) = B(x_m)$, and x_m is between two mean values of A and B (if x_m is not unique, any value of x_m can be used). Then,* MIN *can be defined as follows:*

$$\text{MIN}(A, B)(z) = \begin{cases} (A \cup B)(z), & \textit{if } z < x_m, \\ (A \cap B)(z), & \textit{if } z \geq x_m. \end{cases} \tag{3.3}$$

The second theorem can be used to compute MAX.

Theorem 3.2.2 *Assume that A, B, and x_m are as in Theorem 3.2.1. Then,* MAX *can be defined as follows:*

$$\text{MAX}(A, B)(z) = \begin{cases} (A \cap B)(z), & \textit{if } z < x_m, \\ (A \cup B)(z), & \textit{if } z \geq x_m. \end{cases} \tag{3.4}$$

Example 3.2.3 Consider the following fuzzy numbers:

$$A(x) = \begin{cases} 0, & \text{if } x < -2 \text{ or } x > 4, \\ (x+2)/3, & \text{if} -2 \leq x \leq 1, \\ (4-x)/3, & \text{if } 1 < x \leq 4, \end{cases} \qquad \text{and}$$

$$B(y) = \begin{cases} 0, & \text{if } y < 1 \text{ or } y > 3, \\ y-1, & \text{if } 1 \leq y \leq 2, \\ 3-y, & \text{if } 2 < y \leq 3. \end{cases}$$

These numbers are depicted in Figure 3.7. Assume that $x_m = 7/4 = 1.75$. Obviously, $A(1.75) = B(1.75)$. Also, note that $A(2.5) = B(2.5)$. Let us compute the minimum of the two numbers. First, we have to examine the possible values when $z < 7/4$. Clearly,

$$\text{MIN}(A, B)(z) = (A \cup B)(z) = 0$$

for all $z < -2$. Similarly, $\text{MIN}(A, B)(z) = (A \cap B)(z) = 0$ for all $z > 3$. When $-2 \leq z \leq 1$, then $\text{MIN}(A, B)(z) = (A \cup B)(z) = (z+2)/3$. Next, we need to know what happens when $1 < z < 1.75$. In this case, $\text{MIN}(A, B)(z) = (A \cap B)(z) = (4-z)/3$.

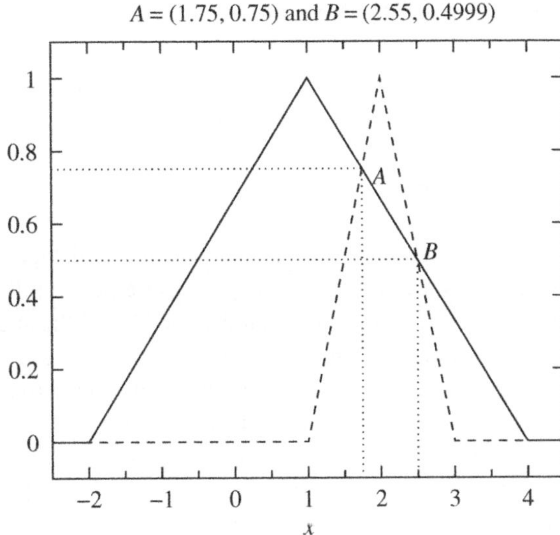

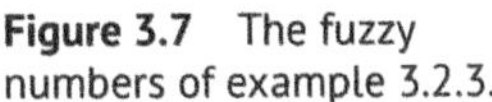
Figure 3.7 The fuzzy numbers of example 3.2.3.

When $1.75 \geq z \leq 2.5$, then $\text{MIN}(A, B)(z) = (A \cup B)(z) = (4 - z)/3$ and when $2.5 < z \leq 3$, then $\text{MIN}(A, B)(z) = (A \cup B)(z) = 3 - z$. In summary,

$$\text{MIN}(A, B)(z) = \begin{cases} 0, & \text{if } z < -2 \text{ or } z > 3, \\ (z + 2)/3, & \text{if} - 2 \leq z \leq 1, \\ (4 - z)/3, & \text{if } 1 < z \leq 2.5, \\ 3 - z, & \text{if } 2.5 < z \leq 3. \end{cases}$$

Dug Hun Hong and Kyung Tae Kim [163] found another easier way to compute the minimum and maximum of many fuzzy numbers at the same time. Their result is based on a theorem that uses the following notation:

$$[A]^\alpha = \begin{cases} \{x|A(x) \geq \alpha\}, & \text{if } 0 < \alpha \leq 1, \\ \text{supp}\, A, & \text{if } \alpha = 0. \end{cases}$$

Theorem 3.2.3 *Assume that A_i, $i = 1, \dots, n$ are fuzzy numbers such that $[A_i]^\alpha = [a_i^1(\alpha), a_i^2(\alpha)]$. Then, the operations* MIN *and* MAX *can be computed from the following formulas:*

$$[\text{MIN}(A_1, \dots, A_n)]^\alpha = [\min_{1 \leq i \leq n} a_i^1(\alpha), \min_{1 \leq i \leq n} a_i^2(\alpha)],$$

$$[\text{MAX}(A_1, \dots, A_n)]^\alpha = [\max_{1 \leq i \leq n} a_i^1(\alpha), \max_{1 \leq i \leq n} a_i^2(\alpha)].$$

3.3 Linguistic Variables

The concept of a *linguistic variable*[4] was introduced by Zadeh [311–313]. According to Zadeh, a linguistic variable is a special kind of variable whose values are not numbers but words or, more generally, sentences in a natural language (e.g. English or Greek). For instance, the temperature of a room is a linguistic variable whose *linguistic values* include the terms "freezing," "very cold," "cold," "cool," "mild," "moderate," "warm," "very warm," and "hot." Other examples of linguistic variables are the age of people, where possible linguistic values include the terms "young," "old," and "middle-aged," and the speed of a car, where possible linguistic values include the terms "fast," "slow," and "stationary."

More formally, a linguistic variable is characterized by a quintuple $(V, T(V), U, G, M)$, where V is the name of the variable, $T(V)$ is the set of terms of V, that is, a set of linguistic values of V, which are fuzzy sets on the universe X, G is a syntactic rule for generating the names of values of V, and M is a semantic rule for associating each value with its meaning, that is, the membership function that characterizes the fuzzy set. Figure 3.8 depicts the linguistic variable *temperature*.

Any word like the word "very" that modifies a linguistic value like "cold" is called a *linguistic hedge*. For example, the words "quite," "very very," "not so," etc., all count as linguistic hedges. A linguistic hedge can either intensify or lessen the meaning of a linguistic value. For example, if T_{cold} is the fuzzy set associated

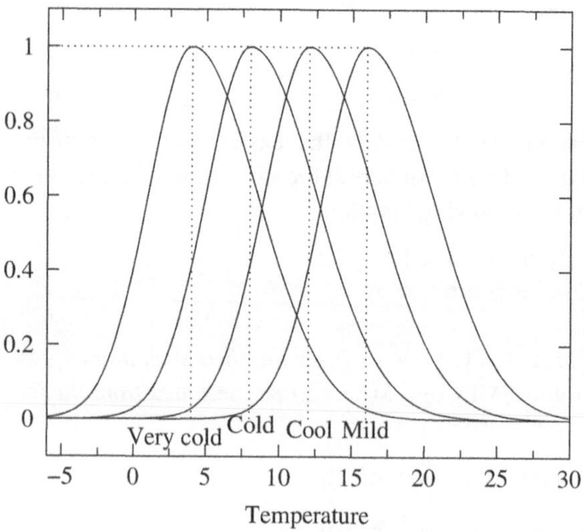

Figure 3.8 Room temperature as a linguistic variable quantified by some linguistic values.

4 This term has also been used in linguistics, where a linguistic variable is a linguistic item that has identifiable variants. For example, the fishing is sometimes pronounced as "*fishin*." The final sound of this word is the linguistic variable (see Ref. [295] for a full discussion).

with the linguistic value "cold," then $T^{1/2}_{\text{cold}}$ could be the fuzzy set associated with the linguistic value "very cold." Similarly, if T_{hot} is the fuzzy set associated with the linguistic value "hot," then T^2_{hot} could be the fuzzy set associated with the linguistic value "very hot." The linguistic hedges and the atomic linguistic variable set (i.e. the "basic" words that characterize a linguistic variable) are put together to create the linguistic values. And this is exactly a possible syntactic rule for generating linguistic values.

3.4 Fuzzy Equations[5]

A fuzzy *equation* is one where both the unknown variables and the coefficients are fuzzy numbers. For example, the equation

$$A \cdot X + B = C, \tag{3.5}$$

where A, B, and C are triangular fuzzy numbers, is the simplest possible fuzzy equation. It is rather tempting to try to solve this equation using techniques we use to solve ordinary algebraic equations. However, this is not possible because $B - B \neq 0$ and so $A \cdot X \neq C - B$! For example, if $B = \text{tfn}(2, 1, 3)$, then $B - B = \text{tfn}(0, -2, 2) \neq 0$, as can be easily verified using the method described in Section 3.2.4.

3.4.1 Solving the Fuzzy Equation $A \cdot X + B = C$

In what follows, we present three methods to solve Eq. (3.5).

3.4.1.1 The Classical Method

This method can be used to compute the solution to an equation, when a solution exists. Assume that ${}^{\alpha}A = [a_1(\alpha), a_2(\alpha)]$, ${}^{\alpha}B = [b_1(\alpha), b_2(\alpha)]$, ${}^{\alpha}C = [c_1(\alpha), c_2(\alpha)]$, and ${}^{\alpha}X_C = [x_1(\alpha), x_2(\alpha)]$, $\alpha \in [0, 1]$. Then, we replace the variables in Eq. (3.5) with the α-cuts:

$$[a_1(\alpha), a_2(\alpha)] \cdot [x_1(\alpha), x_2(\alpha)] + [b_1(\alpha), b_2(\alpha)] = [c_1(\alpha), c_2(\alpha)]. \tag{3.6}$$

Next, we need to solve this equation for $x_1(\alpha)$ and $x_2(\alpha)$ using interval arithmetic (see Section 3.2.1). When the intervals $[x_1(\alpha), x_2(\alpha)]$ define the α-cuts of a fuzzy number, then we get the solution to Eq. (3.5). Note that $x_1(\alpha)$ and $x_2(\alpha)$ specify α-cuts of a fuzzy number when

(i) $x_1(\alpha)$ and $x_2(\alpha)$ are continuous;

5 The exposition that follows is based on [44, 48].

(ii) $x_1(\alpha)$ is monotonically increasing for $0 \leq \alpha \leq 1$;
(iii) $x_2(\alpha)$ is monotonically decreasing for $0 \leq \alpha \leq 1$; and
(iv) $x_1(\alpha) \leq x_2(\alpha)$.

There is no guarantee that this procedure will yield a solution to Eq. (3.5), but if it does produce a solution, then this will satisfy the initial equation.

Example 3.4.1 Suppose that $A = \text{tfn}(2,1,1)$, $B = \text{tfn}(-2,1,1)$, and $C = \text{tfn}(-4,1,1)$. Then, ${}^\alpha A = [1+\alpha, 3-\alpha]$, ${}^\alpha B = [-3+\alpha, -1-\alpha]$, and ${}^\alpha C = [-5+\alpha, -3-\alpha]$. After performing interval arithmetic, Eq. (3.5) becomes

$$[a_1(\alpha)x_1(\alpha) + b_1(\alpha), a_2(\alpha)x_2(\alpha) + b_2(\alpha)] = [c_1(\alpha), c_2(\alpha)],$$

or

$$x_1(\alpha) = -\frac{2}{1+\alpha} \quad \text{and} \quad x_2(\alpha) = -\frac{2}{3-\alpha},$$

after substituting for the known quantities and solving for the unknown. Function $x_1(\alpha)$ is increasing (its derivative is positive) and function $x_2(\alpha)$ is decreasing (its derivative is negative), while $x_1(1) = x_2(1) = -1$. To find to which fuzzy numbers these α-cuts correspond, we simply solve the equations

$$x_1 = -\frac{2}{1+y_1} \quad \text{and} \quad x_2 = -\frac{2}{3-y_2}$$

for y_1 and y_2, that is,

$$y_1 = -\frac{2}{x_1} - 1 \quad \text{and} \quad y_2 = \frac{2}{x_2} + 3.$$

Next, we find for which values of x_1 and x_2, y_1 and y_2 are equal to zero, respectively. We see that for $x_1 = -2$, $y_1 = 0$ and for $x_2 = -2/3$, $y_2 = 0$. Thus, the fuzzy number that solves equation $A \cdot X + B = C$ for the fuzzy numbers $A = \text{tfn}(2,1,1)$, $B = \text{tfn}(-2,1,1)$, and $C = \text{tfn}(-4,1,1)$ is

$$X_C(x) = \begin{cases} 0, & \text{if } x < -2, \\ -\frac{2}{x} - 1, & \text{if } -2 \leq x \leq -1, \\ \frac{2}{x} + 3, & \text{if } -1 \leq x \leq -\frac{2}{3}, \\ 0, & \text{if } -\frac{2}{3} < x. \end{cases}$$

Although we managed to solve this equation using this method, most equations cannot be solved using this technique. Fortunately, there are two more methods which can produce approximate solutions to Eq. (3.5).

3.4.1.2 The Extension Principle Method

As the name of this method suggests, this method uses the extension principle to solve equation $A \cdot X + B = C$. The method is based on a procedure that is used to extend any crisp function $h : [a, b] \to \mathbb{R}$ to a *fuzzy function* $H(X) = Z$. According to this procedure, the crisp function h is extended to its fuzzy counterpart H as follows:

$$Z(z) = \bigvee_{\substack{z=h(x)\\ a\leq x\leq b}} X(x).$$

This equation defines the membership function of Z for any triangular fuzzy number X in $[a, b]$. Also, if h is continuous, then there is a way to compute the α-cuts of Z. Assume that ${}^{\alpha}Z[z_1(\alpha), z_2(\alpha]$. Then,

$$z_1(\alpha) = \min\{h(x)|x \in {}^{\alpha}X\}, \tag{3.7}$$

$$z_2(\alpha) = \max\{h(x)|x \in {}^{\alpha}X\}. \tag{3.8}$$

If we have a crisp function with two independent variables, then we assume that $z = h(x, y)$, where $x \in [a_1, b_1]$ and $y \in [a_2, b_2]$. Then, we extend h to $H(X, Y) = Z$ as follows:

$$Z(z) = \bigvee_{z=h(x,y)} \min[X(x), Y(y)].$$

Provided h is continuous, we can compute the α-cuts with the following equations:

$$z_1(\alpha) = \min\{h(x, y)|x \in {}^{\alpha}X \text{ and } y \in {}^{\alpha}Y\}, \tag{3.9}$$

$$z_2(\alpha) = \max\{h(x, y)|x \in {}^{\alpha}X \text{ and } y \in {}^{\alpha}Y\}. \tag{3.10}$$

As an exercise, explain how one can fuzzify a crisp function with four independent variables.

The second method by which we try to solve Eq. (3.5), assumes that the crisp solution is a function of three independent variables. Therefore, all that we have to do is to fuzzify the "function" $(c - b)/a$ using the function fuzzification procedure we just described. Clearly, the solution we are looking for is the fuzzy number $(C - B)/A$, where zero does not belong to the support of A. The fuzzy number $(C - B)/A$ can be computed using the following equation:

$$X_E(x) = \bigvee_{x=(c-b)/a} \min[A(a), B(b), C(c)]. \tag{3.11}$$

Since $(c - b)/a$ is continuous, we can compute the α-cuts ${}^{\alpha}X = [x_1(\alpha), x_2(\alpha)]$, where

$$x_1(\alpha) = \min\{(c - b)/a|a \in {}^{\alpha}A, b \in {}^{\alpha}B, \text{ and } c \in {}^{\alpha}C\}, \tag{3.12}$$

$$x_2(\alpha) = \max\{(c - b)/a|a \in {}^{\alpha}A, b \in {}^{\alpha}B, \text{ and } c \in {}^{\alpha}C\}. \tag{3.13}$$

Clearly, the α-cuts will be ${}^{\alpha}X_E[x_1(\alpha), x_2(\alpha)]$. The solution will be a triangular fuzzy number. However, there is no guarantee that the computed solution will satisfy the initial equation. If X_C does not exist, then the solution of the equation is X_E!

3.4.1.3 The α-Cuts Method

A third method to solve equation $A \cdot X + B = C$ is to use α-cuts and interval arithmetic. In particular, the solution of the equation is assumed to be

$$^{\alpha}X_I = \frac{^{\alpha}C - {}^{\alpha}B}{^{\alpha}A}.$$

This equation can be simplified into

$$[x_1(\alpha), x_2(\alpha)] = \frac{[c_1(\alpha), c_2(\alpha)] - [b_1(\alpha), b_2(\alpha)]}{[a_1(\alpha), a_2(\alpha)]}, \tag{3.14}$$

or

$$x_1(\alpha) = \frac{c_1(\alpha) - b_2(\alpha)}{a_2(\alpha)} \quad \text{and} \quad x_2(\alpha) = \frac{c_2(\alpha) - b_1(\alpha)}{a_1(\alpha)},$$

if $a_1(\alpha) > 0$ for all α. This method always yields the solution

$$X_I = \bigcup_{0 \le \alpha \le 1} [x_1(\alpha), x_2(\alpha)],$$

but, again, there is no guarantee that the computed solution will satisfy the initial equation. If X_C does not exist and it is difficult to get X_E, then we can use X_I as an approximate solution.

Example 3.4.2 Suppose that $A = \text{tfn}(9, 1, 1)$, $B = \text{tfn}(-2, 1, 1)$, and $C = \text{tfn}(5, 2, 2)$. Then, $^{\alpha}A = [8 + \alpha, 10 - \alpha]$, $^{\alpha}B = [-3 + \alpha, -1 - \alpha]$, and $^{\alpha}C = [3 + 2\alpha, 7 - 2\alpha]$. Unfortunately, as the reader can verify, X_C does not exist, so we first need to try to find X_E. In order to find X_E, we need to evaluate Eqs. (3.12) and (3.13). After finding the partial derivatives of $(c - b)/a$, we conclude that $(c - b)/a$ is increasing in c and decreasing in both b and a. Thus,

$$x_{E1}(\alpha) = \frac{c_1(\alpha) - b_2(\alpha)}{a_2(\alpha)} = \frac{4 + 2\alpha}{3 - \alpha},$$

$$x_{E2}(\alpha) = \frac{c_2(\alpha) - b_1(\alpha)}{a_1(\alpha)} = \frac{8 - 2\alpha}{1 + \alpha}.$$

In order to compute X_I, we note that all intervals in Eq. (3.14) are positive. This simply means that $X_E = X_I$.

3.4.2 Solving the Fuzzy Equation $A \cdot X^2 + B \cdot X + C = D$

The fuzzy quadratic equation has the following form:

$$A \cdot X^2 + B \cdot X + C = D. \tag{3.15}$$

For triangular fuzzy numbers A, B, C, and D, the solution of this equation will be also a triangular fuzzy number. The fuzzy quadratic equation does not have the form $A \cdot X^2 + B \cdot X + C = 0$ just because the left-hand side of this equation can never be exactly equal to zero. If we allow complex solutions, then the crisp equation $ax^2 + bx + c = d$ has two solutions, which implies that the fuzzy equation might have solutions that are fuzzy complex numbers, nonetheless we are not interested in fuzzy complex solutions. As in the case of the equation $A \cdot X + B = C$, there are three methods to solve Eq. (3.15).

3.4.2.1 The Classical Method

Assume that ${}^\alpha A = [a_1(\alpha), a_2(\alpha)]$, ${}^\alpha B = [b_1(\alpha), b_2(\alpha)]$, ${}^\alpha C = [c_1(\alpha), c_2(\alpha)]$, ${}^\alpha D = [d_1(\alpha), d_2(\alpha)]$, and ${}^\alpha X = [x_1(\alpha), x_2(\alpha)]$. Then, we substitute these α-cuts into Eq. (3.15) and solve for $x_1(\alpha)$ and $x_2(\alpha)$. In order to proceed, we need to know whether $A > 0$, $B > 0$, and $X > 0$. Suppose that all these numbers are positive. Then,

$$a_1(\alpha)(x_1(\alpha))^2 + b_1(\alpha) + c_1(\alpha) = d_1(\alpha) \quad \text{and}$$
$$a_2(\alpha)(x_2(\alpha))^2 + b_2(\alpha) + c_2(\alpha) = d_2(\alpha).$$

The solution X_C exists if the α-cuts $[x_1(\alpha), x_2(\alpha)]$, where

$$\begin{aligned} x_1(\alpha) &= \frac{-b_1(\alpha) + \sqrt{(b_1(\alpha))^2 - 4a_1(\alpha)(c_1(\alpha) - d_1(\alpha))}}{2a_1(\alpha)} \\ &= S_1(a_1(\alpha), b_1(\alpha), c_1(\alpha), d_1(\alpha)), \\ x_2(\alpha) &= \frac{-b_2(\alpha) - \sqrt{(b_2(\alpha))^2 - 4a_2(\alpha)(c_2(\alpha) - d_2(\alpha))}}{2a_2(\alpha)} \\ &= S_2(a_2(\alpha), b_2(\alpha), c_2(\alpha), d_2(\alpha)) \end{aligned}$$

are of a triangular fuzzy number. This means that $\partial x_1/\partial\alpha > 0$ and $\partial x_2/\partial\alpha < 0$, for $\alpha \in (0,1)$ and $x_1(1) = x_2(1)$. In addition, the solutions must be real numbers, so this means that

$$(b_i(\alpha))^2 - 4a_i(\alpha)(c_i(\alpha) - d_i(\alpha)) \geq 0, \ \ i = 1,2 \text{ and } \alpha \in [0,1].$$

Naturally, if $A < 0$ and $B < 0$ or $A < 0$ and $B > 0$, we may get different results, provided all conditions are met.

3.4.2.2 The Extension Principle Method

This solution fuzzifies the quantities S_1 and S_2 and the α-cut of X_E, when we are working with S_1, is $[x_1(\alpha), x_2(\alpha)]$, where

$$x_1(\alpha) = \min\{S_1(a,b,c,d) | a \in {}^\alpha A, b \in {}^\alpha B, c \in {}^\alpha C, \ \text{and} \ d \in {}^\alpha D\},$$
$$x_2(\alpha) = \max\{S_1(a,b,c,d) | a \in {}^\alpha A, b \in {}^\alpha B, c \in {}^\alpha C, \ \text{and} \ d \in {}^\alpha D\},$$

for $\alpha \in [0,1]$.

3.4.2.3 The α-Cuts Method

This method is employed when it is difficult to compute the min and max in the previous equations. The solution X_I is computed by substitution of A, B, C, and D into S_1 or S_2. Here, we work with S_1, and we assume that the α-cut of X_I is $[x_1(\alpha), x_2(\alpha)]$, where

$$x_1(\alpha) = \frac{-b_2(\alpha) + \sqrt{(b_1(\alpha))^2 - 4a_2(\alpha)(c_2(\alpha) - d_1(\alpha))}}{2a_2(\alpha)} \quad \text{and}$$

$$x_2(\alpha) = \frac{-b_1(\alpha) + \sqrt{(b_2(\alpha))^2 - 4a_1(\alpha)(c_1(\alpha) - d_2(\alpha))}}{2a_1(\alpha)},$$

$\alpha \in [0, 1]$.

3.5 Fuzzy Inequalities

A fuzzy inequality is an expression like $A \cdot X + B \leq C$ or like $A \cdot X + B < C$, where A, B, and C are triangular fuzzy numbers. However, the problem here is, what do the expressions $N \leq M$ and $N < M$ really mean? Unfortunately, there is no unique definition and this means that a possible solution will depend on how we choose to define these two relational operators.

A number of different definitions is presented in [44], but it seems there is no standard definition. Here is a simple definition:

$$[\![N \leq M]\!] = \bigvee_{x \leq y} \min[N(x), M(y)].$$

Based on this, we agree that $N < M$ if $[\![N \leq M]\!] = 1$, but $[\![M \leq N]\!] < \theta$, where θ is a fixed number, such that $\theta \in \mathbb{Q} \cap [0, 1]$. Let us say that $\theta = 0.75$. Then, $N < M$ if $[\![N \leq M]\!] = 1$ and $[\![M \leq N]\!] < 0.75$. We write $N \approx M$ when both $M < N$ and $N < M$ are not true. Moreover, $N \leq M$ means that $N < M$ or that $N \approx M$.

In order to solve the inequality $A \cdot X + B \leq C$, we first try to compute the number $E = A \cdot X + B$. Suppose we are going to use α-cuts and interval arithmetic to compute E. Then, the solution for X to $E \leq C$ or $E < C$ depends on the definition of "$\leq$." However, since there is no standard definition but only proposals, there is no reason to further discuss possible solutions.

3.6 Constructing Fuzzy Numbers

We have shown how to deal with fuzzy numbers, but we have said nothing about how one can actually construct them from real-world data. Chi-Bin Cheng [73] presented a relatively simple method that can be used to construct a triangular fuzzy number. In particular, he explained how one can construct a triangular fuzzy number $\text{tfn}(\overline{x}, e_l, e_r)$ from the grades given to Q, which can be an object, a performance, etc., by a group of experts. We can assume that each expert graded Q with a number in the range from 0 to G. Moreover, $g_1, \dots, g_n$ are the scores that n different experts gave to Q. In addition, we require that $g_i \neq g_j$ for at least one pair of grades g_i and g_j.

The first thing we would like to compute is the number *overlinex*. For this, we build the $n \times n$ matrix $D = [d_{ij}]$, where each $d_{ij} = |g_i - g_j|$. This matrix holds the distances between various g_i, and it is used to locate $\overline{x}$. The average of the relative distances, for each g_i, is given by $\overline{d}_i = \sum_{j=1}^{n} d_{ij}/(n-1)$. This average distance is used to measure the proximity of g_i to *overlinex*. Next, we want to determine the degree of importance of each g_i. So, we build an $n \times n$ pair-wise comparison matrix $P = [p_{ij}]$, where

$$p_{ij} = \frac{\overline{d}_j}{\overline{d}_i}.$$

Because P is obtained from a comparison of distances, it turns out that it is perfectly consistent. Assume that w_i is the true degree of importance of g_i. Then, because of the consistency of P,

$$p_{ij} = \frac{w_i}{w_j}.$$

Suppose that w is a column vector of w_i, where $1 \le i \le n$. Then,

$$Pw = nw,$$

which means that n is an eigenvalue of P and w is the corresponding eigenvector. It holds that

$$\sum_{i=1}^{n} w_i = 1,$$

and we conclude that

$$w_j = \frac{1}{\sum_{i=1}^{n} p_{ij}}, \quad j = 1, \dots, n.$$

From this we can finally compute $\overline{x}$:

$$\boxed{\overline{x} = \sum_{i=1}^{n} w_i g_i.}$$

Now we need to compute e_l and e_r.

Definition 3.6.1 The mean deviation of a triangular fuzzy number $A = \text{tfn}(\overline{x}, e_\text{l}, e_\text{r})$ is

$$\sigma = \frac{\int_a^b |x - \overline{x}| \cdot A(x)\, dx}{\int_a^b A(x)\, dx},$$

where $a = \overline{x} - e_\text{l}$ and $b = \overline{x} + e_\text{r}$.

This last equation can be written as follows:

$$\sigma = \frac{(\overline{x} - a)^2 + (b - \overline{x})^2}{3(b - a)}.$$

Also, let η be

$$\eta = \frac{\overline{x} - a}{b - \overline{x}}.$$

These last two equations can be solved to yield

$$\boxed{a = \overline{x} - \frac{3(1+\eta)\eta\sigma}{1+\eta^2} \quad \text{and} \quad b = \overline{x} + \frac{3(1+\eta)\sigma}{1+\eta^2}.}$$

Obviously, $e_\text{l} = \overline{x} - a$ and $e_\text{r} = b - \overline{x}$.

We use the average deviation that is calculated from the sample scores to approximate the value of σ:

$$\boxed{\sigma \approx \sum_{i=1}^{n} w_i |g_i - \overline{x}|.}$$

The quantity η can be computed approximately as follows. Assume that g^l is the weighted average of the scores that are less than $\overline{x}$ and g^r the weighted average of the scores that are greater than $\overline{x}$. Also, let

$$I = \{1, \ldots, n\}, \quad A = \{i | g_i < \overline{x} \text{ and } i \in I\}, \quad \text{and} \quad B = \{i | g_i > \overline{x} \text{ and } i \in I\}.$$

Next, we compute g^l and g^r:

$$g^l = \frac{\sum_{i \in A} w_i g_i}{\sum_{i \in A} w_i} \quad \text{and} \quad g^r = \frac{\sum_{i \in B} w_i g_i}{\sum_{i \in B} w_i}.$$

Finally, we can approximately compute η by

$$\boxed{\eta \approx \frac{\overline{x} - g^l}{g^r - \overline{x}}.}$$

Assume that $g_i = g_j = g^c, i \neq j$, for all i and j. Then, the method described cannot be used since this condition violates the assumptions of the method. However, the membership function of the corresponding fuzzy number can be constructed easily:

$$A(x) = \begin{cases} 1, & \text{if } x = g^c, \\ 0, & \text{otherwise.} \end{cases}$$

The scores given by the experts are between 0 and G, therefore, the support of a fuzzy number constructed from these scores cannot be outside this range. Thus, the triangular fuzzy number is defined as follows:

$$A(x) = \begin{cases} 0, & \text{if } x \leq \overline{x} - e_l, \\ 1 + (x - \overline{x})/e_l, & \text{if } \overline{x} - e_l < x < \overline{x} \text{ and } x \geq 0, \\ 1 - (x - \overline{x})/e_r, & \text{if } \overline{x} \leq x < \overline{x} + e_r \text{ and } x \leq G, \\ 0, & \text{if } x \geq \overline{x} + e_r. \end{cases}$$

3.7 Applications of Fuzzy Numbers

There are many nontrivial applications of fuzzy numbers and Michael Hanss's monograph [159] describes some very interesting applications. In this section, we present a few applications of fuzzy numbers so as to demonstrate their usefulness.

3.7.1 Simulation of the Human Glucose Metabolism

It is an undeniable fact that diabetes mellitus type I can seriously affect the quality of a patient's life. Since diabetes is the result of a problematic human glucose metabolism, it is of paramount importance to know the main characteristics of glucose metabolism. Naturally, we first need to develop a mathematical model of the metabolism and then use it in simulations so to check its usefulness. Michael Hanss and Oliver Nehls [160] examine such a model and find ways to improve the model by introducing fuzzy numbers. First, let us briefly present the model and then we can see how fuzzy numbers can be introduced.

Generally speaking, the human glucose metabolism model for patients with diabetes mellitus type I is divided into two parts: (i) the part that describes the inflow $I_{ex}(t)$ of insulin into blood as a result of a subcutaneous insulin injection, and (ii) the part that describes the inflow $G_{ex}(t)$ of glucose into blood as a result of food consumption. The second part is divided into two submodels: (i) the submodel that describes metabolisms in the stomach, and (ii) the submodel that describes the metabolism in the intestine. The outputs of these models are combined in a simplified model to predict the amount of in-blood glucose $G_b(t_k)$ at time t_k.

When a patient injects insulin, it appears in two modifications in the subcutaneous depot. These are described by a hemisphere with radial coordinate r: as dimer insulin with concentration $c_{\mathrm{d}}(r,t)$ and as hexamer insulin with concentration $c_{\mathrm{h}}(r,t)$. The uptake of insulin into the blood is only affected by dimer insulin. However, the injected external insulin is a solution of pure hexamer insulin. The following equations describe the model:

$$\frac{\partial c_{\mathrm{h}}(r,t)}{\partial t} = -P(c_{\mathrm{h}} - Qc_{\mathrm{d}}^3) + D\nabla^2 c_{\mathrm{h}},$$

$$\frac{\partial c_{\mathrm{d}}(r,t)}{\partial t} = P(c_{\mathrm{h}} - Qc_{\mathrm{d}}^3) + D\nabla^2 c_{\mathrm{d}} - Bc_{\mathrm{d}},$$

$$I_{\mathrm{ex}}(t) = 2\pi B \int_{r_{\min}}^{r_{\max}} r^2 c_{\mathrm{d}}\, dr,$$

where

$$\nabla^2 = \frac{\partial^2}{\partial r} + \frac{2}{r}\frac{\partial}{\partial r},$$

and the model parameters

$$D = 8.4 \times 10^{-5}\ \mathrm{cm^2/min}, \qquad P = 0.5\ \mathrm{min^{-1}},$$

$$Q = 9.3 \times 10^{-3}\ \mathrm{ml^2/mg^2}, \qquad r_{\min} = 1.5\ \mathrm{cm},$$

$$B = 7 \times 10^{-3}\ \mathrm{min^{-1}} \ldots 13 \times 10^{-3}\ \mathrm{min^{-1}}, \qquad r_{\max} = 6\ \mathrm{cm},$$

and the initial and boundary conditions

$$c_{\mathrm{h}}(r,0) = \begin{cases} 4.0\ \mathrm{mg/ml}, & \text{if } r = r_{\min} \\ 0, & \text{if } r > r_{\min} \end{cases}, \qquad \frac{\partial c_{\mathrm{h}}}{\partial r}(r_{\max}, t) = 0,$$

$$c_{\mathrm{d}}(r,0) = 0, \qquad \frac{\partial c_{\mathrm{d}}}{\partial r}(r_{\max}, t) = 0.$$

The model for the concentration $c_{\mathrm{c}}^S(t)$ of carbohydrates in the stomach of volume $V(t)$ is given by

$$\frac{d}{dt}(c_{\mathrm{c}}^S(t) \cdot V(t)) = -\alpha V c_{\mathrm{c}}^S,$$

$$\frac{dV(t)}{dt} = -\alpha V + q_{\mathrm{b}} = -q(t) + q_{\mathrm{b}},$$

with the initial conditions

$$c_{\mathrm{c}}^S(0) = \frac{m_{c_0}}{V(0)} \quad \text{and} \quad V(0) = V_{\text{empty}} + (1 + f_{\text{sec}})V_{\text{meal}}$$

and the parameters

$$\alpha = \frac{f_{\text{gas}} \ln 2}{V_{\text{meal}}(0.1797 - 0.167e^{-0.2389\kappa})},$$

$$\kappa = \frac{0.0167 m_{c_0} + 0.0167 m_{p_0} + 0.0377 m_{f_0}}{V_{\text{meal}}},$$

$$V_{\text{empty}} = 50 \text{ ml}, \qquad q_{\text{b}} = 0.4861 \text{ ml/min},$$
$$f_{\text{sec}} = 1.0, \qquad f_{\text{gas}} = 0.5 \ldots 0.75.$$

The input parameters m_{c_0}, m_{p_0}, and m_{f_0} designate the amount of carbohydrates, proteins, and fat in the ingested meal.

The model for the concentration $c_{\text{c}}^I(z,t)$ of carbohydrates in the intestine of radius $r(z,t)$ and length l is given by

$$\frac{\partial}{\partial t}(c_{\text{c}}^I(z,t)\pi r^2) = -v\frac{\partial}{\partial z}(c_{\text{c}}^I \pi r^2) - g_{\text{ex}}(c_{\text{c}}^I),$$
$$\frac{\partial r(z,t)}{\partial t} = -v\frac{\partial r}{\partial z},$$
$$g_{\text{ex}}(c_{\text{c}}^I(z,t)) = \frac{\rho c_{\text{c}}^I}{k + c_{\text{c}}^I},$$
$$G_{\text{ex}}(t) = \int_0^l g_{\text{ex}}(z,t)dz,$$

with the initial and boundary conditions

$$c_{\text{c}}^I(0,t) = \frac{q(t)}{q(t) + f_v V_{\text{meal}}\, e^{-\alpha r}} c_{\text{c}}^S(t), \qquad c_{\text{c}}^I(z,0) = 0,$$
$$r(0,t) = \sqrt{\frac{q(t)}{\pi v}}, \qquad r(z,0) = \sqrt{\frac{q_{\text{b}}}{\pi v}} = r_0,$$

and the parameters

$$f_v = 0.5, \qquad v = 1 \text{ cm/min}, \qquad l = 150 \text{ cm},$$
$$r_0 = 0.4 \text{ cm}, \qquad \rho = 16.6 \text{ mg/min/cm}, \qquad k = 27.72 \text{ mg/ml}.$$

The simplified model for the amount of in-blood glucose $G_{\text{b}}(t)$ is described below:

$$G_{\text{b}}(t_{k+1}) = G_{\text{b}}(t_K) + \frac{1}{a(t_k)}\int_{t_k}^{t_k+\Delta t} I_{\text{ex}}\, dt + \frac{1}{b(t_k)}\int_{t_k}^{t_k+\Delta t} G_{\text{ex}}\, dt + \frac{1}{c(t_k)}\Delta t,$$
$$G_{\text{b}}(0) = G_{\text{b}_0}, \quad t_{k+1} = t_k + \Delta t.$$

The sensitivity parameters $a(t)$, $b(t)$, and $c(t)$ can be considered as constants for a multiple N of the time interval Δt. Typically, the time interval is chosen to be 1 min and the sensitivity parameters are considered as constant for about one hour, that is, $N = 60$.

From the description so far, it is obvious that the parameters B and f_{gas} have values that lie within a specific range, which means that they are vague values by definition. In addition, it is next to impossible to predetermine the amount of carbohydrates m_{c_0}. So the model needs at least three fuzzy numbers that are represented by quasi-Gaussian fuzzy numbers:

$$\overset{*}{B} = \text{gfn}(11.8 \times 10^{-3}\ \text{min}^{-1}, 1.6 \times 10^{-3}\ \text{min}^{-1}, 0.4 \times 10^{-3}\ \text{min}^{-1}),$$

$$\overset{*}{f_{\text{gas}}} = \text{gfn}(0.64, 0.03, 0.02),$$

$$\overset{*}{m_{c_0}} = \text{gfn}\left(5\ \text{bu}, \frac{1}{3}\ \text{bu}, \frac{1}{3}\ \text{bu}\right).$$

The various values have been chosen based on the data presented in the original model, while 1 bu = 1 bread unit and the nutritional content of carbohydrates in the ingested food is usually an integer multiple of the bread unit. Finally, the initial condition for the in-blood glucose is set to

$$G_\text{b}(t = 0) = G_{\text{b}_0} = 73\ \text{mgdl}^{-1}.$$

3.7.2 Estimation of an Ongoing Project's Completion Time

For any project it is a good planning strategy to try to anticipate all possible cases that may delay its realization. However, no matter how good we plan a project, it is quite possible that some unexpected things may occur that might eventually delay the realization of the project. Therefore, we need a tool that can be used to analyze a situation and make some sort of predictions. The "obvious" solution is to use probability theory, as noted by Dorota Kuchta [184]. However, it seems that this approach is not useful since it assumes that we can verify certain hypotheses about the probability distributions of activity duration times. Clearly, if we know these times in advance, then we do not need probability theory. As an alternative approach to the solution of this problem, Kuchta suggested the use of fuzzy numbers since they make it easy to describe several criteria that influence the actual duration of a project's activities.

Kuchta's fuzzy numbers are a special form of triangular fuzzy numbers. In particular, she defines a fuzzy number A as a triplet (z_1, z_2, z_3) whose analytic form is as follows:

$$A(x) = \begin{cases} 0, & \text{when } x \leq z_1 - z_2, \\ \frac{x}{z_2} + 1 + \frac{z_1}{z_3}, & \text{when } z_1 \geq x > z_1 - z_2, \\ 1, & \text{when } x = z_1, \\ -\frac{x}{z_3} + 1 + \frac{z_1}{z_3}, & \text{when } z_1 \leq x \leq z_1 + z_3, \\ 0, & \text{when } x \geq z_1 + z_3. \end{cases}$$

Here z_1 is called the mean value, while the variability measures z_2 and z_3 measure the uncertainty linked to the assumption that the unknown magnitude z will be equal to z_1.

3.7.2.1 Model of a Project

Each project should be understood as a set of activities

$$\mathfrak{A} = \{A_1, A_2, \dots, A_I\}.$$

Clearly, the members of this set may have dependencies between them (e.g. A_i should happen before A_j, or A_k and A_l use the same resources, etc.). At the beginning, we provide an estimation of each activity's duration, but when the project is implemented, the mean value of an activity's duration may depend on a number of factors. Such factors are the weather, the mood in the activity team, the skills of the activity team, the attitude of certain stakeholders, etc. Unfortunately, most of these factors cannot be measured, although they may strongly influence the duration of an activity. The set of all these factors will be

$$\mathfrak{F} = \{\alpha_1, \dots, \alpha_s, \dots, \alpha_S\}.$$

For each a_i, we will denote by $\alpha_i(t)$ the impact of α_i on the estimation of the mean values of the durations of project activities at time $t \in [0, T]$, where T is a point beyond which the project cannot go (Kuchta calls it *time horizon*) and 0 is the planning phase of the project. Apart from these factors, it is quite possible to have factors that affect the uncertainty (variability) in the estimation of an activity's duration. The set of these "other" factors will be written as follows:

$$\mathfrak{O} = \{\beta_1, \dots, \beta_r, \dots, \beta_R\}.$$

It is quite possible that $\mathfrak{O}$ is in an one-to-one correspondence with $\mathfrak{F}$. However, this does not imply that in all cases, the sets are in an one-to-one correspondence. Also, all $\beta_r(t)$ will represent the impact of the corresponding factors on the uncertainty of the estimates of the duration of the activities of the project at a given moment t. These two sets are clearly different, and the elements of the first one affect the duration of an activity and can be used to determine the most possible value of the duration. The elements of the second set affect the variability of the estimate around the mean value. For instance, in construction projects, the weather plays a decisive role in the determination of the duration of certain activities and may affect the mean value of the estimated completion time entirely (forecasts of long rainy periods or long sunny periods often affect the mean values in different ways). However, other factors like technical problems or the experience of the team members, do not have such a strong impact on the completion of a project but affect the precise determination of the completion time. Therefore, one should take into account their variability in both directions.

For each activity A_i, $D_i(t)$ will be the estimate of the duration of this activity at a moment t before this activity has been finished:

$$D_i(t) = (f_i^1(\alpha_{s_i}(t)), f_i^2(\beta_{p_i}(t)), f_i^3(\alpha_{r_i}(t))), \quad i = 1, \dots, I, \tag{3.16}$$

where f_i^1, f_i^2, and f_i^3 are invertible functions from and into the set of nonnegative real numbers, $s_i \in \{1, \dots, S\}$, and p_i and r_i are not necessarily different indices from the set $\{1, \dots, R\}$. From this equation, it is clear that all three parameters, that is, the mean value of the estimates as well as its variability measures, depend on exactly one parameter. Although this may seem like a limitation, in most real-life cases, one major factor can be selected: one for the mean value and one for each of the two variability measures. Also, according to Eq. (3.16) the duration of each activity is vague.

When realizing a project, one should be able to update the estimates of the duration of activities which have not been started yet, and thus to update the estimate of the total duration time of the project. For this we need a set of selected control moments

$$\mathfrak{T} = \{t_1, \dots, t_j, \dots, t_J\},$$

where $t_j \geq 0$ and $t_J \leq T$. Depending on the nature of the project, the intervals $[t_{j-1}, t_j]$ for $j = 1, \dots, J$ might be smaller if the project is risky, or they might be bigger if the project is not risky. At a moment $t_j, j \geq 0$, $C(t_j)$ will denote the estimated total completion time of the project at this given moment. $C(t_j)$ is actually the maximum of the estimated lengths of all the paths in the project network, using the actual completion time of the activities which have been completed at moment t_j, and using the estimated duration time of the activities that have not been completed in the form of fuzzy numbers derived from Eq. (3.16):

$$D_i(t_j) = (f_i^1(\alpha_{s_i}(t_j)), f_i^2(\beta_{p_i}(t_j)), f_i^3(\alpha_{r_i}(t_j))), \quad i = 1, \dots, I,\ j = 1, \dots, J.$$

In order to get a reliable and informative estimate $C(t_j)$ at each control moment t_j, it is necessary to have the best possible estimates $D_i(t_j)$ of the durations of those activities A_i that have not been completed at t_j. Kuchta has proposed an algorithm for updating the estimates $D_i(t_j)$, but we will not describe it here. Our purpose was to show the use of fuzzy numbers in specific problems and not to show how specific problems can be solved completely.

Exercises

3.1 Using Eq. (3.1) draw the triangular fuzzy numbers tfn(8, 4, 4) and tfn(9, 1, 5).

3.2 Evaluate the following in interval arithmetic:

a. $[0, 0]/[0, 0]$;
b. $[-1, 1]/[0, 0]$;
c. $[-1, 1]/[-1, 1]$;
d. $[1, 2]/[2, 1]$;
e. $[0, 0]/[-1, 1]$.

3.3 Evaluate in interval arithmetic each of the following expressions: $1/(2-X)+1/(1+X)$ and X^2-2X+5 for X equal to each of the intervals $[0,1]$, $[0.5,1]$, and $[1,2]$.

3.4 Assume that $A=\text{tfn}(6,1,2)$ and $B=\text{tfn}(7,3,2.5)$. Find the fuzzy numbers $A\cdot B, A+B+9$, and A^2+B^2 using the method described in Section 3.2.2.

3.5 Redo the previous exercise using the method described in Section 3.2.4.

3.6 Complete Example 3.2.3 by calculating $\text{MAX}(A,B)(z)$.

3.7 Solve equation $A\cdot X+B=C$ using the classical method when $A=\text{tfn}(2,1,1)$, $B=\text{tfn}(-3,1,1)$, and $C=\text{tfn}(4,1,1)$.

3.8 Verify that the equation $A\cdot X+B=C$ has no classical solution when $A=\text{tfn}(9,1,1)$, $B=\text{tfn}(-2,1,1)$, and $C=\text{tfn}(5,2,2)$.

3.9 Assume that five experts assign the following five scores:

$$g_1=3,\quad g_2=7,\quad g_3=8,\quad g_4=8,\quad \text{and}\quad g_5=9.$$

Construct the corresponding triangular fuzzy number.

4

Fuzzy Relations

Fuzzy relations are to (crisp) relations what fuzzy sets are to (crisp) sets. In particular, elements of some sets are related to a certain degree just like elements of some universe belong to a fuzzy set to some degree. Fuzzy relations have been introduced and further studied by Zadeh [305, 307]. Fuzzy relations are important tools that are used in fuzzy modeling, fuzzy diagnosis, and fuzzy control, which explains why it is useful to have a good understanding of fuzzy relations and their properties.

4.1 Crisp Relations

The concept of a relation is a basic concept in both mathematics and in everyday life. For example, when one has a set of keywords and a set of documents, she can create an association between the keywords and the documents, and this association is a relation. Fuzzy relations extend crisp relations just like fuzzy sets extend crisp sets. Thus, if the keywords w_1 and w_2 characterize document D_1, then in a fuzzy theoretic setting one can say that these keywords characterize the document to some degree. In order to better understand the ideas about fuzzy relations, it is necessary to have a good understanding of crisp relations. In what follows, we will briefly present the relevant ideas. Readers willing to have a more detailed presentation can consult any book that covers relations (e.g. [248] is a very good choice).

In the previous paragraph, it was hinted that a relation describes an association between two sets, more specifically:

Definition 4.1.1 Assume that A and B are sets. Then, a binary relation from A to B is a subset of $A \times B$.

In other words, a binary relation from A to B is a set R of ordered pairs (a, b), where $a \in A$ and $b \in B$. Usually, we write $a\ R\ b$ to denote that $(a, b) \in R$ and $a\ \not R\ b$

A Modern Introduction to Fuzzy Mathematics, First Edition.
Apostolos Syropoulos and Theophanes Grammenos.

to denote that $(a, b) \notin R$. A special kind of relation is a *relation in a set A*, which is a relation from A to A. For example, the relation

$$<= \{(x, y)|x, y \text{ are real numbers and } x \text{ is less than } y\}$$

is a simple example of a relation in $\mathbb{R}$.

Suppose that S is a binary relation. Then, the set dom(S) of all elements a such that for some b, $a\ S\ b$ is called the *domain* of S, that is,

$$\text{dom}(S) = \{a|\text{there is a } b \text{ such that } (a, b) \in S\}.$$

Similarly, the set cod(S) of all elements b such that for a, $a\ S\ b$, is called the *codomain* of S, that is,

$$\text{cod}(S) = \{b|\text{there is an } a \text{ such that } (a, b) \in S\}.$$

4.1.1 Properties of Relations

In Section 2.1.1, we presented some of the properties that a relation may have, but here we will present them in their general form:

Definition 4.1.2 A relation R in a set A is called

reflexive if $(a, a) \in R$ for every element $a \in A$;
symmetric if $(b, a) \in R$ whenever $(a, b) \in R$, for all $a, b \in A$;
antisymmetric if $(a, b) \in R$ and $(b, a) \in R$, then $a = b$, for all $a, b \in A$;
transitive if whenever $(a, b) \in R$ and $(b, c) \in R$, then $(a, c) \in R$, for all $a, b, c \in A$.

The relation $R_1 = \{(a, b) \mid a \leq b \text{ and } a, b \in \mathbb{Z}\}$, where $\mathbb{Z}$ is the set of integers, is reflexive, because $a \leq a$ for every integer a. It is also antisymmetric because $a \leq b$ and $b \leq a$ imply that $a = b$. And R_1 is transitive, because $a \leq b$ and $b \leq c$ imply that $a \leq c$.

4.1.2 New Relations from Old Ones

A relation from A to B is a subset of $A \times B$, therefore, it makes sense to talk about the union, the intersection, and the difference of two relations. In particular, if R and S denote two relations, then

union is the relation $R \cup S$ such that $a(R \cup S)b$ is equivalent to the statement that either $a\ R\ b$ or $a\ S\ b$ is true;

intersection is the relation $R \cap S$ such that $a(R \cap S)b$ is equivalent to the statement that both $a\ R\ b$ and $a\ S\ b$ are true; and

difference is the relation $R - S$ such that $a(R - S)b$ is equivalent to the statement that both $a\ R\ b$ and $a \not S\ b$ are true.

In addition, relations can be composed:

Definition 4.1.3 Suppose that R is a relation from a set A to a set B and S is a relation from B to a set C. Then, the composite of R and S is the relation consisting of ordered pairs (a, c), where $a \in A$, $c \in C$, and for which there exists an element $b \in B$ such that $(a, b) \in R$ and $(b, c) \in S$. The composite of R and S is written as $S \circ R$.

Example 4.1.1 Assume that $A = \{x, y, z\}$ and that R is a relation in A defined by the set

$$\{(x, y), (x, z), (z, y)\}.$$

Then, $R \circ R = \{(x, y)\}$.

Powers of a relation are compositions of the relation itself. Assume that R is a relation in the set A. Then, $R \circ R = R^2$, $R \circ R \circ R = R \circ R^2 = R^3, \dots, R^{n+1} = R \circ R^n$.

4.1.3 Representing Relations Using Matrices

A binary relation between finite sets can be represented using a matrix. The elements of the matrix are either the number 1 or the number 0, depending on whether the address of an element belongs or does not belong to the relation, respectively. More precisely, if $A = \{a_1, a_2, \dots, a_m\}$ and $B = \{b_1, b_2, \dots, b_n\}$, then a relation $R \subset A \times B$ can be represented by the matrix $\mathbf{M}_r = [m_{ij}]$:

$$m_{ij} = \begin{cases} 1, & \text{if } (a_i, b_j) \in R, \\ 0, & \text{if } (a_i, b_j) \notin R. \end{cases}$$

Example 4.1.2 Assume that $A = \{1, 2, 3\}$ and $B = \{1, 2\}$ are two sets and a relation R from A to B is defined by the set $\{(2, 1), (3, 1), (3, 2)\}$. Then, the following matrix is the matrix of R:

$$\begin{array}{cc} & \begin{array}{cc} 1 & 2 \end{array} \\ \mathbf{M}_R = & \begin{bmatrix} 0 & 0 \\ 1 & 0 \\ 1 & 1 \end{bmatrix} \begin{array}{c} 1 \\ 2 \\ 3 \end{array} \end{array}.$$

4.1.4 Representing Relations Using Directed Graphs

Directed graphs or *digraphs* are used to provide a pictorial representation of relations. Assume that S is a relation in a set $A = \{a_1, a_2, \dots, a_n\}$. Then, the elements of A are represented by points or little circles called *nodes* or *vertices*. A node corresponding to a_k is labeled a_k. If $a_i \; S \; a_j$, then we connect nodes a_i and

a_j by means of an arrow in the direction from a_i to a_j. If $a_i\ S\ a_j$ and $a_j\ S\ a_i$, then we can draw either two arrows or a line segment or an arc with two arrowheads attached at the end of the line segment or the arc, respectively.

Example 4.1.3 If $R = \{(1,3),(1,4),(2,1),(2,2),(2,3),(3,1),(3,3),(4,1),(4,3)\}$ is a relation in $\{1,2,3,4\}$, then the following directed graph represents R:

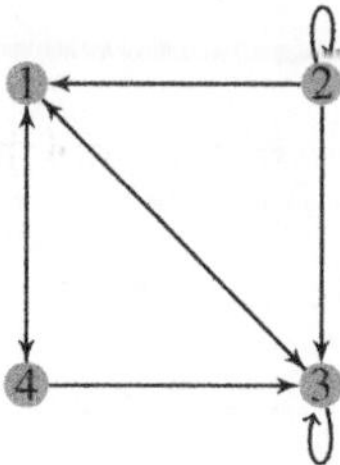

4.1.5 Transitive Closure of a Relation

Suppose that we have a relation R in a set A of sites, where each site is a data center. The diagram that follows shows the sites that are directly connected with a one-way telephone line:

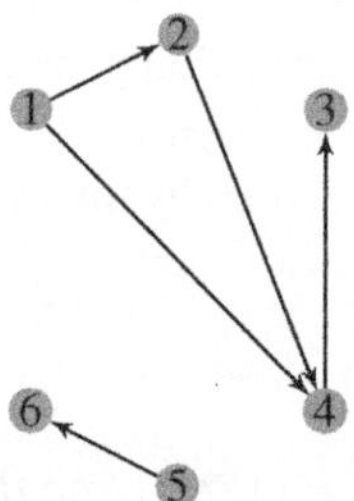

An important question is: How can we determine if there is a path from any vertex A to another vertex B? In other words, we ask if there is a direct or indirect link consisting of one or more telephone lines from one data center to another? Unfortunately, we cannot use R to answer this question, simply because it is not transitive. In order to answer this question, we need to construct the transitive closure of R:

Definition 4.1.4 Assume that A is a finite set and R is a relation in A. Then, the relation $R^+ = R \cup R^2 \cup R^3 \cup \cdots$ in A is the *transitive closure* of R in A.

Remark 4.1.1 R^+ is the smallest transitive relation containing R.

4.1.6 Equivalence Relations

Equivalence relations generalize the notion of equality.

Definition 4.1.5 A relation on a set A is called an equivalence relation if it is reflexive, symmetric, and transitive.

If R is an equivalence relation and $a\ R\ b$, then a and b are called equivalent. Usually, we denote that a and b are equivalent by $a \sim b$.

Example 4.1.4 Suppose that W is the set of all words. Then, $w_1, w_2 \in W$ are equivalent if they have the same number of letters or if the first n letters of each word are identical.

Definition 4.1.6 Assume that S is an equivalence relation on a set A. Then, the set of all elements that are related to an element $a \in A$ is called the equivalence class of a. This set is written as $[a]_S$. In other words, the equivalence class of the element a is

$$[a]_S = \{s | a\ S\ s\}.$$

Any element $b \in [a]_S$ is called a representative of the equivalence class $[a]_S$.

Assume that S is an equivalence relation on a set A. Then,

$$A = \bigcup_{a \in A} [a]_S.$$

Also, $[a]_R \cap [b]_R = \emptyset$, when $[a]_R \neq [b]_R$. Thus, the equivalence classes form a *partition* of A since they break A into disjoint subsets.

4.2 Fuzzy Relations

As we have already mentioned, a fuzzy relation is a fuzzy extension of the notion of a relation. In fact, one extends the matrix representation of relations by allowing elements of the matrix to take values from $[0, 1]$.

Definition 4.2.1 Assume that A_1, A_2,..., A_n are n crisp sets. Then, a mapping $R : A_1 \times A_2 \times \cdots \times A_n \to [0, 1]$ is called a fuzzy relation. The number $R(a_1, a_2, \ldots, a_n) \in [0, 1]$ can be interpreted as the degree of relationship between a_1, a_2,..., and a_n.

When $n = 2$, the mapping $R : A_1 \times A_2 \to [0, 1]$ is called a *fuzzy binary relation*. Clearly, the number $R(a_1, a_2) \in [0, 1]$ can be interpreted as the degree of relationship between a_1 and a_2.

Example 4.2.1 Assume that the following sets

$$D = \{D_1, D_2, D_3, D_4\} \quad \text{and} \quad W = \{w_1, w_2, w_3\}$$

represent documents and keywords, respectively. Then, a fuzzy (binary) relation in D and W will be a mapping $R : D \times W \to [0, 1]$. The *membership matrix* of a fuzzy relation, which is the analog of a crisp relation matrix, is shown as follows:

$$R = \begin{array}{c} \begin{matrix} w_1 & w_2 & w_3 \end{matrix} \\ \begin{bmatrix} 1 & 0 & 0.6 \\ 0.8 & 1 & 0 \\ 0 & 1 & 0 \\ 0.8 & 0 & 1 \end{bmatrix} \end{array} \begin{matrix} d_1 \\ d_2 \\ d_3 \\ d_4 \end{matrix}.$$

Figure 4.1 is a diagrammatic representation of R. Note that not all membership degrees have been specified in the diagram. This kind of diagrams are known as *sagittal* diagrams.

Given a fuzzy binary relation $S : A \times B \to [0, 1]$, then its *domain* is a fuzzy set on A whose membership function is defined by

$$\text{dom}(S)(a) = \bigvee_{b \in B} S(a, b), \text{ for all } a \in A.$$

Similarly, the *codomain* of S is a fuzzy set on B whose membership function is defined by

$$\text{cod}(S)(b) = \bigvee_{a \in A} S(a, b), \text{ for all } b \in B.$$

Also, the *height* of S is defined by

$$h(S) = \bigvee_{b \in B} \bigvee_{a \in A} S(a, b).$$

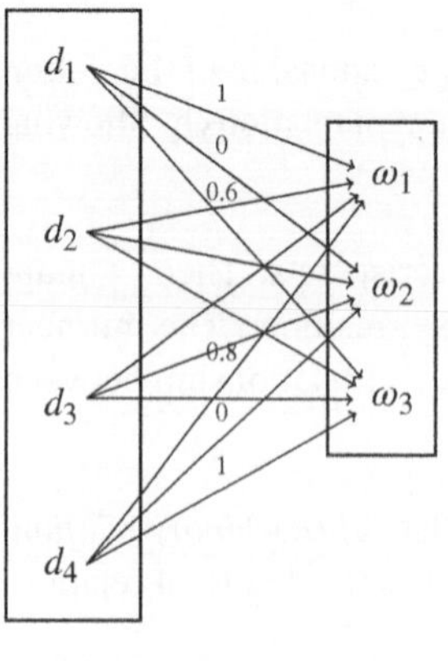

Figure 4.1 A sagittal diagram of the fuzzy relation in Example 4.2.1.

4.3 Cartesian Product, Projections, and Cylindrical Extension

The notions of Cartesian product, projections, and cylindrical extension of fuzzy sets are used to produce fuzzy relations from ordinary fuzzy sets.

4.3.1 Cartesian Product

Assume that $A_1, A_2, \dots, A_n$ are fuzzy sets whose universes are the crisp sets $X_1, X_2, \dots, X_n$, respectively. Then, their *Cartesian product* $A_1 \times A_2 \times \cdots \times A_n$ is a fuzzy relation P in $X_1 \times X_2 \times \cdots \times X_n$ such that

$$P(x_1, x_2, \dots, x_n) = \min[A_1(x_1), A_2(x_2), \dots, A_n(x_n)],$$

for all $x_i \in X_i$, $i = 1, \dots, n$. In other words, the Cartesian product of n fuzzy sets is used to define an n-ary fuzzy relation. More generally, we can use a t-norm $*$ to define the Cartesian product as follows:

$$P(x_1, x_2, \dots, x_n) = A_1(x_1) * A_2(x_2) * \cdots * A_n(x_n),$$

for all $x_i \in X_i$, $i = 1, \dots, n$.

4.3.2 Projection of Fuzzy Relations

Let Q be a fuzzy relation in X_1, X_2,..., X_n. Then, its *projection* on $Z = X_i \times X_j \times \cdots \times X_k$, where $I = \{i, j, \cdots, k\} \subset N = \{1, 2, \dots, n\}$, is a fuzzy relation Q_Z defined as follows:

$$Q_Z(x_i, x_j, \dots, x_k) = \mathrm{Proj}_Z Q(x_1, x_2, \dots, x_n) = \bigvee_{x_l, x_m, \dots, x_p} R(x_1, x_2, \dots, x_n),$$

where $J = \{x_l, x_m, \dots, x_p\} \subset N$ and $I \cup J = N$ and $I \cap J = \emptyset$.

Example 4.3.1 Assume that S is a fuzzy binary relation:

$$S = \int_{A \times B} \frac{S(a, b)}{(a, b)}.$$

Then, its first projection is the following fuzzy set:

$$R_A = \mathrm{Proj}_A S(a, b) = \int_A \frac{R_A(a)}{a}, \quad \text{where} \quad R_A(a) = \bigvee_{b \in B} A(a, b).$$

Similarly, its second projection is the fuzzy set R_2 defined as follows:

$$R_B = \mathrm{Proj}_B S(a, b) = \int_A \frac{R_B(b)}{b}, \quad \text{where} \quad R_B(b) = \bigvee_{a \in A} A(a, b).$$

Example 4.3.2 Consider the fuzzy relation from Example 4.3.1. Then R_W is the following fuzzy set:

$$\begin{aligned} R_W &= \frac{\max(1, 0.8, 0, 0.8)}{w_1} + \frac{\max(0, 1, 1, 0)}{w_2} + \frac{\max(0.6, 0, 0, 1)}{w_3} \\ &= \frac{1}{w_1} + \frac{1}{w_2} + \frac{1}{w_3}. \end{aligned}$$

Clearly, R_W is a crisp set. Similarly, we can compute R_D.

4.3.3 Cylindrical Extension

Let $A : X \to [0, 1]$ be a fuzzy subset of a crisp set X. Then, a *cylindrical extension* of A on $X \times Y$ is a fuzzy relation cyl A defined as follows:

$$\text{cyl}\ A(x, y) = A(x), \quad \text{for all } x \in X \text{ and } y \in Y.$$

Example 4.3.3 Suppose that $X = \{a_1, a_2, a_3, a_4\}$, $Y = \{b_1, b_2, b_3\}$, and

$$A = \frac{0.3}{a_1} + \frac{0.4}{a_2} + \frac{0.6}{a_3} + \frac{0.8}{a_4}.$$

Then,

$$\text{cyl}\ A(x, y) = \begin{array}{c} \\ a_1 \\ a_2 \\ a_3 \\ a_4 \end{array} \begin{array}{c} \begin{array}{ccc} b_1 & b_2 & b_3 \end{array} \\ \begin{bmatrix} 0.3 & 0.3 & 0.3 \\ 0.4 & 0.4 & 0.4 \\ 0.6 & 0.6 & 0.6 \\ 0.8 & 0.8 & 0.8 \end{bmatrix} \end{array}.$$

More generally, if S is a fuzzy relation in $X_{j_1} \times \cdots \times X_{j_k}$, where $\{j_1, \ldots, j_k\}$ is a subsequence of $\{1, 2, \ldots, n\}$, the cylindrical extension of S on $X_1 \times \cdots \times X_n$ is a fuzzy relation cyl S in $X_1 \times \cdots \times X_n$ such that

$$\text{cyl}\ S(x_1, \ldots, x_n) = S(x_{j_1}, \ldots, x_{j_k}).$$

4.4 New Fuzzy Relations from Old Ones

Assume that R and S are two binary fuzzy relations in A and B. Then, the basic operations between these relations in their most general form are defined as follows:

Union If $Q = R \cup S$, then $Q(a, b) = R(a, b) \bigstar S(a, b)$, for all $a, b \in A \times B$, where $\bigstar$ is a t-conorm;

Intersection If $Q = R \cap S$, then $Q(a, b) = R(a, b) * S(a, b)$, for all $a, b \in A \times B$, where $*$ is a t-norm ;

Complement $S^{\complement}(x,y) = 1 - S(x,y)$, for all $a, b \in A \times B$;
Transpose $S^{\mathrm{T}}(b,a) = S(a,b)$, for all $a, b \in A \times B$.

Also, a fuzzy relation P in A and B can be represented by its *resolution form*:

$$P = \bigcup_{\alpha \in [0,1]} \alpha^{\alpha} P,$$

where ${}^{\alpha}P = \{(a,b) | P(a,b) \geq \alpha\}$ is the α-cut of P, which is actually a crisp relation. Alternatively, if one is interested in membership degrees, the previous equality leads to the following one:

$$P(a,b) = \bigvee_{\alpha \in [0,1]} \{\min[\alpha, P(a,b)]\}.$$

Naturally, the most important operation between fuzzy relations is *composition*:

Definition 4.4.1 Assume that $P : A \times B \to [0,1]$ and $Q : B \times C \to [0,1]$ are two fuzzy relations. Then, the *max–min* composition $R = P \circ Q$ is a fuzzy relation in A and C defined by

$$R(a,c) = \max_{b \in B} \min[P(a,b), Q(b,c)].$$

If we replace min with another t-norm, then we get a more general definition of composition.

We can perform compositions of binary fuzzy relations by using their membership matrices. Suppose that $P = [p_{ik}]$, $Q = [q_{kj}]$, and $R = [r_{ij}]$ are the membership matrices of three fuzzy relations such that $R = P \circ Q$. Now, this can be written as

$$[r_{ij}] = [p_{ik}] \circ [q_{kj}],$$

where

$$r_{ij} = \max_{k} \min(p_{ik}, q_{kj}). \tag{4.1}$$

In other words, we are performing matrix "multiplication", but we use the min and max operators instead of the multiplication and addition operators, respectively. The following relation composition is a typical example that shows the use of Eq. 4.1 in practice:

$$P \circ Q = \begin{bmatrix} 0.3 & 0.5 & 0.8 \\ 0 & 0.7 & 1 \\ 0.4 & 0 & 0.5 \end{bmatrix} \circ \begin{bmatrix} 0.9 & 0 & 0.7 & 0.7 \\ 0.3 & 0.2 & 0 & 0.9 \\ 1 & 0 & 0.5 & 0.5 \end{bmatrix} = \begin{bmatrix} 0.8 & 0.2 & 0.5 & 0.5 \\ 1 & 0.2 & 0.5 & 0.7 \\ 0.5 & 0 & 0.5 & 0.5 \end{bmatrix} = R.$$

Here is how we computed r_{11}:

$$\begin{aligned} r_{11} &= \max\{\min(0.3, 0.9), \min(0.5, 0.3), \min(0.8, 1)\} \\ &= \max\{0.3, 0.5, 0.8\} \\ &= 0.8 \end{aligned}$$

Proposition 4.4.1 *The max–min composition is associative, that is,*

$$R \circ (S \circ Q) = (R \circ S) \circ Q,$$

where R, S, and Q are fuzzy relations in A and B, B and C, and C and D, respectively.

Proof: It is enough to prove that

$$(R \circ (S \circ Q))(a, d) = ((R \circ S) \circ Q)(a, d), \text{ for all } (a, d) \in A \times D.$$

Clearly,

$$(S \circ Q)(b, d) = \max_{c \in C} \min[S(b, c), Q(c, d)],$$

$$(R \circ S)(a, c) = \max_{b \in B} \min[R(a, b), S(b, c)].$$

Now,

$$\begin{aligned}(R \circ (S \circ Q))(a, d) &= \max_{b \in B} \min[R(a, b), \max_{c \in C} \min[S(b, c), Q(c, d)]] \\ &= \max_{b \in B} \min[\max_{c \in C} \min[S(b, c), Q(c, d)], R(a, b)]\end{aligned}$$

and for some $c_k \in C$

$$= \max_{b \in B} \min[\min[S(b, c_k), Q(c_k, d)], R(a, b)]$$

and for some $b_l \in B$

$$= \min[\min[S(b_l, c_k), Q(c_k, d)], R(a, b_l)].$$

If we set $x = S(b_l, c_k)$, $y = Q(c_k, d)$, and $z = R(a, b_l)$, then

$$(R \circ (S \circ Q))(a, d) = \min[\min[x, y], z] = \min[x, y, z].$$

Similarly, we can show that $((R \circ S) \circ Q)(a, d) = \min[x, y, z]$, which proves the associativity of the max–min composition. □

Proposition 4.4.2 *For any two fuzzy relations R and S in A and B, and a third relation Q in B and C, we have*

$$(R \cup S) \circ Q = (R \circ Q) \cup (S \circ Q).$$

Proof:

$$\begin{aligned}((R \cup S) \circ Q)(a, b) &= \max_{b \in B} \min[(R \cup S)(a, b), Q(b, c)] \\ &= \max_{b \in B} \min[\max(R(a, b), S(a, b)), Q(b, c)] \\ &= \max_{b \in B} \max[\min(R(a, b), Q(b, c)), \min(S(a, b), Q(b, c))] \\ &\leq \max\{\max_{b \in B}[\min(R(a, b), Q(b, c))], \max_{b \in B}[\min(S(a, b), Q(b, c))]\}.\end{aligned}$$

In other words, $(R \cup S) \circ Q \subseteq (R \circ Q) \cup (S \circ Q)$. On the other hand, $R \circ Q \subseteq (R \cup S) \circ Q$ and $S \circ Q \subseteq (R \cup S) \circ Q$ simply because $R \subseteq (R \cup S)$ and $S \subseteq (R \cup S)$, respectively. Therefore,

$$(R \circ Q) \cup (S \circ Q) \subseteq (R \cup S) \circ Q.$$

And since $(R \cup S) \circ Q \subseteq (R \circ Q) \cup (S \circ Q)$, the equality follows. □

In a way, the min–max fuzzy relation composition is the dual of the max–min composition operation. The min–max composition is defined as follows:

Definition 4.4.2 Assume that $P : A \times B \to [0,1]$ and $Q : B \times C \to [0,1]$ are two fuzzy relations. Then, the *min–max* composition [24, 228] $R = P \bullet Q$ is a fuzzy relation in A and C defined by

$$R(a,c) = \min_{b \in B} \max[P(a,b), Q(b,c)].$$

Not so surprisingly, we can perform compositions of binary fuzzy relations by using their membership matrices: The principle is the same, but we just interchange the use of min and max. The two composition operators are similar in other ways:

Proposition 4.4.3 *The min–max composition is associative, that is,*

$$R \bullet (S \bullet Q) = (R \bullet S) \bullet Q,$$

where R, S, and Q are fuzzy relations in A and B, B and C, and C and D, respectively.

Proposition 4.4.4 *For any two fuzzy relations R and S in A and B and a third relation Q in B and C, we have*

$$(R \cup S) \bullet Q = (R \bullet Q) \cup (S \bullet Q).$$

The proofs of these propositions are left as exercises to the reader.

Naturally, we can define even more general composition operators by using a t-norm instead of min:

Definition 4.4.3 Assume that $P : A \times B \to [0,1]$ and $Q : B \times C \to [0,1]$ are two fuzzy relations. Then, the *max-t-norm* composition $R = P \circ_* Q$ is a fuzzy relation in A and C defined by

$$R(a,c) = \max_{b \in B} [P(a,b) * Q(b,c)].$$

4.5 Fuzzy Binary Relations on a Set

Given some crisp set A, a fuzzy binary relation in A is a fuzzy relation $R : A \times A \to [0,1]$. These fuzzy relations can be "visualized" using a membership matrix or a

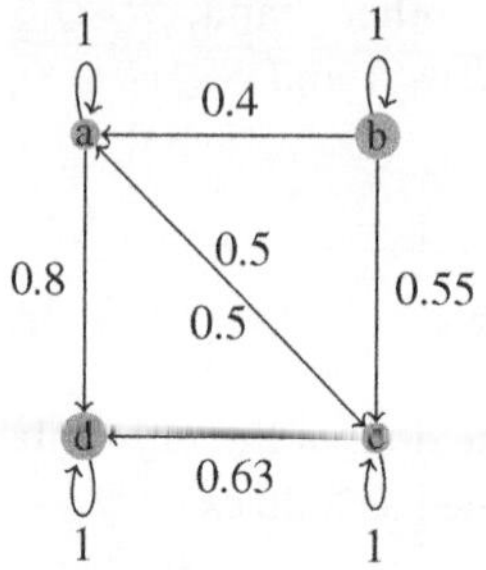

Figure 4.2 A directed graph that depicts a fuzzy relation.

directed graph. In particular, each edge of such a graph is annotated with a number that denotes the degree of relationship between the elements that correspond to the tail and the head of the arrow. Figure 4.2 depicts a graph that represents a fuzzy relation.

Fuzzy relations, like their crisp counterparts, can have a number of properties. Thus, a relation $R : A \times A \to [0, 1]$ is

reflexive if $R(a, a) = 1$, for all $a \in A$;
irreflexive if $R(a, a) = 0$, for all $a \in A$;
symmetric if $R(a, b) = R(b, a)$, for all $a, b \in A$;
antisymmetric if $R(a, b) > 0$ and $R(b, a) > 0$, then $a = b$, for all $a, b \in A$;
transitive if $R \circ R \subseteq R$ or, equivalently, if

$$R(a, c) \geq \bigvee_{b \in A} [R(a, b) * R(b, c)].$$

One can replace the t-norm $*$ with min to get the so-called sup–min transitive.

For example, it is not difficult to see that the fuzzy relation shown in Figure 4.2 is reflexive, but not symmetric. A relaxed version of reflexivity can be defined as follows:

Definition 4.5.1 Suppose that $\epsilon \in [0, 1]$. Then, R is ϵ-reflexive if $R(a, a) \geq \epsilon$.

Also, a relation R is *locally reflexive* when, for any $(a_1, a_2) \in A \times A$,

$$\max[R(a_1, a_2), R(a_2, a_1)] \leq R(a_1, a_1).$$

4.5.1 Transitive Closure

Suppose that R is a fuzzy relation on a finite set A, such that $\text{card}(A) = k$. Then, according to a result in [87] (see also the discussion in [238]), R has a $*$-transitive closure R^+ given by

$$R^+ = R \cup R^2 \cup \cdots \cup R^k,$$

where $R^i = R \circ R^{i-1}$ is a power of R. In general, it is easy to compute the transitive closure of some fuzzy relation R. Assume that $[r_{ij}]$ is the membership matrix of R. Then, Floyd's $*$-transitive closure algorithm is used to compute the membership matrix of R^+:

```
for(i=1; i<=n; i++)
  for(j=1; j<=n; j++)
    for(k=1; k<=n; k++)
      r_jk = max(r_jk, r_ji * r_ik);
```

Note that the algorithm gradually transforms the input (i.e. the membership matrix of R) into the output (i.e. the membership matrix of R^+).

Given a fuzzy relation R, the supremum of its max–min powers is given by

$$R^{\infty}(x,y) = \bigvee \{R^i(x,y) | i \geq 1\}.$$

An algorithm that can be used to compute R^{∞} has been presented in [29].

4.5.2 Similarity Relations

Zadeh [307] introduced the concept of a similarity relation.[1] This concept is essentially a generalization of the concept of an equivalence relation.

Definition 4.5.2 A fuzzy binary relation in A that is reflexive, symmetric, and transitive, is a *similarity* relation in A.

Example 4.5.1 Assume that $X = \{a, b, c, d, e, f\}$. Then, the fuzzy relation $S : X \times X \to [0, 1]$, whose membership matrix is shown below, is a similarity relation.

$$S = \begin{array}{c} \begin{matrix} a & b & c & d & e & f \end{matrix} \\ \begin{bmatrix} 1 & 0.2 & 1 & 0.6 & 0.2 & 0.6 \\ 0.2 & 1 & 0.2 & 0.2 & 0.8 & 0.2 \\ 1 & 0.2 & 1 & 0.6 & 0.2 & 0.6 \\ 0.6 & 0.2 & 0.6 & 1 & 0.2 & 0.8 \\ 0.2 & 0.8 & 0.2 & 0.2 & 1 & 0.2 \\ 0.6 & 0.2 & 0.6 & 0.8 & 0.2 & 1 \end{bmatrix} \end{array} \begin{matrix} a \\ b \\ c \\ d \\ e \\ f \end{matrix}$$

The fuzzy relation S was originally discussed in [307].

4.5.3 Proximity Relations

Years after Zadeh introduced similarity relations, it was discovered that they are an important special case of the more general concept of a proximity relation [229].

---

1 Sometimes similarity relations are called *fuzzy equivalence relations*.

A fuzzy relation that is reflexive and symmetric is a fuzzy proximity relation. In the literature, fuzzy proximity relations are also known as fuzzy *compatibility* relations or as *tolerance* relations. In what follows, we will briefly present the theory of proximity relations.

Definition 4.5.3 Assume that A is a fuzzy set defined on a universe X. Then, a family of fuzzy sets $\Sigma = \{P_i\}_{i \in J}$ is a *fuzzy covering* of A if

$$A = \bigcup_{i \in J} P_i.$$

Clearly, $P_i \subset A$ for all $i \in J$.

Definition 4.5.4 Suppose that A is a fuzzy set whose universe is the set X. Then, R is a *proximity relation on A* if

$$R(x, y) = R(y, x),$$
$$R(x, y) \leq \min[R(x, x), R(y, y)],$$
$$R(x, x) = A(x),$$

for all x and y in X.

Given any fuzzy covering $\Sigma = \{P_i\}_{i \in J}$ of A, we can define a fuzzy binary relation R_Σ as follows:

$$R_\Sigma(x, y) = \max_{i \in J} \min[P_i(x), P_i(y)]. \tag{4.2}$$

Lemma 4.5.1 *R_Σ is a proximity relation on A.*

Proof: From Eq. (4.2), we can obviously deduce that $R_\Sigma(x, y) = R_\Sigma(y, x)$ and that $R_\Sigma(x, x) = A(x)$. Since each P_i is a subset of A, we have

$$R_\Sigma(x, y) = \max_{i \in J} \min[P_i(x), P_i(y)] \leq \min[A(x), A(y)] = \min[R_\Sigma(x, x), R_\Sigma(y, y)].$$

□

Classes of fuzzy binary relations have a central role in the theory of proximity and similarity relations.

Definition 4.5.5 Assume that R is a fuzzy binary relation on X. Then, a fuzzy set K is a *pre-class* of R if $\min[K(x), K(y)] \leq R(x, y)$ for all $x, y \in X$. Maximal preclasses of R with respect to the inclusion relation $\subset$ are called *classes* of R. The set of all classes of R is denoted by Σ_R.

Theorem 4.5.1 *Suppose that R is a proximity relation on a fuzzy set A. Then, there exists a covering Σ of A such that*

$$R = R_\Sigma.$$

Proof: We consider the family $\{K_{\{a,b\}}\}$ of fuzzy sets defined as follows:

$$K_{\{a,b\}} = \begin{cases} R(a,b), & \text{if } x \in \{a,b\} \\ 0, & \text{otherwise,} \end{cases}$$

for all a and b in X. The following equalities and inequalities can be easily proved by using the criteria of Definition 4.5.4:

$$\min[K_{\{a,b\}}(a), K_{\{a,b\}}(b)] = R(a,b) \leq R(a,a),$$
$$\min[K_{\{a,b\}}(a), K_{\{a,b\}}(b)] = R(a,b),$$
$$\min[K_{\{a,b\}}(a), K_{\{a,b\}}(b)] = R(a,b) = R(b,a),$$
$$\min[K_{\{a,b\}}(a), K_{\{a,b\}}(b)] = R(a,b) \leq R(b,b),$$
$$\min[K_{\{a,b\}}(a), K_{\{a,b\}}(b)] = 0 \text{ if } x, y \notin \{a,b\}.$$

For this reason, $\{K_{\{a,b\}}\}$ is a preclass of R. Assume that Σ is a set of classes such that for any $\{a,b\}$, there is a class in Σ that contains $\{K_{\{a,b\}}\}$. Then, for any $K \in \Sigma$, $\min[K(a), K(b)] \leq R(a,b)$, and there is a K' such that $\min[K'(a), K'(b)] = R(a,b)$. Indeed, suppose that $K \in \Sigma$ is a class that contains $\{K_{\{a,b\}}\}$. Then,

$$R(a,b) = \min[K_{\{a,b\}}(a), K_{\{a,b\}}(b)] \leq \min[K(a), K(b)] \leq R(a,b).$$

From this, we can deduce that

$$\max_{K\in\Sigma} \min[K(a), K(b)] = R(a,b) \text{ for all } a, b \in X.$$

Obviously, R is a proximity relation on A and so we have from the previous formula

$$\max_{K\in\Sigma} K(a) = A(a) \text{ for all } a \in X,$$

that is, Σ is a cover of A and $R_\Sigma = R$. □

Just like equivalence relations have equivalence classes, similarity relations have similarity classes:

Definition 4.5.6 Assume that Q is a similarity relation on X. Then, for any $a \in X$, a *similarity class* of a is a fuzzy set $Q[a]$ on X defined as follows:

$$Q[a](x) = S(a,x) \text{ for all } x \in X.$$

Theorem 4.5.2 *A fuzzy set Q is a similarity class of a similarity relation S if and only if it is a class of S considered as a proximity relation.*

Proof: Let us start by noting that any similarity class of Q is a preclass of Q. Indeed,

$$\begin{aligned}\min[Q[a](x), Q[a](y)] &= \min[Q(a,x), Q(a,y)]\\ &= \min[Q(x,a), Q(y,a)] \le Q(x,y),\end{aligned}$$

because Q is a symmetric and transitive fuzzy binary relation.[2]

Suppose that P is a class of Q and that a is an element of X such that $P(x) \le P(a)$ for all x. Since X is a finite set, the element a does exist. It holds that

$$P(x) = \min[P(a), P(x)] \le Q(a,x) = Q[a](x).$$

Because $Q[a]$ is a preclass, we conclude that $P = Q[a]$.

Assume now that $Q[a]$ is a similarity class of Q and P is a class of Q containing $Q[a]$. Then, $P(a) = 1$, since Q is a reflexive relation and $P(x) \ge Q[a](x)$ for all $x \in X$. We have

$$Q[a](x) \le P(x) = \min[P(a), P(x)] \le Q(a,x) = Q[a](x)$$

for all $x \in X$. Therefore, $Q[a] = P$. □

The following result shows that $Q[a] = Q[b]$ can be true for different elements a and b.

Proposition 4.5.1 $Q[a] = Q[b]$ *if and only if* $Q(a,b) = 1$.

Proof: Let $Q[a] = Q[b]$. Then, $Q(a,b) = Q[a](b) = Q[b](b) = Q(b,b) = 1$. Conversely, let $Q(a,b) = 1$. Then, because of transitivity à la Ovchinnikov,

$$Q[a](x) = Q(a,x) \ge \min[Q(a,b), Q(b,x)] = Q(b,x) = S[b](x).$$

Similarly, $Q[b](x) \ge Q[a](x)$. □

Having defined similarity classes, the next logical thing is to define the fuzzy equivalent of a partition:

Definition 4.5.7 A fuzzy cover Σ of X is a *fuzzy partition* of X if there is a similarity relation Q on X such that Σ is the set of all distinct similarity classes of Σ.

Suppose that P is a similarity relation on X. Then, for any $\alpha \in [0,1]$, ${}^{\alpha}P$ is an equivalence relation on X, and clearly $a\ {}^{\alpha}P\ b$ if and only if $P(a,b) \ge \alpha$. For a

2 Sergei Ovchinnikov [229] defines transitivity of a fuzzy relation Q as follows:

$$\min[Q(a,b), Q(b,c)] \le Q(a,c).$$

However, he notes that if Q is reflexive, then his definition is equivalent to the one we have given above.

given $a \in X$

$$^{\alpha}P[a] = \{x|P[a] \geq \alpha\} = \{x|P(a,x) \geq \alpha\} = \{x|a\,^{\alpha}P\,b\}.$$

Therefore, α-cuts of similarity classes of P are equivalence classes of the α-cut of P.

4.6 Fuzzy Orders

A *fuzzy order* relation is a fuzzy transitive relation [307]. If a fuzzy relation is reflexive, transitive, and antisymmetric, then it is a *fuzzy partial order* relation. For example, the following membership matrix

$$S = \begin{array}{c} \begin{matrix} a & b & c & d & e & f \end{matrix} \\ \begin{bmatrix} 1 & 0.8 & 0.2 & 0.6 & 0.6 & 0.4 \\ 0 & 1 & 0 & 0 & 0.6 & 0 \\ 0 & 0 & 1 & 0 & 0.5 & 0 \\ 0 & 0 & 0 & 1 & 0.6 & 0.4 \\ 0 & 0 & 0 & 0 & 1 & 0 \\ 0 & 0 & 0 & 0 & 0 & 1 \end{bmatrix} \end{array} \begin{matrix} a \\ b \\ c \\ d \\ e \\ f \end{matrix}$$

is the membership matrix of a fuzzy partial order relation. The pair (X, S), where X is a crisp set and S is a fuzzy partial order in X, is called a *fuzzy partially ordered set* or just *fuzzy poset*.

Proposition 4.6.1 *Suppose that the expression*

$$P = \bigcup_{\alpha\in[0,1]} \alpha\,^{\alpha}P$$

is the resolution form of a fuzzy partial order relation in X. Then, each $^{\alpha}P$ is a partial order in X. On the other hand, if all $^{\alpha}P$, $0 < \alpha \leq 1$, are a nested sequence of distinct partial orders in X, where $\alpha_1 > \alpha_2$ means that $^{\alpha_1}P \subset {}^{\alpha_2}P$ and vice versa, α^1P is nonempty, and $\mathrm{dom}^{\alpha}P = \mathrm{dom}^1P$, *then for any choice of α's in* $(0, 1]$ *that includes $\alpha = 1$, P is a fuzzy partial order relation in X.*

Assume that P is a fuzzy partial order relation. Then, each $x \in X$ is associated with two fuzzy sets: the *dominating* class, written as $P_{\geq}[x]$, where

$$P_{\geq}[x](y) = P(x,y),\ y \in X,$$

and the *dominated* class, written as $P_{\leq}[x]$, where

$$P_{\leq}[x](y) = P(y,x),\ y \in X.$$

We say that x is *undominated* if and only if $P(x,y) = 0$, for all $y \neq x$, and x is *undominating* if and only if $P(y,x) = 0$, for all $y \neq x$. Clearly, if P is a fuzzy partial order in $X = \{x_1, \dots, x_n\}$, the sets of undominated and undominating elements of X are nonempty. Given a nonfuzzy subset A of a crisp set X, then its fuzzy upper-bound $U_{\phi(A)}$ is defined as follows:

$$U_{\phi(A)} = \bigcap_{x_i \in A} P_{\geq}[x_i].$$

For a nonfuzzy partial order, this definition is actually the definition of an upper-bound. Also, for a nonfuzzy subset A of a crisp set X, its fuzzy lower-bound $L_{\phi(A)}$ is defined as follows:

$$L_{\phi(A)} = \bigcup_{x_i \in A} P_{\leq}[x_i].$$

Zadeh did not define fuzzy lattices, nevertheless, their definition followed a few years later [63].

Definition 4.6.1 Assume that X is a fuzzy partially ordered set and A a fuzzy subset of X. Then, A is a fuzzy lattice in X if every pair of elements in X has a fuzzy lower-bound L_ϕ and a fuzzy upper- bound U_ϕ (where both L_ϕ and U_ϕ are fuzzy subsets of X) such that

- $\max\{U_\phi\}(x) \geq A(x)$, for all $x \in X$; and
- $\min\{L_\phi\}(x) \geq A(x)$, for all $x \in X$.

Example 4.6.1 Suppose that $X = \{\zeta_1, \zeta_2, \zeta_3\}$ and P is a fuzzy partial order in X defined as follows:

$$P = \begin{array}{c} \\ \\ \\ \end{array}\!\!\begin{array}{c} \begin{matrix} \zeta_1 & \zeta_2 & \zeta_3 \end{matrix} \\ \begin{bmatrix} 1 & 1 & 0.8 \\ 0 & 1 & 0 \\ 0 & 0.7 & 1 \end{bmatrix} \end{array} \begin{matrix} \\ \zeta_1 \\ \zeta_2 \\ \zeta_3 \end{matrix}$$

Also, assume that $A = \{\zeta_1/0, \zeta_2/0, \zeta_3/0.6\}$. Then,

$$\begin{aligned}
U_{\phi(\zeta_1,\zeta_2)} &= \{\zeta_1/1, \zeta_2/0, \zeta_3/0\},\\
L_{\phi(\zeta_1,\zeta_2)} &= \{\zeta_1/1, \zeta_2/1, \zeta_3/0.8\},\\
U_{\phi(\zeta_1,\zeta_3)} &= \{\zeta_1/0.8, \zeta_2/0, \zeta_3/0\},\\
L_{\phi(\zeta_1,\zeta_3)} &= \{\zeta_1/1, \zeta_2/1, \zeta_3/1\},\\
U_{\phi(\zeta_2,\zeta_3)} &= \{\zeta_1/0, \zeta_2/0, \zeta_3/0.7\},\\
L_{\phi(\zeta_2,\zeta_3)} &= \{\zeta_1/0, \zeta_2/1, \zeta_3/1\}.
\end{aligned}$$

Thus,

$$\max\{U_\phi\} = \{\zeta_1/1, \zeta_2/0, \zeta_3/0.6\},$$

and

$$\max\{L_\phi\} = \{\zeta_1/0, \zeta_2/1, \zeta_3/0.8\}.$$

Clearly, for all $\zeta_i \in X$, $i = 1, 2, 3$, it holds that

$$\max\{U_\phi\}(\zeta_i) \geq A(\zeta_i),$$
$$\min\{L_\phi\}(\zeta_i) \geq A(\zeta_i).$$

This means that A is a fuzzy lattice in X.

A fuzzy partial order relation S in X is a *fuzzy total order* if and only if $S(x, y) > 0$ or $A(y, x) > 0$ for all $x, y \in X$. A fuzzy *preorder* S is a fuzzy relation in Y that is reflexive and transitive. The resolution form of S is

$$S = \bigcup_{\alpha\in[0,1]} \alpha^{\alpha}S, 0 < \alpha \leq 1,$$

where the α-cuts ${}^{\alpha}S$ are nonfuzzy preorders.

A fuzzy *linear order* L is a fuzzy antisymmetric, transitive relation in some set X. In addition, for all $x, y \in X$, that are not equal, either $L(x, y) > 0$ or $L(y, x) > 0$.

4.7 Elements of Fuzzy Graph Theory

Fuzzy graphs have been introduced by Arnold Kaufmann [174]. However, a more elaborate characterization of fuzzy graphs was provided by Azriel Rosenfeld [250]. Both characterizations are based on the remark that any binary relation R in a set S can be regarded as defining a graph whose set of vertices and set of edges are the set S and the relation R, respectively. Based on this, Rosenfeld introduced fuzzy graphs by assuming that any binary fuzzy relation $R : S \times S \to [0, 1]$ can be thought of as defining a *fuzzy graph*. Quite naturally, later on, the theory of fuzzy graphs was further developed (e.g. [221] is a good description of some more recent results). Before proceeding with a formal definition of fuzzy graphs, it is useful to recall some notions from (crisp) graph theory (for a thorough introduction to graphs see Ref. [248]). Readers familiar with these notions can safely skip the next subsection.

4.7.1 Graphs and Hypergraphs

In Section 4.1, we introduced graphs but not so formally. A graph consists of two sets V and E, where V is a finite nonempty set of *vertices* and E is a set of pairs of vertices. Usually, we write $G = (V, E)$ to represent a graph and $V(G)$ and $E(G)$ to represent the sets of vertices and edges of G. If either V or E are not finite sets, the graph is called an *infinite graph*, whereas a graph with a finite vertex set and a finite edge set is called a *finite graph*. There are two kinds of graphs: *undirected* and *directed* graphs. In an undirected graph, the pairs (v_1, v_2) and (v_2, v_1) represent the same edge and, in comparison, these pairs represent different edges in a directed graph. In addition, for the edge (v_1, v_2) of a directed graph, v_1 is its *tail* and v_2 its *head*. A *subgraph* of G is a graph G' such that $V(G') \subseteq V(G)$ and $E(G') \subseteq E(G)$. A path from vertex v_p to vertex v_q in graph G is a sequence of vertices $v_p, v_{i_1}, \dots, v_{i_n}, v_q$ such that $(v_p, v_{i_1}), (v_{i_1}, v_{i_2}), \dots, (v_{i_n}, v_q)$ are edges in $E(G)$. The path is a cycle, if it begins and ends at the same vertex. A path or cycle is simple, if it does not contain the same edge more than once.

Trees and forests are special kinds of graphs.

Definition 4.7.1 A tree is a connected undirected graph with no simple cycles.

The following result helps us to easily identify trees.

Theorem 4.7.1 *An undirected graph is a tree if and only if there is a unique simple path between any two of its vertices.*

A graph that contains no simple cycles and it is not necessarily connected is called a *forest*. A forest has the property that each of its connected components is a tree.

A hypergraph is a generalization of the concept of a graph. The definition that follows is from [28]:

Definition 4.7.2 Suppose that $X = \{x_1, x_2, \dots, x_n\}$ is a finite set and $\mathscr{E} = \{E_i | i \in I\}$ is a family of subsets of X. The family $\mathscr{E}$ is said to be a *hypergraph on X* if

(i) $E_i \neq \emptyset$ for all $i \in I$; and
(ii) $\bigcup_{i \in I} E_i = X$.

The pair $\mathscr{H} = (X, \mathscr{E})$ is called a *hypergraph*. The number $n = \text{card}X$ is called the *order* of the hypergraph. The elements $x_1, \dots, x_n$ are called the *vertices* and the sets $E_1, \dots, E_m$ are called the *edges*. Thus, the big difference between a graph and a hypergraph is that the edges of a hypergraph can be determined by one or more vertices, while the edges of a graph are determined always by two vertices. Also, a hypergraph is simple if $E_i \cap E_j = \emptyset$ for all E_i and E_j. Figure 4.3 shows a nonsimple hypergraph.

A hypergraph $\mathscr{H} = (\mathscr{X}, \mathscr{E})$ can be visualized by its *incidence matrix*, A, that contains m rows representing the edges of $\mathscr{H}$ and n columns that represent the vertices of $\mathscr{H}$, such that

$$A_{ij} = \begin{cases} 1, & \text{if } x_j \in E_i \\ 0, & \text{if } x_j \notin E_i. \end{cases}$$

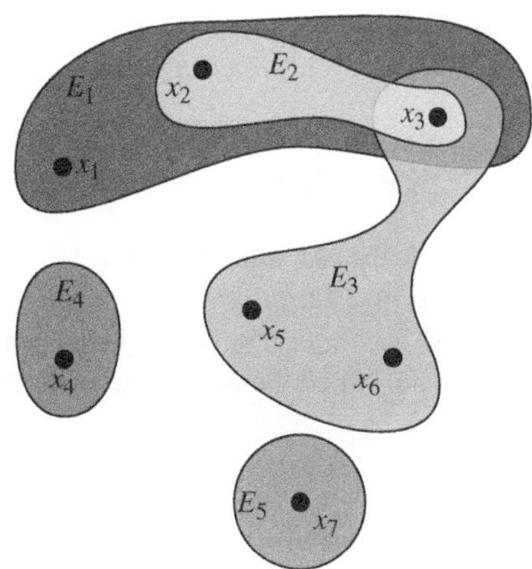

Figure 4.3 A typical hypergraph.

4.7.2 Fuzzy Graphs

Definition 4.7.3 A fuzzy graph $G = (S, V, E)$ is a triple, where S is a nonempty set and $V : S \to [0, 1]$ and $E : S \times S \to [0, 1]$ are a fuzzy set and a fuzzy relation, respectively, such that $E(x, y) \leq V(x) \wedge V(y)$, for all $x, y \in S$.

Remark 4.7.1 Some authors (e.g. see Ref. [190]) define a fuzzy graph $G = (V, E)$ to be a set of vertices and a fuzzy set of edges, respectively.

Example 4.7.1 Assume that $S = \{a, b, c\}$. Also, assume that $V(a) = 0.6$, $V(b) = 0.9$, and $V(c) = 0.75$. Furthermore, assume that $E(a, b) = 0.6$, $E(b, c) = 0.6$, and $E(a, c) = 0.5$. Then, $E(x, y) \leq V(x) \wedge V(y)$, for all $x, y \in S$. Thus, the triple (S, V, E) is a fuzzy graph.

A fuzzy graph $G' = (S, V', E')$ is a partial fuzzy subgraph of $G = (S, V, E)$ if $V'(x) \leq V(x)$, for all $x \in S$, and $E'(x, y) \leq E(x, y)$, for all $x, y \in S$. In other words, G' is a partial fuzzy subgraph of G if $V' \subseteq V$ and $E' \subseteq E$. The fuzzy graph $H = (P, W, F)$ is called a fuzzy subgraph of G induced by P if $P \subseteq V$, $W(x) = V(x)$, for all $x \in P$, and $F(x, y) = E(x, y)$, for all $x, y \in P$. Assume that $0 \leq t \leq 1$, and that

$$V_t = \{x | x \in S \text{ and } V(x) \geq t\},$$
$$E_t = \{(x, y) | (x, y) \in S \times S \text{ and } E(x, y) \geq t\}.$$

Then, $E_t \subseteq V_t \times V_t$ and (S, V_t, E_t) is a fuzzy graph.

Proposition 4.7.1 *If* $0 \leq u \leq v \leq 1$, *then* (S, V_v, E_v) *is a subgraph of* (S, V_u, E_u).

Proposition 4.7.2 *If (S, V', E') is a fuzzy subgraph of (S, V, E), then for any t, $0 \leq t \leq 1$, (S, V'_t, E'_t) is a subgraph of (S, V, E).*

A partial fuzzy subgraph (S, V', E') of a fuzzy graph (S, V, E) *spans* the latter if $V = V'$, and so the former is called a spanning fuzzy subgraph of (S, V, E).

4.7.2.1 Paths and Connectedness

A path ρ in a fuzzy graph is a sequence of distinct vertices $x_0, x_1, \dots, x_n$ such that $E(x_{i-1}, x_i) > 0$, $1 \leq i \leq n$. The length of ρ is equal to n. The consecutive pairs (x_{i-1}, x_i) are called the arcs of the path. The *plausibility degree* of ρ is

$$\bigwedge_{i=1}^{n} E(x_{i-1}, x_i),$$

that is, the plausibility degree of a path is equal to the plausibility degree of the least plausible arc of the path. For "paths" that have length equal to 0, their plausibility degrees are equal to $V(x_0)$. Also, a path ρ' is called a cycle if $x_0 = x_n$ and $n \geq 3$. The diameter of $x, y \in S$, written diam(x, y), is the length of the longest path joining x and y. It was shown in [29] that if diam$(x, y) = 1$, then $E^\infty(x, y) = E(x, y)$. Also, it was shown that if diam$(x, y) = k$, then $E^\infty(x, y) = \bigvee\{E^i(x, y) | i = 1, 2, \dots, k\}$.

Proposition 4.7.3 *If (S, W, F) is a partial fuzzy subgraph of (S, V, E), then $F^\infty \subseteq E^\infty$.*

Definition 4.7.4 Assume that $\rho = x_0, \dots, x_n$ is a path of some fuzzy graph (S, V, E). Then, the μ-length of ρ is defined as follows:

$$\ell(\rho) = \sum_{i=1}^{n} \frac{1}{E(x_{i-1}, x_i)}.$$

If $n = 0$, then $\ell(\rho) = 0$.

Obviously, for $n \geq 1$, it holds that $\ell(\rho) \geq 1$. For any two vertices x and y, their μ-distance $\delta(x, y)$ is the smallest μ-length of any path from x to y.

Proposition 4.7.4 *$\delta(x, y)$ is a metric.*

Proof: A *metric* defines a distance on a set S. In addition, a metric should have as values only nonnegative real numbers. Clearly, δ is such a function. Also, if δ is a metric, then the following should hold true:

(i) $\delta(x, y) = 0$ if and only if $x = y$;
(ii) $\delta(x, y) + \delta(y, z) \geq \rho(x, z)$;
(iii) $\delta(x, y) = \delta(y, x)$.

Obviously, $\delta(x, y) = 0$ if and only if $x = y$, because $\ell(\rho) = 0$ when the length of ρ is equal to 0. Also, $\delta(x, y) = \delta(y, x)$ since the reversal of a path is a path and E is symmetric. Finally, $\delta(x,z) \leq \delta(x,y) + \delta(y,z)$ because the concatenation of a path from x to y and a path from y to z is a path from x to z, and ℓ is additive for the concatenation of paths. □

4.7.2.2 Bridges and Cut Vertices

Assume that $G = (S, V, E)$ is a fuzzy graph. Also, assume that a and b are two distinct vertices and that G' is the partial fuzzy subgraph of G obtained by deleting the edge (a, b). In other words, $G' = (S, V, E')$, where $E'(a, b) = 0$ and $E'(x,y) = E(x,y)$ for all $x, y \in S$ such that $x \neq a$ and $y \neq b$. Then, we say that (a, b) is a *bridge* in G if $E'^{\infty}(u, v) < E^{\infty}(u, v)$ for some u and v. This simply means that the effect of deleting the edge (a, b) is the reduction of the plausibility degree of connectedness between some pair of vertices.

Theorem 4.7.2 *The following statements are equivalent:*

(i) (a, b) *is a bridge;*
(ii) $E'^{\infty}(a, b) < E(a, b)$;
(iii) (a, b) *is not the least plausible edge of any cycle.*

Proof: (ii) ⇒ (i) If (a, b) is not a fuzzy bridge, then $E'^{\infty}(a, b) = E^{\infty}(a, b) \geq E(a, b)$.

(i) ⇒ (iii) If (a, b) is the least plausible edge of a cycle, then any path ρ containing edge (a, b) can be converted into a path not containing (a, b) that is at least as plausible, by using the rest of the cycle as a path from a to b. Thus, (a, b) cannot be a bridge.

(iii) ⇒ (ii) If $E'^{\infty}(a, b) \geq E(a, b)$, then there is a path from a to b not containing the edge (a, b), which has plausibility greater or equal to $E(a, b)$, and this path together with (a, b) forms a cycle of which (a, b) is the least plausible edge. □

Definition 4.7.5 An edge w of a fuzzy graph $G = (S, V, E)$ is called a *cutvertex* if $E'^{\infty}(u, v) < E^{\infty}(u, v)$ for some u and v such that $u \neq w \neq v$.

Definition 4.7.6 A fuzzy graph G is called *block* if it has no cutvertices.

If between every two vertices a and b of G there exist two most plausible paths that are disjoint (except for a and b themselves), G is a block.

4.7.2.3 Fuzzy Trees and Fuzzy Forests

Fuzzy trees and fuzzy forests are special kinds of fuzzy graphs. A fuzzy graph is a fuzzy forest, if the graph consisting of its nonzero edges (i.e. edges whose plausibility degree is not equal to zero) is a forest, and it is a fuzzy tree if this graph is also connected. More precisely, a fuzzy graph $G = (S, V, E)$ is a fuzzy forest if it has a partial fuzzy spanning subgraph $F = (S, V, F)$ which is a forest, where for all edges (x, y) not in F (i.e. such that $F(x, y) = 0$), we have $E(x, y) < F^{\infty}(x, y)$. This simply means that if (x, y) is in G, but it is not in F, there is a path in F between x and y whose plausibility degree is greater than $E(x, y)$. It is not difficult to see that a forest is a fuzzy forest. If G is connected, then F is also connected because any edge of a path in G is either in F, or can be diverted through F. In this case, we call G a fuzzy tree. Figure 4.4 shows the difference between fuzzy simple graphs, fuzzy forests, and fuzzy trees.

The following result characterizes fuzzy forests:

Theorem 4.7.3 *A fuzzy graph G is a fuzzy forest if and only if in any cycle of G, there is an edge (a, b) such that $E(a, b) < E'^{\infty}(a, b)$, where $G' = (S, V, E')$ is the partial fuzzy subgraph obtained by removing the edge (a, b) from G.*

Proof: Assume that (a, b) is an edge that belongs to a cycle and has the property of the theorem and for which $E(a, b)$ is smallest. (Obviously, if G has no cycles, it is a forest and we are done.) If we remove (a, b), the new partial fuzzy subgraph satisfies the path property of a fuzzy forest. If there are still cycles in this graph, we can repeat the process. At each stage of this process, no previously removed edge is more plausible than the edge being currently removed. Thus, the path guaranteed by the property of the theorem contains only edges that have not yet been removed. When no cycles remain, the new partial fuzzy subgraph is a forest H. Suppose that (a, b) is not an edge of H. Then, (a, b) is one of the edges that we removed in the process of constructing H, and there is a path from a to b that is stronger than $E(a, b)$ and that does not contain (a, b) nor any of the edges deleted prior to it. If this path contains edges that were removed later, it can be diverted around them using a path of still more plausible edges; if any of these were removed later, the path can be further diverted; and so on. This process eventually stabilizes with a path consisting entirely of edges of H. Thus, G is a fuzzy forest.

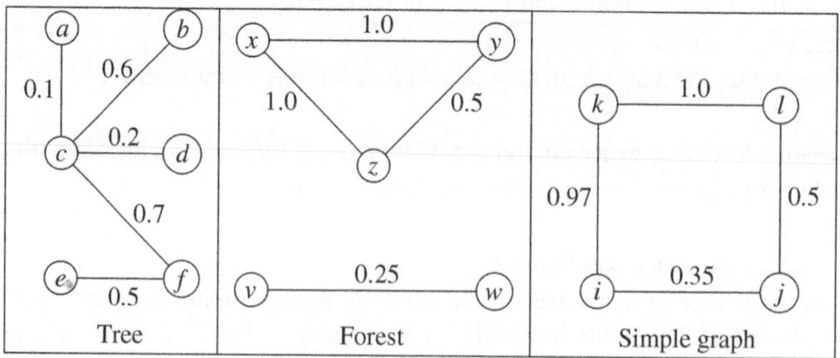

Figure 4.4 Types of fuzzy graphs.

On the other hand, if G is a fuzzy forest and ρ is any cycle, then some edge (a, b) of ρ is not in F. Thus, by definition of a fuzzy forest, we have $E(a, b) < F(a, b) < E'^{\infty}(a, b)$. □

Proposition 4.7.5 *If there is at most one most plausible path between any two vertices of G, then G must be a fuzzy forest.*

Proof: Assume that G is not a fuzzy forest. Then, by the previous theorem there is a cycle ρ in G such that $E(a, b) \geq E'(a, b)$ for all edges (a, b) of ρ. This implies that (a, b) is a most plausible path from a to b. If we select (a, b) to be a least plausible edge of ρ, this means that the rest of ρ is also a most plausible path from a to b which is clearly a contradiction. □

The converse of the previous proposition is not true. The fuzzy graph G can be a fuzzy forest and still have multiple most plausible paths between edges. This happens, because the plausibility of a path is that of its least plausible edge and, as long as this edge lies in H, there is little constraint on the other edges. For example, the fuzzy graph

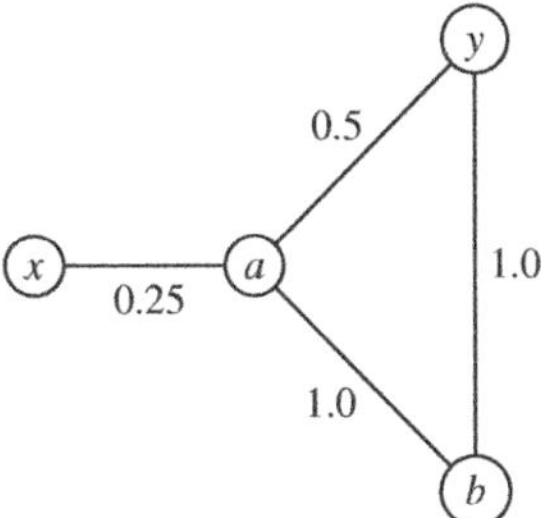

is a fuzzy forest and H consists of all edges except (a, y). The most plausible paths between x and y have plausibility degree 0.25, because the plausibility degree of (x, a) is 0.25. Note that (x, a, b, y) and (x, a, y) are such paths, where the former is contained in H but the latter is not.

Proposition 4.7.6 *If G is a fuzzy forest, the edges of F are just the bridges of G.*

Proof: An edge (a, b) that is not in F is certainly not a bridge because $F(a, b) < H^{\infty}(a, b) \leq E'^{\infty}(a, b)$. On the other hand, assume that (a, b) is an edge of F. Then, if it were not a bridge, there would be a path ρ from a to b not containing (a, b) whose plausibility degree would be greater or equal to $E(a, b)$. This path must contain edges that are not in F, because F is a forest and has no cycles. However, by definition, any such edge (u_i, v_i) can be replaced by a path ρ_i in F whose plausibility degree is greater than $E(u, v)$. The path ρ_i cannot contain (a, b) because all its edges are strictly more plausible than $E(u, v) \geq E(a, b)$. Therefore, by substituting each (u_i, v_i) by ρ_i, we can build a path in H from a to b that does not contain (a, b), giving us a cycle in H, which is a contradiction. □

4.7.3 Fuzzy Hypergraphs[3]

A fuzzy hypergraph is an extension of the notion of a hypergraph. Fuzzy hypergraphs have been introduced by William L. Craine [81] in his doctoral dissertation.

Definition 4.7.7 Suppose that X is a finite set and $\mathscr{E}$ is a finite family of nontrivial fuzzy subsets of X such that

$$X = \bigcup_{E \in \mathscr{E}} \operatorname{supp}(E).$$

The pair $\mathscr{H} = (X, \mathscr{E})$ is called a *fuzzy hypergraph on* X and $\mathscr{E}$ is the set of edges of H. Each element of $\mathscr{E}$ is a fuzzy edge.

The height of $\mathscr{H}$, $h(\mathscr{H})$, is defined as follows:

$$h(\mathscr{H}) = \bigvee \{h(E) | E \in \mathscr{E}\},$$

where $h(E)$ denotes the height of A.

Definition 4.7.8 A fuzzy hypergraph $\mathscr{H} = (X, \mathscr{E})$ is called simple when $\mathscr{E}$ has no repeated fuzzy edges and whenever $E_i, E_j \in \mathscr{E}$ and $E_i \subseteq E_j$, then $E_i = E_j$.

Definition 4.7.9 A fuzzy hypergraph $\mathscr{H} = (X, \mathscr{E})$ is called support simple if $E_i, E_j \in \mathscr{E}$, $E_i \subseteq E_j$, and $\operatorname{supp} E_i = \operatorname{supp} E_j$, then $E_i = E_j$.

Definition 4.7.10 Suppose that $\mathscr{H} = (X, \mathscr{E})$ is a fuzzy hypergraph. Assume that $t \in [0, 1]$ and that

$${}^t\mathbf{E} = \{{}^tE \neq \emptyset | E \in \mathscr{E}\} \quad \text{and} \quad {}^tX = \bigcup_{E \in \mathscr{E}} {}^tE.$$

If ${}^t\mathbf{E} \neq \emptyset$, then the nonfuzzy hypergraph

$${}^tH = ({}^tX, {}^t\mathbf{E})$$

is the t-level hypergraph of $\mathscr{H}$.

The t-level of a fuzzy hypergraph is essentially its α-cut.

---

3 The presentation and the example in this section have been borrowed from [221].

Assume that $\mathscr{A}$ and $\mathscr{B}$ are two families of sets such that for each set $A \in \mathscr{A}$, there is at least one set $B \in \mathscr{B}$ that contains A. In this case, $\mathscr{B}$ *absorbs* $\mathscr{A}$, and we write $\mathscr{A} \sqsubseteq \mathscr{B}$. If $\mathscr{A} \sqsubseteq \mathscr{B}$ and $\mathscr{A} \neq \mathscr{B}$, we write $\mathscr{A} \sqsubset \mathscr{B}$.

Definition 4.7.11 Assume that $\mathscr{H} = (X, \mathscr{E})$ is a fuzzy hypergraph, and for $0 < t \leq h(\mathscr{H})$, assume that ${}^tH = ({}^tX, {}^t\mathbf{E})$ is the t-level hypergraph of $\mathscr{H}$. The sequence of real numbers $\{r_1, r_2, \dots, r_n\}$, $0 < r_n < \cdots < r_1 = h(\mathscr{H})$ that has the following properties:

(i) if $r_{i+1} < s \leq r_i$, then ${}^s\mathbf{E} = {}^{r_i}\mathbf{E}$, and
(ii) ${}^{r_i}\mathbf{E} \sqsubseteq {}^{r_{i+1}}\mathbf{E}$,

is called the *fundamental sequence of* $\mathscr{H}$, and is written as $\mathbf{F}(\mathscr{H})$. The set of r_i-level hypergraphs $\{{}^{r_1}H, {}^{r_2}H, \dots, {}^{r_n}H\}$ is the core set of $\mathscr{H}$ and is written as $\mathbf{C}(\mathscr{H})$.

Fuzzy hypergraphs can be visualized by incidence matrices.

Example 4.7.2 We will examine the fuzzy hypergraph $\mathscr{H} = (X, \mathscr{E})$, where $X = \{a, b, c, d\}$ and $\mathscr{E} = \{E_1, E_2, E_3, E_4, E_5\}$. The hypergraph is represented by the following incidence matrix:

$$\begin{array}{c} \\ a \\ b \\ c \\ d \end{array} \begin{array}{c} \begin{array}{ccccc} E_1 & E_2 & E_3 & E_4 & E_5 \end{array} \\ \begin{bmatrix} 0.7 & 0.9 & 0 & 0 & 0.4 \\ 0.7 & 0.9 & 0.9 & 0.7 & 0 \\ 0 & 0 & 0.9 & 0.7 & 0.4 \\ 0 & 0.4 & 0 & 0.4 & 0.4 \end{bmatrix} \end{array}$$

From this matrix we conclude, for example, that $E_3 : X \to [0, 1]$ satisfies $E_3(a) = 0$, $E_3(b) = 0.9$, $E_3(c) = 0.9$, and $E_3(d) = 0$. Obviously, $h(\mathscr{H}) = 0.9$ and $r_1 = 0.9$. Also,

$${}^{0.9}\mathbf{E} = \{\{a, b\}, \{b, c\}\} = {}^{0.7}\mathbf{E}$$

and

$${}^{0.4}\mathbf{E} = \{\{a, b\}, \{a, b, d\}, \{b, c\}, \{b, c, d\}, \{a, c, d\}\}.$$

Clearly, when $0.4 < t \leq 0.9$, then ${}^t\mathbf{E} = \{\{a, b\}, \{b, c\}\}$. Similarly, when $0 < t \leq 0.4$, then ${}^t\mathbf{E} = \{\{a, b\}, \{a, b, d\}, \{b, c\}, \{b, c, d\}, \{a, c, d\}\}$. Also, note that ${}^{0.9}\mathbf{E} \sqsubseteq {}^{0.4}\mathbf{E}$ and ${}^{0.9}\mathbf{E} \neq {}^{0.4}\mathbf{E}$. From this one concludes that the fundamental sequence is $\mathbf{F}(\mathscr{H}) = \{r_1 = 0.9, r_2 = 0.4\}$ and the set of core hypergraphs is $\mathbf{C}(\mathscr{H}) = \{H_1 = (X_1, \mathbf{E}_1) = {}^{0.9}H, H_2 = (X_2, \mathbf{E}_2) = {}^{0.4}H\}$, where $X_1 = \{a, b, c\}$, $\mathbf{E}_1 = {}^{0.9}\mathbf{E}$, $X_2 = \{a, b, c, d\}$, and $\mathbf{E}_2 = {}^{0.4}\mathbf{E}$. The reader is invited to check whether $\mathscr{H}$ is simple and/or support simple.

4.8* Fuzzy Category Theory

Categories, which were invented by Samuel Eilenberg and Saunders Mac Lane, form a very high-level abstract mathematical theory that unifies all branches of mathematics. The standard reference on category theory is Mac Lane's book-length introduction to categories [202], nevertheless [189] is a highly accessible introduction to category theory. Category theory plays a central role in modern mathematics, theoretical computer science, and, in addition, it is used in mathematical physics (e.g. see Refs. [124, 154]), software engineering [123], etc. However, what makes category theory even more interesting is that it is an alternative to set theory as a foundation for mathematics. Indeed, as Mac Lane [201] pointed out

> It is now possible to develop almost all of ordinary mathematics in a well-pointed topos [i.e. an ordinary category that has some additional properties] with choice and a natural number object. The development would seem unfamiliar; it has nowhere been carried out yet in great detail. However, this possibility does demonstrate one point of philosophical interest: The foundation of Mathematics on the basis of set theory (ZFC) is by no means the only possible one!

Roughly, a category is a universe that includes all mathematical objects of a particular form together with maps between them. These maps must obey a few basic principles. In particular,

Definition 4.8.1 A category $\mathscr{C}$ is made up of

(i) a collection of things that are called $\mathscr{C}$-objects;
(ii) a collection of "bridges" between $\mathscr{C}$-objects that are called $\mathscr{C}$-arrows;
(iii) each arrow f has as *domain* the object dom f and as *codomain* the object cod$f$. If $A = \text{dom}f$ and $B = \text{cod}f$, then we write $f : A \to B$;
(iv) an operation that assigns to each pair (g,f) of $\mathscr{C}$-arrows, such that dom$g$ = cod$f$, a $\mathscr{C}$-arrow $g \circ f$, such that $\text{dom}(g \circ f) = \text{dom}f$ and $\text{cod}(g \circ f) = \text{cod}g$, that is, $g \circ f : \text{dom}f \to \text{cod}g$. In addition, given the arrows $f : A \to B$, $g : B \to C$, and $h : C \to D$, then $h \circ (g \circ f) = (h \circ g) \circ f$;
(v) for each $\mathscr{C}$-object A there is a $\mathscr{C}$-arrow $\text{id}_A : A \to A$ called the *identity* arrow, such that for any $f : A \to B$ and $g : B \to C$, $\text{id}_B \circ f = f$ and $g \circ \text{id}_B = g$.

Example 4.8.1 The collection of all sets and functions between them forms the category that is traditionally denoted by **Set**.

4.8.1 Commutative Diagrams

In category theory *commutative diagrams* play, the role equations play in algebra. In the simplest case, a commutative diagram can be identified with two different paths starting from the same object A and ending with the same object B in which the composition of the arrows that make up the first path and the composition of the arrows of the second path yield two arrows that have the same effect (i.e. when applied to the same object(s), they yield the same result). For example, the identity law can be expressed by the following commutative diagrams:

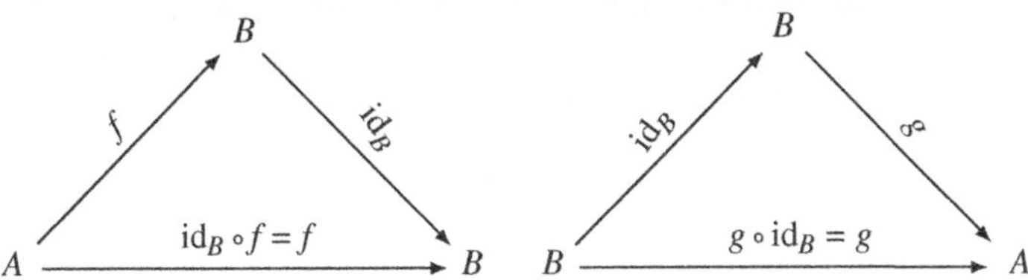

4.8.2 Categories of Fuzzy Structures

It is not difficult to define categories of fuzzy structures. For instance, Carol Walker [291] and Siegfried Gottwald [146] have defined such categories. However, some other categories of fuzzy structures emerged from efforts to define fuzzy models of *linear logic* [136]. In particular, Michael Barr [23], Basil Papadopoulos, and Apostolos Syropoulos [231, 232, 272] have introduced such categories.

Typically, one defines a general category of fuzzy structures (preferably fuzzy sets) and then finds a way to embed them into a categorical model of linear logic. The categories **SET**(L) of L-fuzzy sets (e.g. see Refs. [116, 232, 269]) are a simple example of a category of fuzzy sets:

Definition 4.8.2 Let L be a frame. Category **SET**(L) has as objects Goguen sets, that is, pairs (S, σ), where S is a set and $\sigma \in L^S$. Category **SET**(L) has as arrows maps between Goguen sets: Given two Goguen sets (S, σ) and (T, τ), a map $f : (S, \sigma) \to (T, \tau)$ is a function $f : S \to T$ such that $\sigma \leq f^{\leftarrow}(\tau)$.

Clearly, composition of maps is function composition, which is associative, and the identity arrows are the identity functions. Now that we have defined a category of fuzzy structures, we can embed this category into one that has known properties. Here, we will show how we can embed it into *Dialectica* and *Chu* categories.

In 1958, the Austrian logician Kurt Friedrich Gödel [156] published in the journal *Dialectica* an interpretation of intuitionistic arithmetic (i.e. the intuitionistic analog of Peano arithmetic known as Heyting arithmetic) in a quantifier-free theory of functionals of finite type, which has come to be known as *Dialectica Interpretation*. The set of types is generated inductively by the following rules:

(i) 0 is a type;
(ii) if σ and τ are types, then so is $\sigma \to \tau$; and
(iii) if σ and τ are types, then so is $\sigma \times \tau$.

Objects of type 0 are the natural numbers $0, 1, 2, \ldots$. Also, objects of a type $(\sigma \to \tau) \to \rho$ take functions as arguments and are usually called functionals.

Valeria de Paiva [90] presented a categorical version of the Dialectica interpretation. In her thesis, she presented two categories with arrows that correspond to the Dialectica interpretation of implication. Later on she presented one more categorical version of the Dialectica interpretation [91]. The term *Dialectica space* refers to objects of *Dialectica* categories. The Dialectica categories are models of linear logic. This is particularly interesting, as categorical semantics model derivations (i.e. proofs) are not just used to deduce whether a theorem is true or not. Another widely known categorical model of linear logic is based on the Chu construct described by Po-Hsiang Chu in an appendix of [22].

A *Chu space* over an alphabet Σ (i.e. an arbitrary set whose structure is of no importance) is a triplet (X, r, A), where X and A are arbitrary sets and $r : X \times A \to \Sigma$ is a function. Function r relates the elements of X with the elements of A. For example, suppose that $\Sigma = \{0, 1\}$ and that A stands for the set of open subsets of X. Then, $r(x, a) = 1$ if x belongs to the open subset a, else $r(x, a) = 0$. Following a similar way of thinking, one can represent any relational structure (e.g. groups, vector spaces, categories, etc.,) as a Chu space. Assume that $\mathscr{A} = (X, r, A)$ and $\mathscr{B} = (Y, s, B)$ are two Chu spaces. Then, a transformation from $\mathscr{A}$ to $\mathscr{B}$ is just a pair of functions $(f, \overline{f})$, where $f : X \to Y$ and $\overline{f} : B \to A$, such that

$$s(f(x), b) = r(x, \overline{f}(b)), \text{ for all } x \in X \text{ and all } b \in B, \tag{4.3}$$

or, as a commutative diagram:

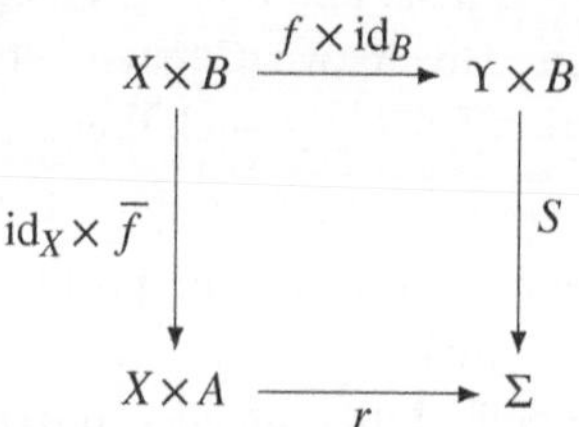

This condition is called the *adjointness condition*. We can build a Chu category **Chu**(Σ) with objects all Chu spaces and with arrows pairs of functions that fulfill the adjointness condition. Arrow composition is the usual composition of functions pairwise. For any Chu space (X, r, A), it is easy to verify that the identity arrow is the pair of functions $(\mathrm{id}_X, \mathrm{id}_A)$.

4.8.3 Embedding Fuzzy Categories to Chu Categories

We consider a subcategory of **SET**(L) that has the following property: for each pair of objects (S, σ) and (T, τ), a function $f : S \to T$ is an arrow between them if and only if $\sigma = f^{\leftarrow}(\tau)$. We call this subcategory **SET**$(L)_=$. Although the restriction imposed on the arrows of **SET**$(L)_=$ may seem too strong, the following proposition shows that there are enough arrows in **SET**$(L)_=$.

Proposition 4.8.1 *Suppose that (S, σ) and (T, τ) are Goguen sets and that $f : S \to T$ is an injective function satisfying $f^{\rightarrow}(\sigma) = \tau$. Then, $f : (S, \sigma) \to (T, \tau)$ is an arrow such that $\sigma = f^{\leftarrow}(\tau)$.*

This proposition is immediate from the following remarks:

Remark 4.8.1 If $f : L \to M$, $g : L \leftarrow M$ are isotone maps (i.e. maps that are monotone increasing and therefore order-preserving) between posets, then $f \dashv g$ implies that $f \circ g \circ f = f$. In addition, if f is injective, then $g \circ f = \mathrm{id}_L$.

Remark 4.8.2 If $f : X \to Y$ is injective, then $f^{\rightarrow}$ is injective, which implies that $f^{\leftarrow} \circ f^{\rightarrow} = \mathrm{id}_{L^X}$.

A functor is a map between categories and an embedding is realized by showing that there is a functor between two categories. Thus, one can say that an embedding is a functor.

Definition 4.8.3 A *functor* from category $\mathscr{C}$ to category $\mathscr{D}$ is a function that assigns

(i) to each $\mathscr{C}$-object A, a $\mathscr{D}$-object $F(A)$;
(ii) to each $\mathscr{C}$-arrow $f : A \to B$ a $\mathscr{D}$-arrow $F(f) : F(A) \to F(B)$ such that
   (a) $F(\mathrm{id}_A) = \mathrm{id}_{F(A)}$, that is, the identity arrow on A is assigned the identity on $F(A)$;
   (b) $(g \circ f) = F(g) \circ F(f)$ if $g \circ f$ is defined.

Let us proceed with the definition of the functor $\mathfrak{F}$ from the subcategory **SET**$(L)_=$, for some fixed L, to the category **Chu**(L). We first define the object part:

Definition 4.8.4 (Object Part) Let (S, σ) be an object of **SET**$(L)_=$. Then, functor $\mathfrak{F}$ maps it to the Chu space $(S, r, \{\sigma\})$, where $r(s, \sigma) = \sigma(s)$.

The following result is a direct consequence of the previous definition:

Corollary 4.8.1 *Functor $\mathfrak{F}$ is injective on objects.*

We now define the morphism part of the functor:

Definition 4.8.5 (**Arrow Part**) Suppose that $\mathfrak{F}(S,\sigma) = (S, r, \{\sigma\})$ and $\mathfrak{F}(T,\tau) = (T, s, \{\tau\})$. Moreover, suppose that $f : (S,\sigma) \to (T,\tau)$ is an arrow of $\mathbf{SET}(L)_{=}$. Then $\mathfrak{F}(f) = (f, g)$, where $g(\tau) = \sigma$.

The following result is a direct consequence of the above definitions:

Theorem 4.8.1 *Any subcategory* $\mathbf{SET}(L)_{=}$ *fully embeds*[4] *into a category* $\mathbf{Chu}(L)$.

Goguen sets can be represented more naturally as Dialectica spaces. Generally speaking, Dialectica spaces are Chu spaces with more arrows between them. Although there are at least three different families of Dialectica categories, here we are interested only in the categories $\mathbf{Dial}_L\mathbf{Set}$, where L is a *lineale* [92], that is, a monoidal poset with additional structure.

Definition 4.8.6 A monoidal poset is a poset $(L, \leq)$ with a given symmetric monoidal structure $(L, \circ, \mathbf{1})$. That is, a set L equipped with a binary relation $\leq$, together with a monoid structure $(\circ, \mathbf{1})$ consisting of a (order-preserving) multiplication $\circ : L \times L \to L$ and a distinguished object $\mathbf{1}$ of L. We write a monoidal poset as a quadruple $(L, \leq, \circ, \mathbf{1})$.

Operator "$\circ$" is a logical conjunction operator, which is not necessarily idempotent. In addition, $\mathbf{1}$ is not necessarily the top element of L. Suppose now that L is a monoidal poset and a, b are elements of L. Then, if there is an $x \in L$, which is the largest element of L such that $a \circ x \leq b$, this element is denoted by $a \multimap b$.

Definition 4.8.7 A lineale (or close poset) is a monoidal poset such that $a \multimap b$ exists for all $a, b \in L$. We write a lineale as a quintuple $(L, \leq, \circ, \mathbf{1}, \multimap)$.

In addition, it holds that $z \circ x \leq y$ if and only if $z \leq (x \multimap y)$. This means that $\circ$ and $\multimap$ form an *adjoint pair*. Practically, this is a proof of the following:

---

4 Assume that $F : \mathcal{S} \to \mathcal{C}$ is a functor, where $\mathcal{S}$ is a subcategory of $\mathcal{C}$ (i.e. $\mathcal{S}$ consists of a subcollection of the collections of objects and arrows of $\mathcal{C}$). Then, $\mathcal{S}$ is a *full subcategory* if for every pair of $\mathcal{S}$-objects A and B and to every $\mathcal{C}$-arrow $g : F(A) \to F(B)$, there is an $\mathcal{S}$-arrow $f : A \to B$ such that $g = F(f)$.

Corollary 4.8.2 *Given a frame L, the quintuple $(L, \leq, \wedge, \mathbf{1}, \Rightarrow)$, where $\Rightarrow$ is the exponential, is a lineale.*

We are now ready to define the family of categories $\mathbf{Dial}_L\mathbf{Set}$ [91]:

Definition 4.8.8 Let $(L, \leq, \circ, \mathbf{1}, \multimap)$ be a lineale. Then, the objects of a category $\mathbf{Dial}_L\mathbf{Set}$ are triplets (X, r, A), where X and A are arbitrary sets and $r : X \times A \to L$ is a function. Suppose that (X, r, A) and (Y, s, B) are two objects. Then, an arrow between them is a pair (f, g), where $f : X \to Y$ and $g : B \to A$, such that

$$r(x, g(b)) \leq s(f(x), b), \quad \forall x \in X, \forall b \in B.$$

Note that any category $\mathbf{Dial}_L\mathbf{Set}$ is a symmetric monoidal closed category with involution and products, and it is a categorical model of intuitionistic linear logic.

By following a line of arguments similar to those employed for Theorem 4.8.1, we can prove the following theorem:

Theorem 4.8.2 *Any category* $\mathbf{SET}(L)$ *embeds fully into a Dialectica category* $\mathbf{Dial}_L\mathbf{Set}$.

4.8.4 Fuzzy Categories

Not so surprisingly, a category of fuzzy structures is not a fuzzy structure itself. Indeed, Alexander Šostak [265, 266] was the first researcher who had realized this. In order to remedy this situation, Šostak introduced a new structure that mimics the way fuzzy graphs are defined. However, these new structures make use of GL-monoids:

Definition 4.8.9 A GL-monoid is a triple $(L, \leq, \circledast)$, where L is a complete lattice and $\circledast$ is a binary operator such that:

(i) $a \leq b$ implies $a \circledast c \leq b \circledast c$, for all $a, b, c \in L$;
(ii) $a \circledast b = b \circledast a$, for all $a, b \in L$;
(iii) $a \circledast (b \circledast c) = (a \circledast b) \circledast c$, for all $a, b, c \in L$;
(iv) $a \circledast \mathbf{1} = a$, for all $a \in L$;
(v) $a \circledast \mathbf{0} = \mathbf{0}$, for all $a \in L$;
(vi) $a \circledast (b_1 \vee \cdots \vee b_n) = (a \circledast b_1) \vee \cdots \vee (a \circledast b_n)$, for all $a, b_1, \ldots, b_n \in L$;
(vii) when $a \leq b$, then there is a $c \in L$ such that $a = b \circledast c$.

Let us proceed with the definition of L-fuzzy categories:

Definition 4.8.10 An L-fuzzy category (where L is a GL-monoid) is a quintuple

$$\mathscr{C} = (\mathrm{Ob}(\mathscr{C}), \omega, M(\mathscr{C}), \mu, \circ),$$

where

(i) $\mathscr{C}_\perp = (\text{Ob}(\mathscr{C}), M(\mathscr{C}), \circ)$ is a usual (classical) category called the bottom frame of the fuzzy category $\mathscr{C}$;
(ii) $\omega : \text{Ob}(\mathscr{C}) \to L$ is an L-fuzzy subclass of the class of objects $\text{Ob}(\mathscr{C})$ of $\mathscr{C}_\perp$;
(iii) $\mu : M(\mathscr{C}) \to L$ is an L-subclass of the class of morphisms $M(\mathscr{C})$ of $\mathscr{C}_\perp$.

In addition, ω and μ must satisfy the following conditions:

(i) if $f : X \to Y$, then $\mu(f) \leq \omega(X) \wedge \omega(Y)$;
(ii) $\mu(g \circ f) \geq \mu(g) \circledast \mu(f)$ whenever the composition $g \circ f$ is defined;
(iii) if $e_X : X \to X$ is the identity morphism, then $\mu(e_X) = \omega(X)$.

Given an L-fuzzy category $\mathscr{C}$ and $X \in \text{Ob}(\mathscr{C})$, then $\omega(X)$ is the *degree* to which a potential object X of the L-fuzzy category $\mathscr{C}$ is indeed its object. Similarly, if $f \in M(\mathscr{C})$, then $\mu(f)$ is the degree to which the map f is indeed a morphism $\mathscr{C}$.

Definition 4.8.11 Assume that $\mathscr{C} = (\text{Ob}(\mathscr{C}), \omega, M(\mathscr{C}), \mu, \circ)$ is a fuzzy category. Then, an L-fuzzy subcategory of $\mathscr{C}$ is an L-fuzzy category

$$\mathscr{C}' = (\text{Ob}(\mathscr{C}), \omega', M(\mathscr{C}), \mu', \circ),$$

where $\omega' \leq \omega$ and $\mu' \leq \mu$. $\mathscr{C}'$ is a full subcategory of $\mathscr{C}$ if

$$\mu'(f) = \mu(f) \wedge \omega'(X) \wedge \omega'(Y),$$

for all $f \in M_{\mathscr{C}}(X, Y)$ and all $X, Y \in \text{Ob}(\mathscr{C})$.

In the preface of the first edition of his book [202], Mac Lane noted that:

> Since a category consists of arrows, our subject could also be described as learning how to live without elements, using arrows instead. This line of thought, present from the start, comes to a focus in Chapter VIII, which covers the elementary theory of abelian categories and the means to prove all of the diagram lemmas without ever chasing an element around a diagram.

If we are not interested in objects in category theory, why should we care about objects in some sort of fuzzy category theory? And if we do not care about objects, in general, why should we care about their membership degree to a given category? In fact, we can assume that all possible candidates are objects of the category. Thus, fuzzy categories should be simpler than Šostak's fuzzy categories. A first attempt to define such categories was presented by the first author [274].

Definition 4.8.12 A fuzzy category $\mathscr{C}$ comprises

(i) a collection of entities called *objects*;
(ii) another collection of entities called *arrows* or *morphisms*;
(iii) operations assigning to each $\mathscr{C}$-arrow f a $\mathscr{C}$-object $A = \mathrm{dom}f$, its domain, a $\mathscr{C}$-object $B = \mathrm{cod}f$, its codomain, and a plausibility degree $\rho = \mathrm{p}f$. Typically, the plausibility degree is a real number belonging to the unit interval. These operations on f are indicated by displaying f as an arrow starting from A and ending at B with plausibility degree ρ:

$$A \underset{\rho}{\overset{f}{\rightarrow}} B \quad \text{or} \quad f : A \overset{\rho}{\rightarrow} B;$$

(iv) an operation assigning to each pair (g, f) of arrows with $\mathrm{dom}g = \mathrm{cod}f$, an arrow $g \circ f$, the *composite* of f and g, having $\mathrm{dom}(g \circ f) = \mathrm{dom}f$, $\mathrm{cod}(g \circ f) = \mathrm{cod}g$, and $\mathrm{p}(g \circ f) = \min\{\mathrm{p}f, \mathrm{p}g\}$. This operation and the previous three are subject to the *associative law*: Given the configuration

$$A \underset{\rho_1}{\overset{f}{\rightarrow}} B \underset{\rho_2}{\overset{g}{\rightarrow}} C \underset{\rho_3}{\overset{h}{\rightarrow}} D$$

of $\mathscr{C}$-objects and $\mathscr{C}$-arrows, then $h \circ (g \circ f) = (h \circ g) \circ f$;
(v) an *assignment* to each $\mathscr{C}$-object B of a $\mathscr{C}$-arrow $\mathbf{1}_B : B \overset{1}{\longrightarrow} B$, called the *identity arrow on B*, such that the following *identity law* holds true:

$$\mathbf{1}_B \circ f = f \quad \text{and} \quad g \circ \mathbf{1}_B = g$$

for any $\mathscr{C}$-arrows $f : A \overset{\rho_f}{\longrightarrow} B$ and $g : B \overset{\rho_g}{\longrightarrow} A$.

4.9* Fuzzy Vectors[5]

Suppose that $x, y \in \mathbb{R}^n$. Then, the set

$$[x, y] = \{z = \lambda x + (1 - \lambda)y | 0 \leq \lambda \leq 1\}$$

is called a segment with endpoints x and y. A subset S of $\mathbb{R}^n$ is called convex if for any pair $x, y \in S$ it contains the entire segment $[x, y]$. The open sets of the Euclidean topology on $\mathbb{R}^n$ are given by arbitrary unions of the sets $B_r(p)$:

$$B_r(p) = \{\mathbf{x} | \mathbf{x} \in \mathbb{R}^n \text{ and } d(\mathbf{p}, \mathbf{x}) < r\},$$

for all $r > 0$ and $\mathbf{p} \in \mathbb{R}^n$, and the Euclidean metric d. If S is an open set of $\mathbb{R}^n$, then its complement, $\mathbb{R}^n \setminus S$, is closed. Also, a set in $\mathbb{R}^n$ is bounded if and only if it is contained inside some *ball* $x_1^2 + x_2^2 + \cdots + x_n^2 \leq R^2$ of finite radius. In what follows, a fuzzy number is a fuzzy set A on $\mathbb{R}$ if $\mathrm{core}(A) \neq \varnothing$, ${}^{\alpha}A$, where $\alpha \in (0, 1]$, are closed intervals, and $\mathrm{supp}(A)$ is bounded.

5 The short exposition that follows is based on [65, 236].

Assume that $\mathbf{V}$ is a fuzzy relation $\mathbb{R}^n$. Then, $\mathbf{V}$ is an n-dimensional *fuzzy vector* if it has the following properties:

(i) core($\mathbf{V}$) $\neq \emptyset$;
(ii) for all $\alpha \in (0,1]$, ${}^\alpha\mathbf{V}$ are closed and convex subsets of $\mathbb{R}^n$;
(iii) supp($\mathbf{V}$) is bounded.

Suppose that $\mathcal{K}(\mathbb{R}^n)$ is a family of all nonempty compact subsets of $\mathbb{R}^n$ (recall that a compact set K is a set for which every open cover of K contains a finite subcover of K). Then, a fuzzy relation $\mathbf{V}$ in $\mathbb{R}^n$ is *compact* if ${}^\alpha\mathbf{V} \in \mathcal{K}(\mathbb{R}^n)$ for all $\alpha \in [0,1]$. Also, $\mathbf{V}$ is *convex* if ${}^\alpha\mathbf{V}$ is convex for all $\alpha \in [0,1]$.

For any $1 \leq i \leq n$, the ith projection of the n-dimensional vector $\mathbf{V}$ is denoted by $[\mathbf{V}]_i$. The membership function of $[\mathbf{V}]_i$ is given below:

$$[\mathbf{V}]_i(y) = \max\{\mathbf{V}(x_1,x_2,\dots,x_n)|(x_1,x_2,\dots,x_n \in \mathbb{R}^n \text{ and } x_i = y\}.$$

Also, it holds that ${}^\alpha[\mathbf{V}]_i = [{}^\alpha\mathbf{V}]_i$ for all $\alpha \in (0,1]$.

The projections of a fuzzy vector are fuzzy numbers and the Cartesian product of fuzzy numbers is a fuzzy vector, as the following results show:

Theorem 4.9.1 *If $\mathbf{V}$ is an n-dimensional fuzzy vector, then its projections $[\mathbf{V}]_i$, $1 \leq i \leq n$, are fuzzy numbers.*

Proof: Each set core($[\mathbf{V}]_i$) $\neq \emptyset$, where $0 \leq i \leq n$, because core($\mathbf{V}$) $\neq \emptyset$. Also, for any $\alpha \in (0,1]$, the α-cut ${}^\alpha\mathbf{V}$ is a closed bounded convex set and its projections ${}^\alpha[\mathbf{V}]_i = [{}^\alpha\mathbf{V}]_i$ are closed intervals. In addition, supp($\mathbf{V}$) is a bounded set, and supp($[\mathbf{V}]_i$, where $1 \leq i \leq n$, are also bounded sets, which completes the proof. □

Theorem 4.9.2 *Suppose that V_i, where $1 \leq i \leq n$, are fuzzy numbers. Then, the fuzzy set $\mathbf{V} = V_1 \times V_2 \times \cdots \times V_n$ is an n-dimensional fuzzy vector.*

Proof: We have to verify that the fuzzy set $\mathbf{V} = V_1 \times V_2 \times \cdots \times V_n$ has the properties of a fuzzy vector. First, core($\mathbf{V}$) = core(V_1) $\times \cdots \times$ core(V_n) $\neq \emptyset$ if and only if core(V_i) $\neq \emptyset$, where $1 \leq i \leq n$. Second, ${}^\alpha\mathbf{V} = {}^\alpha\mathbf{V}_1 \times \cdots \times {}^\alpha\mathbf{V}_n$ is closed and convex for all $\alpha \in (0,1]$ if and only if each ${}^\alpha\mathbf{V}_i$ is a closed interval. Third, supp($\mathbf{V}$) = supp(V_1) $\times \cdots \times$ supp(V_n) is bounded if and only if each supp(V_i) is bounded. □

Theorem 4.9.3 *Assume that $\mathbf{V} = V_1 \times V_2 \times \cdots \times V_n$, where each V_i is a fuzzy number. Then, $[\mathbf{V}]_i = V_i$ for all $1 \leq i \leq n$.*

Proof: For any i such that $1 \leq i \leq n$, it follows from the definition of the membership function of $[\mathbf{V}]_i$ that for all $y \in \mathbb{R}$ the following holds:

$$[\mathbf{V}]_i(y) = \max\{\min\{V_1(x_1),\dots,V_i(x_i),\dots,V_n(x_n)\}|(x_1,x_2,\dots,x_n) \in \mathbb{R}^n \text{ and } x_i = y\}.$$

Also, $\text{core}(V_j) \neq \emptyset$ for all $1 \leq j \leq n$, and so it is obvious that $[\mathbf{V}]_i(y) = V_i(y)$ for all $y \in \mathbb{R}$ and this completes the proof. □

The following result "explains" how one can perform calculations using fuzzy vectors.

Theorem 4.9.4 *Assume that f_F is the fuzzy extension of a continuous function $f : \mathbb{R}^n \to \mathbb{R}$, where $n \geq 1$. Also, suppose that* **X** *is an n-dimensional fuzzy vector. Then, $f_F(\mathbf{X})$ is a fuzzy number whose α-cuts ${}^{\alpha}f_F(\mathbf{X})$ for all $\alpha \in (0, 1]$ are given by*

$$^{\alpha}f_F(\mathbf{X}) = f(^{\alpha}\mathbf{X}).$$

4.10 Applications

The most striking usage example of fuzzy relations are *fuzzy databases* [132]. A database is a collection of related data that are stored in a computer system and are organized in such a way so that they can be easily accessed, managed, and updated. A database consists of tables that are made of rows and columns. Each row is like a record in Pascal, a struct in C/C++, or a tuple in Haskell, which, in turn, can be viewed as an element of some Cartesian product, that is, a relation. Naturally, the columns of such a table contain the same kind of data (e.g. age of people) and this makes it possible to query the database about these data (e.g. how many people are older than 55 years?). Databases that have this kind of characteristics are called *relational databases*.

In the simplest case, each row of a fuzzy relational database has one more field when compared to a crisp database where a membership degree of this row is stored. Alternatively, one can think that this membership degree is some sort of importance degree or a fulfilment degree of a condition. Furthermore, there are database designs that allow fuzzy values to be stored in fuzzy attributes using fuzzy sets. Of course, a fuzzy database should allow fuzzy queries using fuzzy or non-fuzzy data. For example, the fuzzy query language FSQL [132], an extension of SQL, can be used to perform fuzzy queries. The following is an example of a fuzzy query borrowed from [276]:

```
SELECT City, Inhabitants, CDEG(*)
FROM   Population
WHERE Country = 'Greece'
   AND Inhabitants FGEQ $[200,300,650,800] THOLD .75
   AND Inhabitants IS NOT UNKNOWN
ORDER BY 3 DESC;
```

Function CDEG computes the Compatibility DEGree of each row, which is the fulfillment degree of each row to the fuzzy condition included in the WHERE clause. FGEQ is a fuzzy relational operator (fuzzy greater or equal) and the numbers in the square brackets define a trapezoidal fuzzy number. Also, number 0.75 is the minimum threshold. Since this command uses the FGEQ operator, the last two values of the trapezoid will not be used. This means that if the number of inhabitants is equal to or greater than 300, then the degree will be 1. Clearly, when the number of inhabitants is less than or equal to 200, then the degree will be 0. Here UNKNOWN is a fuzzy variable.

Fuzzy diagnosis is another area where fuzzy relations are used (see Ref. [185] for an overview). A basic problem in medical diagnosis is how to handle *nonstatistical uncertainty*. A typical example of this kind of uncertainty is "high" blood pressure. More specifically, for a normally hypotonic patient "high" blood pressure is not the same as it is for a normally hypertonic patient. In addition, different experts have different opinions about the values of "high" blood pressure. Finally, definitions depend on the medical context (e.g. during anaesthesia "high" blood pressure may be "normal" under *normal* conditions). Although doctors measure blood pressure and know exactly its value, it is better to say whether it is "high" or "low" instead of giving its exact value (e.g. 145 mmHg).

The problem of medical diagnosis has been formalized as follows [185]. Assume that $\mathbf{C} = \{C_1, C_2, \dots, C_M\}$ is a set of M possible diagnoses for some medical problem. Then, the elements of $\mathbf{C}$ can be a list of disorders, types of blood cells, etc., and the set itself is called set of class labels. Suppose that $\mathbf{x} \in \mathbb{R}^n$ is the description of an object (e.g. a patient, a cell). Then, the components of $\mathbf{x}$ encode various features such as clinical measurements and findings, details from the patient's history, test results, image parameters like gray-level intensity etc. A crisp classifier is any mapping

$$D : \mathbb{R}^n \to \mathbf{C}$$

This means that D maps the various features of a patient to a single class label which is interpreted as a *diagnosis*.

In the case of a fuzzy diagnosis, we are making use of a classifier that relies on fuzzy sets to perform its task. These fuzzy sets are employed in varius ways and in different stages of the classifier design. The most obvious choices are

- **Fuzzy inputs**: We are no longer using the various input values (e.g. measurements) but their "fuzzy" versions. For example, if the value of a blood pressure mesurement is 145 mmHg, then we use a triple $(0.0, 0.4, 0.6)$, where $\text{low}_{\text{bp}}(145) = 0.0$, $\text{medium}_{\text{bp}}(145) = 0.4$, and $\text{high}_{\text{bp}}(145) = 0.6$. The three fuzzy sets low_{bp}, $\text{medium}_{\text{bp}}$, and high_{bp} characterize low, medium, and high blood pressure.

- **Fuzzy reasoning**: The classifier is implemented atop some fuzzy inference system.
- **Fuzzy classes**: It is a fact the some of the disorders can occur simultaneously in the same patient but with varying degrees. However, in the crisp case, we assume that this cannot happen. Thus, each patient can be assumed to belong to different classes.

A fuzzy classifier realizes the mapping

$$\mathcal{D} : \mathbb{R}^n \to [0,1]^M,$$

that is, $\mathcal{D}(\mathbf{x}) = (C_1(\mathbf{x}), \ldots, C_M(\mathbf{x}))$, where $C_i(\mathbf{x})$ is the degree to which $\mathbf{x}$ belongs in class C_i. There are several, different interpretations of this degree and here are the most reasonable ones:

- Typicality of case $\mathbf{x}$ with respect to diagnosis C_i.
- Severity of disorder C_i in $\mathbf{x}$.
- Support for the hypothesis that C_i is the real diagnosis for $\mathbf{x}$ based on available evidence.
- Probability that C_i is the real diagnosis for $\mathbf{x}$.

It is possible to impose more restrictions to $\mathcal{D}(\mathbf{x})$ so to get a single class label from $\mathbf{C}$. The following gives the most "supported" class:

$$\mathcal{D}(\mathbf{x}) = C_j \in \mathbf{C} \Longleftrightarrow C_j(\mathbf{x}) = \max_i C_i(\mathbf{x}).$$

Exercises

4.1 Consider the relation $R_2 = \{(a,b) | (a = b \text{ or } a = -b) \text{ and } a, b \in \mathbb{Z}\}$. Which properties of relations does the R_2 relation have?

4.2 Generalize the concept of cylindrical extension to multidimensional cases.

4.3 Assume that

$$\begin{bmatrix} 0.2 & 0 & 0.5 \\ 0.4 & 0.2 & 1 \\ 0.5 & 0.6 & 0.5 \end{bmatrix} \quad \text{and} \quad \begin{bmatrix} 0.4 & 0.5 & 0 \\ 0.6 & 1 & 0.4 \\ 1 & 0.8 & 0.5 \end{bmatrix}$$

are the membership matrices of two fuzzy relations. Find their max–min and their min–max composition.

4.4 Write down the membership matrix of the relation depicted in Figure 4.2.

4.5 Suppose that $Q : A \times A \rightarrow [0, 1]$ is a fuzzy binary relation on $A = \{a, b, c\}$ whose membership matrix is

$$Q = \begin{bmatrix} 0.7 & 0.8 & 0.3 \\ 0.2 & 0.6 & 0.9 \\ 0.4 & 0 & 0.1 \end{bmatrix}.$$

Find its transitive closure using min–max composition.

4.6 Write down the incidence matrix of the hypergraph shown in Figure 4.3 on page 91.

4.7 Verify that an α-cut of a similarity relation is an equivalence relation.

4.8 Verify that $B = \{\zeta_1/0.5, \zeta_2/0, \zeta_3/0.9\}$ is not a fuzzy lattice for X and P as defined in Example 4.6.1.

5

Possibility Theory

The English economist George Lennox Sharman Shackle [256] proposed that uncertainty is related to surprise. For example, when Jane says that an event can happen, then she will not be surprised if this event will actually happen. In general, when we say that an event can "happen", we actually mean that it is a "possible" event. Since there are degrees of surprise, it is absolutely reasonable to talk about degrees of possibility. Thus, in a way, Shackle laid the theoretical basis for a theory of possibility. Indeed, Zadeh [314] used fuzzy sets to define a mathematical *theory of possibility*. Later on, others tried to describe possibilities using different tools that culminated to what is now known as possibility theory.

5.1 Fuzzy Restrictions and Possibility Theory

Although Zadeh was quite aware of the measure-theoretic axiomatization of probability[1] (see Ref. [257] for a discussion of Kolmogorov's work), still he opted to define the notion of *possibility* in terms of *fuzzy restrictions* [309]. In Zadeh's own words, a fuzzy restriction is "a fuzzy relation which acts as an elastic constraint on the values that may be assigned to a variable." Here the term "elastic" does not have any special meaning, and it just means that the constraint is not rigid. Let us first give an informal description of fuzzy restrictions.

Consider the fuzzy proposition *Emma is young* and let Age(Emma) denote the value of variable Emma. The value is a whole number that belongs to the interval [0,100]. Using this interval as a universe, we define a fuzzy set Young for the attribute *young*. For example, it makes sense to have Young(28) = 0.7. The assertion *Emma is young* says something about Emma's age and thus is a restriction on the possible values of Emma's age. However, it says nothing about the group of

1 In fact, Zadeh used measure theory to provide "[a]dditional insight into the distinction between probability and possibility" [314, p. 9].

A Modern Introduction to Fuzzy Mathematics, First Edition.
Apostolos Syropoulos and Theophanes Grammenos.

people to which Emma possible belongs. The following equation

$$R(\text{Age}(\text{Emma})) = \text{Young}$$

denotes the *restriction* on the age of Emma. And here the restriction $R(\text{Age}(\text{Emma}))$ is a fuzzy one, since Young is a fuzzy set.

Assume that F denotes a fuzzy set. Then, the following sentences

Eila is brunette.
Avery is tall.

are instances of the proposition:

$$p \stackrel{\text{def}}{=} X \text{ is } F.$$

In general, the interpretation of "X is F" will be characterized by a *relational assignment equation*. More specifically, we have

Definition 5.1.1 The meaning of the proposition

$$p \stackrel{\text{def}}{=} X \text{ is } F,$$

where X is a name of an object and F is a label of a fuzzy set of a universe of U, is expressed by the relational assignment equation

$$R(A(X)) = F,$$

where A is an attribute which is implied by X and F, and R denotes a fuzzy restriction on $A(X)$ to which the value F is assigned by this equation.

Example 5.1.1 In the sentence "Eila is brunette," the implied attribute is color(Hair), and the relational assignment equation takes the form

$$R(\text{Color}(\text{Hair}(\text{Eila}))) = \text{brunette}.$$

Example 5.1.2 Suppose that the fuzzy set "young" is defined by

$$Y(u) = 1 - S(u; 20, 30, 40),$$

where

$$S(u; \alpha, \beta, \gamma) = \begin{cases} 0, & \text{when } u \leq \alpha, \\ 2\left(\dfrac{u-\alpha}{\gamma-\alpha}\right)^2, & \text{when } \alpha \leq u \leq \beta, \\ 1 - 2\left(\dfrac{u-\alpha}{\gamma-\alpha}\right)^2, & \text{when } \beta \leq u \leq \gamma, \\ 1, & \text{when } u \geq \gamma. \end{cases}$$

Note that the semicolon separates arguments from parameters, thus, u is the argument and α, β, and γ are the parameters. Assume that Emma is 28 years old. Then, $Y(28) \approx 0.7$. Then, we interpret 0.7 as the degree of *compatibility* of 28 with the concept labeled young. In addition, one may argue that the proposition "Emma is young" transforms the meaning of 0.7 from a degree of compatibility to the degree of possibility that Emma is 28. Thus, the possibility that Emma is young is 0.7. More generally, the compatibility of a value of u with young is transformed into the possibility of that value of u given "Emma is young."

The previous example is used to describe the general notion of possibility.

Definition 5.1.2 Suppose that F is a fuzzy subset of a universe U, where $F(u)$ is assumed to be the compatibility of u with the concept that F corresponds to. Suppose that X is a variable whose values are drawn from U. Also, we view F as a fuzzy restriction, $R(X)$, associated with X. Then, the proposition "X is F," which is represented by

$$R(X) = F$$

associates a *possibility distribution* Π_X with X which we propose to be equal to $R(X)$, that is,

$$\Pi_X = R(X).$$

Correspondingly, the *possibility distribution function associated with* X is written as π_X, and its values are the corresponding membership degrees of F, that is,

$$\pi_X(u) \stackrel{\text{def}}{=} F(u).$$

Therefore, $\pi_X(u)$, the possibility that $X = u$, is proposed to be equal to $F(u)$.

5.2 Possibility and Necessity Measures

In order to introduce possibility measures and necessity measures, we need to know what a *measure* is. All the definitions that are presented in this section are borrowed from [294]. Assume that X is a nonempty set, that $\mathbf{C}$ is a nonempty class of subsets of X, and that $\mu : \mathbf{C} \to [0, \infty]$ is a nonnegative, extended real valued set function defined on $\mathbf{C}$.

Definition 5.2.1 A set E in $\mathbf{C}$ is called the *null set* (with respect to μ) if and only if $\mu(E) = 0$.

Definition 5.2.2 μ is *additive* if and only if

$$\mu(E \cup F) = \mu(E) + \mu(F),$$

where $E, F, E \cup F \in \mathbf{C}$ and $E \cap F = \emptyset$.

Definition 5.2.3 μ is *finitely additive* if and only if

$$\mu\left(\bigcup_{i=1}^{n} E_i\right) = \sum_{i=1}^{n} \mu(E_i)$$

for any finite, disjoint class $\{E_1, E_2, \dots, E_n\}$ of sets in $\mathbf{C}$ whose union is also in $\mathbf{C}$.

Definition 5.2.4 μ is *countably additive* if and only if

$$\mu\left(\bigcup_{i=1}^{\infty} E_i\right) = \sum_{i=1}^{\infty} \mu(E_i)$$

for any disjoint class $\{E_1, E_2, \dots, E_n\}$ of sets in $\mathbf{C}$ whose union is also in $\mathbf{C}$.

Definition 5.2.5 μ is *subtractive* if and only if

$$\mu(F - E) = \mu(F) - \mu(E)$$

for $E, F, E \subset F, F \setminus E \in \mathbf{C}$ and $\mu(E) < \infty$.

Theorem 5.2.1 *If μ is additive, then it is subtractive.*

Definition 5.2.6 μ is called a *measure* on $\mathbf{C}$ if and only if it is countably additive, and there exists $E \in \mathbf{C}$ such that $\mu(E) < \infty$.

Definition 5.2.7 Assume that μ is a measure on $\mathbf{C}$. Then, μ is *complete* if and only if $E \in \mathbf{C}$, $F \subset E$, and $\mu(E) = 0$ imply that $F \in \mathbf{C}$.

Definition 5.2.8 Suppose that μ is a measure on $\mathbf{C}$. Then, μ is *monotone* if and only if

(i) $\mu(\emptyset) = 0$ when $\emptyset \in \mathbf{C}$;
(ii) $\mu(E) \leq \mu(F)$ when $E, F \in \mathbf{C}$ and $E \subset F$.

Definition 5.2.9 A monotone measure μ is called *maxitive* on $\mathbf{C}$ if and only if

$$\mu\left(\bigcup_{t \in T} E_t\right) = \bigvee_{t \in T} \mu(E_t)$$

for any subclass $\{E_t | t \in T\}$ of **C** whose union is in **C**, where T is an arbitrary index set.

If **C** is a finite class, then the previous requirement is replaced by

$$\mu(E_1 \cup E_2) = \max[\mu(E_1), \mu(E_2)],$$

where $E_1, E_2 \in \mathbf{C}$ and $E_1 \cup E_2 \in \mathbf{C}$.

Definition 5.2.10 A monotone measure μ is called a *generalized possibility measure* on **C** if and only if it is maxitive on **C**, and there exists $E \in \mathbf{C}$ such that $\mu(E) < \infty$.

In what follows, the letter π will be used to denote a generalized possibility measure.

Definition 5.2.11 A monotone measure μ on $(X, \mathbf{C})$, where $\mathbf{C} \subseteq \mathscr{P}(X)$, is normalized if $X \in \mathbf{C}$ and $\mu(X) = 1$.

Definition 5.2.12 If a generalized possibility measure π defined on $\mathscr{P}(X)$ is normalized, it is called a possibility measure.

Definition 5.2.13 If π is a possibility measure on $\mathscr{P}(X)$, then the set function ν defined as

$$\nu(E) = 1 - \pi(\overline{E}) \text{ for any } E \in \mathscr{P}(X)$$

is called a necessity measure on $\mathscr{P}(X)$.

5.3 Possibility Theory

A theory that is based on possibility and necessity measures is usually called a possibility theory [111]. Roughly, possibility theory is the reinterpretation of results about possibility and necessity measures in terms of events, and the degree to which they surprise us. Thus, we start with a set of events that are supposed to be subsets of a reference set Ω, which will be called the "event that does not surprise us at all" (practically, an event that will happen). The empty set is identified with the "most surprising event" (practically, even that is impossible). Each event $A \subseteq \Omega$ is associated with a real number $g(A)$. This number is not random, but it is computed and/or estimated by someone who happens to have knowledge of the context in which the event occurs. The number $g(A)$ is a measure of the confidence one has that this particular event will happen. Typically, $g(A)$

increases as confidence increases. In addition, if A is an absolutely possible event, then $g(A) = 1$, but if A is an impossible event, then $g(A) = 0$. In particular,

$$g(\emptyset) = 0 \quad \text{and} \quad g(\Omega) = 1.$$

However, $g(A) = 1$ or $g(A) = 0$ does not necessarily mean that A is an absolutely possible or impossible event. The following axiom is necessary in order to ensure that function g is coherent:

$$A \subseteq B \Rightarrow g(A) \leq g(B)$$

This axiom means that if the event A implies the event B, we should be sure that our confidence that B will happen should be at least as much as our confidence that A will happen. When Ω is an infinite reference set, it is possible to introduce axioms of continuity as follows: For every nested sequence $(A_n)_n$ of sets $A_0 \subseteq A_1 \subseteq \cdots A_n \subseteq \cdots$, or $A_0 \supseteq A_1 \supseteq \cdots A_n \supseteq \cdots$, we have

$$\lim_{n\to+\infty} g(A_n) = g\left(\lim_{n\to+\infty} A_n\right).$$

Given two events $A, B \subseteq \Omega$, for the conjunction and disjunction of events the following inequalities hold:

$$g(A \text{ and } B) = P(A \cap B) \leq \min(g(A), g(B)),$$
$$g(A \text{ or } B) = P(A \cup B) \geq \max(g(A), g(B)).$$

A possibility measure Π is a limiting case of confidence measures for which:

$$\Pi(A \cup B) = \max(\Pi(A), \Pi(B)), \quad \forall A, B.$$

Also, if A and A^{C} are two contradictory events (A^{C} is the complement of A in Ω), then

$$\max(\Pi(A), \Pi(A^{\mathrm{C}})) = 1,$$

which means that of two contradictory events, at least one is completely possible. In addition, if an event is considered to be possible, this does not mean that the contrary event is completely impossible.

If the set Ω is finite, then any possibility measure Π can be defined in terms of its values on the singletons of Ω:

$$\Pi(A) = \bigvee \{\pi(\omega) | \omega \in A\}, \quad \forall A,$$

where $\pi(\omega) = \Pi(\{\omega\})$ and $\pi : \Omega \to [0, 1]$ is called *possibility distribution*. Mapping π is *normalized*, meaning that there is an ω' such that

$$\pi(\omega') = 1.$$

Of course, this happens because $\Pi(\Omega) = 1$.

Necessity measures, which are denoted by N, are another form of a limiting case of confidence measures:

$$N(A \cap B) = \min(N(A), N(B)), \quad \forall A, B.$$

In addition, we demand that

$$\Pi(A) = 1 - N(A^{\mathsf{C}}), \quad \forall A.$$

A direct consequence of this requirement is that

$$N(A) = \bigwedge\{1 - \pi(\omega) | \omega \notin A\}.$$

Necessity measures satisfy the relation

$$\min(N(A), N(A^{\mathsf{C}})) = 0,$$

which means that two contrary events cannot be necessary at all at the same time. Finally,

$$N(\Omega) = 1 \quad \text{and} \quad N(\varnothing) = 0.$$

Conditional possibilities and necessities are something special, and a comprehensive review of conditional possibilities is presented in [76]. Roughly, these kinds of possibilities and necessities try to answer the question: Now that the event A has happened, what is the chance of event B? In general, conditional possibilities and necessities are mostly presented as a derived notion of the unconditional ones. For instance, given a possibility Π on a Boolean algebra $\mathbf{B}$ (see Section 2.9) and a t-norm $*$, for every $H \in \mathbf{B} \setminus \varnothing$, a t-conditional possibility $\Pi(\cdot|H)$ on E is defined as any solution of the equation

$$\Pi(E \wedge H) = x * \Pi(H).$$

Depending on the choice of the t-norm the pairs $(\Pi(E \wedge H), \Pi(H))$ may not lead to a unique solution. For instance, if the t-norm is the min function, the possible solutions of the previous equation are

$$\begin{aligned} \Pi(A|B) = \Pi(A \wedge B) \quad & \text{if } \Pi(A \wedge B) < \Pi(B), \\ \Pi(A \wedge B) \le \Pi(A|B) \le 1 \quad & \text{if } \Pi(A \wedge B) = \Pi(B). \end{aligned}$$

In general, we do not demand that an arbitrary solution corresponds to a normalized possibility. However, the following definition avoids this problem:

$$\Pi(A|B) = \begin{cases} \Pi(A \wedge B), & \text{if } \Pi(A \wedge B) < \Pi(B), \\ 1, & \text{if } \Pi(A \wedge B) = \Pi(B). \end{cases}$$

One major problem with this definition is that, for incompatible A and B (i.e. $A \wedge B = 0$) and $\Pi(B) = 0$, we have that $\Pi(A|B) = 1$, while it is natural to expect that $\Pi(A|B) = 0$! Interestingly, if we force $\Pi(A|B) = \Pi(A \wedge B|B) = \Pi(\varnothing|B) = 0$, then $\Pi(\cdot|B)$ is not a possibility. The definition that follows solves this problem [37]:

Definition 5.3.1 Assume that $\mathbf{E} = \mathbf{B} \times \mathbf{H}$ is a set of conditional events $E|H$ such that $\mathbf{B}$ is a Boolean algebra and $\mathbf{H}$ is an additive set (i.e. closed with respect to $\wedge$), with $\mathbf{H} \subset \mathbf{B}$ and $\emptyset \notin \mathbf{H}$. Also, assume that $*$ is a t-norm. Then, a function $\Pi : \mathbf{E} \to [0, 1]$ is a $*$-conditional possibility if it satisfies the following properties:

(i) $\Pi(E|H) = \Pi(E \wedge H|H)$, for all $E \in \mathbf{B}$ and $H \in \mathbf{H}$;
(ii) $\Pi(\cdot|H)$ is a possibility measure, for all $H \in \mathbf{H}$; and
(iii) for all $H, E \wedge H \in \mathbf{H}$, and $E, F \in \mathbf{B}$

$$\Pi(E \wedge F|H) = \Pi(E|H) * \Pi(F|E \wedge H).$$

Recall that a necessity function is the dual function $N : \mathbf{B} \to [0, 1]$ of a possibility Π that satisfies

$$N(E_j \wedge E_j) = \min\{N(E_i), N(E_j)\}.$$

Conditional necessities for any $E|H \in \mathbf{B} \times \mathbf{H}$ have been introduced in [37]:

$$N(E|H) = 1 - \Pi(E^{\mathrm{C}}|H).$$

This definition induces the following one:

Definition 5.3.2 Assume that $\mathbf{E} = \mathbf{B} \times \mathbf{H}$ is a set of conditional events $E|H$ such that $\mathbf{B}$ is a Boolean algebra and $\mathbf{H}$ is an additive set, with $\mathbf{H} \subset \mathbf{B}$ and $\emptyset \notin \mathbf{H}$. Then, a function $N : \mathbf{E} \to [0, 1]$ is a conditional necessity if it satisfies the following properties:

(i) $N(E|H) = N(E \wedge H|H)$, for all $E \in \mathbf{B}$ and $H \in \mathbf{H}$;
(ii) $N(\cdot|H)$ is a necessity measure, for all $H \in \mathbf{H}$; and
(iii) for all $H, E^{\mathrm{C}} \wedge H \in \mathbf{H}$, and $E, F \in \mathbf{B}$

$$N(E \vee F|H) = \max\{N(E|H), N(F|E^{\mathrm{C}} \wedge H)\}.$$

5.4 Possibility Theory and Probability Theory

Roughly speaking, probability is the measure of the likelihood that an event will occur in a random experiment. And this is why people often confuse randomness with probability theory. Typically, a probability is a number that takes values between 0 and 1. Here 0 indicates impossibility and 1 indicates certainty.

Obviously, two events A and B have probabilities p_A and p_B, respectively, and $p_A < p_B$, then A is less likely to occur than B. This brief description explains why possibility theory and probability theory look so similar. Also, this explains why some people superficially assume that these theories are identical. However, there are three different views of the relationship between probability theory and possibility theory. The first view sees the two theories as independent theories that are used to describe different facets of uncertainty. According to the second view, both theories are identical, and one can easily translate possibilities to probabilities and vice versa. And according to the third view, probability theory is just a special case of possibility theory.

The first view has been advocated by Zadeh [317] who, in addition, argued that probability theory is not adequate to deal with uncertainty and imprecision and that the two theories are complementary rather than competitive. In particular, the reasons for the inadequacy of probability theory to deal with uncertainty and imprecision are given below.

(i) Probability theory does not support the concept of a fuzzy event, where examples of such events are a cold day, a strong earthquake, the near future, etc.

(ii) It is not possible to deal with fuzzy quantifiers like many, most, several, and few in probability theory.

(iii) In probability theory, one performs computations with numbers, and it is not possible to perform computations with *fuzzy probabilities* such as likely, unlikely, not very likely, and so forth. Fuzzy probabilities have been introduced by Zadeh [316]:

Assume that A is a fuzzy set defined as follows:

$$A = a_1/u_1 + a_2/u_2 + \cdots + a_n/u_n,$$

where $U = \{u_1, \dots, u_n\}$ is a finite universe. Also, assume that X is a variable that takes the values $u_1, \dots, u_n$ with a uniform probability $(1/n)$. If $A \subset U$, then the nonfuzzy probability of the fuzzy proposition or, equivalently, of the fuzzy event

$$p \stackrel{\text{def}}{=} X \text{ is } A$$

is given by

$$P(p) = \frac{\text{card}(A)}{n}$$

and the fuzzy probability is the fuzzy number

$$FP(p) = \frac{\sum_\alpha \alpha/\text{card}({}^\alpha A)}{n}.$$

(iv) It is not possible to give an estimation of fuzzy probabilities in probability theory. This very simply means that there is no answer to questions like "What is the probability that my car may be stolen?"
(v) Probability theory cannot be used as a meaning-representation language. This means that the sentence "It is not likely that there will be a sharp increase in the price of oil in the near future" has no meaning in probability theory.
(vi) The expressive limits of probability theory is a burden to the analysis of problems in which the data are described in fuzzy terms. For example, the following problem borrowed from [317] explains this point.

> A variable X can take the values small, medium, and large with respective probabilities low, high, and low. What is the expected value of X? What is the probability that X is not large?

Of course, it is an undeniable fact that probability theory has been and continues to be used successfully in areas where the systems are mechanistic, while human reasoning, perceptions, and emotions do not play a significant role. For example, statistical mechanics, quantum mechanics, communication systems, evolutionary programming, are all areas where probability theory is used successfully. However, there are areas such as economics, pattern recognition, group decision analysis, speech and handwriting recognition, expert systems, weather and earthquake forecasting, where probability theory is not that successful. The main reason for this is that dependencies between variables are not well defined, the knowledge of probabilities is imprecise and/or incomplete, the systems are not mechanistic, and human reasoning, perceptions, and emotion do play an important role.

According to George Jiří Klir and Behzad Parviz [177]

> probability theory is a natural tool for formalizing uncertainty in situations where class frequencies are known or where evidence is based on outcomes of large series of independent random experiments. Possibility theory, on the other hand, is a natural tool for formalizing uncertainty that results from information that is both imprecise and fuzzy. When information of both kinds is available, it is useful to have the capability of transforming probabilities to possibilities or vice versa.

A very systematic approach to this "problem" was presented by Dubois et al. [113]. In particular, their solution is based on the fact that a possibility measure Π on a set X (i.e. a maxitive measure) is equivalent to the family $\mathscr{P}(\Pi)$ of probability measures (i.e. measures that have the property described in Definition 5.2.4) such that

$$\mathscr{P}(\Pi) = \{P | \forall A \subseteq X \text{ and } P(A) \leq \Pi(A)\}.$$

Mathematics is an abstract language that has its own rules and principles. What makes this language important is the use of mathematics to describe the world. Thus, probability and possibility theories are mathematical tools that are used to describe the world. However, just because one can use the mathematical properties of these two theories to transform one to the other, it does not mean that it makes sense to make any of these transformations. The following text by Philip Klöcking, which appeared in Philosophy Stack Exchange, explains the difference between possibility and probability and indirectly explains why we cannot transform possibilities to probabilities.

> Possibility means being able to be thought without contradiction at the same time (!). In the sense of probabilities, it means a state (the total number of states = total number of possibilities in Laplace's sense) that can be thought as an outcome without contradiction.
> Probability of an event then simply means a certain number of states (i) thought as causally invoked by or (ii) conceptually thought within an event divided by the total number of states being able to be thought without contradiction.
> This also means: no probability without possibility, but possibility without probability. Therefore, it is (strictly speaking, see below) perfectly possible to become President of the United States for every American (= no contradiction), although it is thought improbable for nearly all of them.
> The problem of thought arises when situations with a probability near or (mathematically) equal 0 are named impossible, although they can be thought without contradiction. Like saying "My neighbor becoming the president is impossible." It often occurs when we are judging heuristically, which we are doing because we are simply not able to say what the probability of an outcome really is (otherwise we would have holy wisdom). It is an equivocation, a slight move of sense ruining the scientific usability of a language. Becoming rid or at least revealing equivocations is basically the main task of philosophy, Husserl would have said.

In addition, one could say that fuzzy set theory is a mathematical theory that describes our vague world. Of course, that our world is vague does not rule out the use of probabilities in the description of certain phenomena. Nevertheless, probabilities are not that fundamental as are possibilities, since the later are very closely related to vagueness.

In conclusion, the idea that probabilities and possibilities describe different facets of uncertainty is weird since we do not know exactly where are the boundaries that separate these facets. But if we find a way to exactly specify these boundaries, then there is no uncertainty! Of course, the idea that possibility and probability theories are interchangeable is dead wrong. Thus, it only makes sense to assume that possibility theory is the most general theory and probability theory is a special case of possibility theory.

5.5 An Unexpected Application of Possibility Theory

As expected, possibility theory has been applied in many areas (e.g. data analysis, database query, diagnosis, argumentations, see Ref. [112] for more details). However, the replacement of probability theory by possibility theory, as suggested by Michael Smithson [263], is particularly interesting. Of course, the reason is that practitioners of psychology have relied almost exclusively on statistical models and methods known as the Neyman–Pearson–Fisher (NPF) framework for the quantitative analysis of human behavior. Although psychologists still use NPF (see Ref. [241] for a very recent overview of the framework), this framework was originally used in military strategic and decision analysis, industrial quality control, and agricultural experimentation. Thus, the NPF was designed for the solution of problems one encounters in these fields and not the human sciences. However, for some reasons, it was fully adopted by psychologists for the solution of their problems. Smithson used the *general linear model* (GLM) in order to assess whether the exclusive reliance on the NPF obstructs a systematic investigation of key theoretical concepts in psychology.

The GLM splits variation in human behavior into two parts: one that is predicted by the instrumental variable(s), which are necessarily one-to-one, and one that is not predicted because there are unobserved variables or random processes. Thus, psychological predictions become an injective "mapping" from states of values of the independent variables to the output variable. Consequently, human agents operate within a specific range, thus making impossible to have any kind of behavior. And the random processes do not make human behavior… random, as they have to be in agreement with the first part. Therefore, the GLM produces a stochastically deterministic view of human behavior, something that does not agree with everyday observations and with other theories. For example, cognitivism requires concepts such as intentionality, choice, or decision, that have no place in a one-to-one predictive model, where randomness is the only source of uncertainty. On the other hand, there are theories in social psychology that have been deeply affected by GLM. For example, cognitive dissonance theory originally assumed that people are mindful actors whose choices are only partially constrained and not even determined by motivations. Subsequently, the theory assumed that all subjects under the same experimental conditions should respond identically. In conclusion, Smithson notes that

> [u]nfortunately, the GLM cannot tell the clinician whether the treatment was sufficient but not necessary, necessary but not sufficient, or weakly contributing to improvement in the clients. A "statistical significant" difference between the means of treatment and control groups could arise from any of those outcomes.

The NPF and all other "statistical paradigms," as Smithson calls them, including the Bayesian approach, rely on probability theory to handle two kinds of uncertainty. The first one is directly related to the model of behavior and is about the uncertainty one encounters when studying what actors will do. The second kind is about the knowledge of what the actors may do. The first kind of uncertainty is best exemplified by the sentence "I can ride my bicycle to work," but, of course, this says nothing about what I will actually do. And it is a fact that nonprobabilistic types of uncertainty play a central role in many psychological theories.

Probability theory has one more drawback: It cannot satisfactorily capture incomplete knowledge. For example, in the simplest case of a binary outcome setup (A and B), probability theory cannot distinguish uncertainty about whether A will happen from ignorance about whether A will happen. Thus, the statement $P(A) = 0.5$ means that either A and B are equally likely or that we have no idea what is the likelihood of A or B.

In general, one can say that whether we are going to use statistical models for the analysis of quantitative data depends on our interpretive and theoretical basis instead of logical or empirical reasons. For example, Figure 5.1 is a scatter plot and Table 5.1 contains information that has an almost identical pattern. In particular, one may notice that high IEPV or animal fat intake is sufficient but not necessary to produce high APV or age-adjusted death rate, respectively. The real question is whether this says something. According to most medical models, animal fat intake and risk of cancer are causal, while some social scientists may not agree that the approval of violence is causally related to the intention to engage in it. Although it might make sense to most social scientists to conclude that people who approve violence may choose to engage in it or not, most medical researchers would not like to claim that countries in which people do not consume much animal fat may opt to have high death-rates from breast cancer. The data in Table 5.1 and the scatter plot say the same thing, logically and empirically. However, our interpretive or theoretical perspective is responsible for assigning causality or intentionality to one or the other. Therefore, Smithson concluded that statistical models cannot be replaced in all circumstances. Of course, this last statement is not entirely true

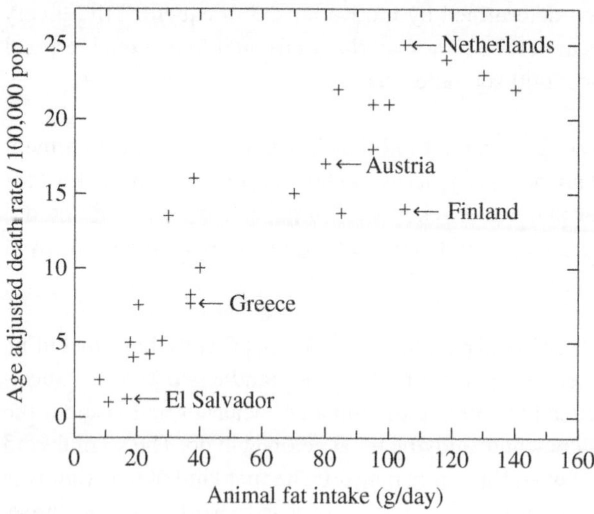

Figure 5.1 A scatter plot. Source: Data from Carroll 1975 [58].

Table 5.1 Intention to engage in political violence (IEPV) and approval of political violence (APV).

IEPV	APV					
	0	**1**	**2**	**3**	**4**	**5**
0	97	89	43	9	2	2
1	5	75	45	17	4	2
2	0	5	39	8	11	2
3	0	0	3	13	8	4
4	0	0	1	1	3	4
5	0	0	0	0	2	5

Source: Data from Muller 1972 [226].

today, as fuzzy statistics is gaining approval in the scientific community (see Ref. [78] for an overview of the fuzzy approach to statistical analysis).

Smithson proposed the use of possibility theory for the working psychologist. His approach is based on the "fact" that "possibility is not a kind of probability, nor can probability represent possibilistic uncertainty" [263, p. 11]. The following inequality is very important in the development of his toolkit:

$$\mathrm{ne}_i \leq pr_i \leq \mathrm{po}_i,$$

where ne_i is the necessity of the ith option, pr_i is probability, and po_i is possibility. This inequality can be used to measure the *freedom of action* and the *relative preference*. In general, when an action is possible, but not necessary, then one is free to choose whether to perform this action or not. Formally, this is expressed as follows:

$$F_i = \mathrm{po}_i - \mathrm{ne}_i. \tag{5.1}$$

Many surveys measure *preference* by calculating the percentage of people who select a specific option. The problem with this approach is that, for all options, the possibility equals to one and the necessity equals to zero. A better approach is the associated preference with the extent to which people freely select an option, then a valid relative preference measure can be defined as follows:

$$S_i = \frac{\mathrm{pr}_i - \mathrm{ne}_i}{\mathrm{po}_i - \mathrm{ne}_i}.$$

Assume that we have a population and that the portions of people for whom the ith and jth options are possible are po_i and po_j, respectively. Then, the question is: What is the joint possibility po(i and j), that is, the portion of people for whom both options are possible? The range of values for joint possibility is

$$\max(0, \mathrm{po}_i + \mathrm{po}_j - 1) \leq \mathrm{po}(i \text{ and } j) \leq \min(\mathrm{po}_i, \mathrm{po}_j).$$

When there is no empirical information about po(i and j), there are two definitions one could adopt. According to the first, $\mathrm{po}(i \text{ and } j) = \min(\mathrm{po}_i, \mathrm{po}_j)$. The second approach uses the inequality above as the definition of the joint possibility.

The relevant interval for po(i and j) is

$$\max(\mathrm{po}_i, \mathrm{po}_j) \leq \mathrm{po}(i \text{ and } j) \leq \min(\mathrm{po}_i + \mathrm{po}_j, 1).$$

If we have adopted that $\mathrm{po}(i \text{ and } j) = \min(\mathrm{po}_i, \mathrm{po}_j)$, then we are forced to deduce that $\mathrm{po}(i \text{ or } j) = \max(\mathrm{po}_i, \mathrm{po}_j)$. Otherwise, the previous inequality becomes the definition. In addition, if we happen to know the value po(i and j), then

$$\mathrm{po}(i \text{ or } j) = \mathrm{po}_i + \mathrm{po}_j) - \mathrm{po}(i \text{ and } j).$$

The possibility that the portion of people for whom option i is possible and have access to option j is expressed by po(i/j). An alternative interpretation of po(i/j) is that it is a measure of the degree to which option i is included by option j. This possibility is defined as follows:

$$\mathrm{po}(i/j) = \frac{\mathrm{po}(i \text{ and } j)}{\mathrm{po}_i}.$$

The concepts presented so far make up Smithson's basic calculus of possibility theory.

When we measure uncertainty with some form of possibility theory, then we need to know how relatively free are the agents of a system to make choices. Naturally, this relative freedom should take under consideration all the constraints of a given system. Eq. (5.1) could be used as a starting point for the definition of relative freedom. Smithson proposed that the *relative amount of freedom enjoyed by the ith individual* is

$$F_i = \sum_{j=1}^{r} \frac{\mathrm{po}_{ij} - \mathrm{ne}_{ij}}{r},$$

where r is the number of conceivable options, po_{ij} and ne_{ij} are the possibility and necessity values for the ith individual on the jth option, respectively. Note that the number of individuals is assumed to be N and that po_{ij} and ne_{ij} can assume only two values: 0 and 1. The marginal possibility and necessity distributions over the r options may be recovered by N:

$$\mathrm{po}_j = \sum_{i=1}^{N} \frac{\mathrm{po}_{ij}}{N} \quad \text{and} \quad \mathrm{ne}_j = \sum_{i=1}^{N} \frac{\mathrm{ne}_{ij}}{N}.$$

The following can be used to measure the relative freedom for part of the population that is free to choose the jth option:

$$F_j = \mathrm{po}_j - \mathrm{ne}_j.$$

What if we want to measure the freedom available to the entire population of N individuals? A straightforward solution is to measure the average F_i or F_j, but this solution does not take under consideration both partitions and permutations. If we take under consideration permutations, then the system freedom F^* is defined as follows:

$$\begin{aligned} F^* &= \frac{\prod_i rF_i}{rN} \\ &= 0 \text{ only if all } F_i = 0, \end{aligned} \tag{5.2}$$

where only nonzero F_i are multiplied together. The third approach ignores permutations and considers only partitions that are equipossible. Without going into the various details, when there are two options, then the relative group freedom (FG) can be computed by the following formula:

$$\mathrm{FG} = (1 - \mathrm{ne}_1 - \mathrm{ne}_2)\max(0, 1 - \mathrm{po}_1 - \mathrm{ne}_2) - \max(0, 1 - \mathrm{po}_2 - \mathrm{ne}_1).$$

When we have three options, then

$$\begin{aligned} \mathrm{FG} = {} & (1 - \mathrm{ne}_1 - \mathrm{ne}_2 - \mathrm{ne}_3)^2 - \max(0, 1 - \mathrm{po}_1 - \mathrm{ne}_2 - \mathrm{ne}_3)^2 \\ & - \max(0, 1 - \mathrm{po}_2 - \mathrm{ne}_1 - \mathrm{ne}_3)^2 - \max(0, 1 - \mathrm{po}_3 - \mathrm{ne}_1 - \mathrm{ne}_2)^2 \\ & + \max(0, 1 - \mathrm{po}_1 - \mathrm{po}_2 - \mathrm{ne}_3)^2 + \max(0, 1 - \mathrm{po}_1 - \mathrm{po}_3 - \mathrm{ne}_2)^2 \\ & + \max(0, 1 - \mathrm{po}_2 - \mathrm{po}_3 - \mathrm{ne}_1)^2. \end{aligned}$$

The general formula for the case of *r* options follows:

$$\mathrm{FG} = \left(1 - \sum_{i=1}^{r} \mathrm{ne}_i\right)^{r-1} - \sum_{i=1}^{r} \max\left(0, 1 - \mathrm{po}_i - \sum_{j \neq i} \mathrm{ne}_j\right)^{r-1}$$
$$+ \sum_{i=1}^{r} \sum_{j>i} \max\left(0, 1 - \mathrm{po}_i - \mathrm{po}_j - \sum_{k \neq i,j} \mathrm{ne}_k\right)^{r-1} - \cdots$$
$$+ (-1)^{r-1} \sum_{i=1}^{r} \max\left(0, 1 - \mathrm{ne}_i - \sum_{j \neq i} \mathrm{po}_j\right)^{r-1}.$$

Let us briefly say what happens when the N individuals are assigned possibilities over a range of values on a continuous variable rather than over a discrete option. Initially, we assume that there is a variable X that can take any value from the closed interval $[d, u]$. Also, we assume that X is the ith individual that is assigned the possibility $\mathrm{po}_i(x) = 1$ for all x in some subrange of $[a_i, b_i]$ and 0 for all y outside this subrange. Then, F_i is defined as

$$F_i = \frac{b_i - a_i}{u - d}.$$

The average F_i and the system freedom F^* are defined as above by replacing r with $u - d$ in (5.2). More generally, we assume that $\mathrm{po}_i(x)$ and $\mathrm{ne}_i(x)$ take values from the closed interval $[0, 1]$. Then, the previous equation is generalized to

$$F_i = \int_d^u \frac{\mathrm{po}_i(x) - \mathrm{ne}_i(x)}{u - d}\, dx. \tag{5.3}$$

If each state x of X is assigned a possibility and/or a necessity value po)(x) and ne(x), respectively, then these values do not refer to individuals. The relative freedom for each individual F_i is identical for all i and is defined by (5.3) without the use of the subscripts in the right-hand terms.

There is one more class of situations where the density functions constrained by possibility and necessity do not have to accumulate to 1. Instead, 1 is considered as a limit on their accumulation value. Typically, we consider M shareholders that may use a resource up to some limit L. Each shareholder's usage of the resource is also limited by her share of the holdings, h_i. Assume that H is the sum of the holdings h_i, w_i is the amount actually used by the ith shareholder, $b_i = w_i/H$, and B is the sum of the b_i. Then,

$$\mathrm{po}(b_i) = \min\left(\frac{h_i}{H}, \frac{L}{H}\right) \quad \text{and} \quad \mathrm{po}(B) = \min(L/H, 1).$$

The value of FG can be calculated by the following formula:

$$\mathrm{FG} = \frac{1}{r!\prod_{i=1}^{r}\mathrm{po}(b_i)}\Bigg[(L/H)^r - \sum_{i=1}^{r}\max(0, L/H - \mathrm{po}(b_i))^r + \sum_{i=1}^{r}\sum_{j>i}\max(0, L/H - \mathrm{po}(b_j) - \mathrm{po}(b_j))^r - \cdots + (-1)^{r-1}\sum_{i=1}^{r}\max(0, L/H - \sum_{j\neq j}\mathrm{po}(b_j))^r\Bigg].$$

We stop here the presentation of Smithson's work. He discussed also how this calculus can be used in psychology, but we feel the discussion is far too specialized.

Exercises

5.1 Write down the truth tables of the following expressions: $((A \Rightarrow B) \wedge (\neg B))$, $((A \Rightarrow (B \Rightarrow C)) \Leftrightarrow (A \Rightarrow B))$, and $(A \vee (\neg A))$.

5.2 Show that $(A \Rightarrow B)$ is logically equivalent to $(\neg(A \wedge (\neg B)))$.

5.3 Prove in BL the following:
- (a) $A \to (B \to (A \& B))$;
- (b) $(A \to B) \to ((A \& C) \to (B \& C))$; and
- (c) $((A \to B) \wedge (A \to C)) \to (A \to (B \wedge C))$.

5.4 Show that $\mathbf{0}$ is equivalent to $A \odot \mathbf{0}$ in BL.

6

Fuzzy Statistics

The uncertainty characterizing decisions and indeed the process of decision-making on the basis of statistical reasoning can be traced, in many cases, to the lack of imprecise information coming from vagueness in the data considered. In this sense, fuzzy set theory comes to the forefront and, as it is plausibly expected, it plays a prominent role. Before inserting fuzziness, we will present a brief review of *random variables* (also known as *stochastic variables*) and their properties as well as the classical statistical notions of *point estimation*, *interval estimation*, *hypothesis testing*, and *regression*. Then, their fuzzy analogues will be introduced.

6.1 Random Variables

A mapping from the set Ω of possible outcomes (sample points) of an experiment to a subset of real numbers $\mathbb{R}$ is a random variable. A rigorous definition of a random variable is the following:

Definition 6.1.1 A (real) measurable function X from a sample space Ω of a probability model to the set of real numbers, $X : \Omega \rightarrow \mathbb{R}$, is called a (real) *random variable.*

Some remarks are in order at this point:

(i) a random variable is not a variable in the usual sense, but a function with the domain Ω and the range $\mathbb{R}$;
(ii) the random variable X may be undefined or infinite for a subset of Ω with zero probability;
(iii) the mapping $X(\omega)$ (i.e. the sample value for a sample point ω) must be such that

$$\{\omega \in \Omega \mid X(\omega) = x\}$$

A Modern Introduction to Fuzzy Mathematics, First Edition.
Apostolos Syropoulos and Theophanes Grammenos.

for $X = x$, where ω is an *event* for a fixed sample value x, for all $x \in \mathbb{R}$. In fact, one can similarly define the events

$$\{\omega \mid X(\omega) \leq x\}$$

for $X \leq x$, or

$$\{\omega \mid X(\omega) > x\}$$

for $X > x$, or even

$$\{\omega \mid x_1 < X(\omega) \leq x_2\}$$

for $x_1 < X \leq x_2$ (see, e.g. [133, 233]).

It is possible to assign probabilities corresponding to the aforesaid events, for example,

$$\Pr(X = x) = \Pr\{\omega \mid X(\omega) = x\},$$

and so on. Now, let us define the *distribution function of a random variable*:

Definition 6.1.2 The *cumulative distribution function* or simply distribution function of a random variable X is defined as

$$D_X(x) = \Pr(X \leq x) = \Pr\{\omega \mid X(\omega) \leq x\}.$$

The cumulative distribution function has to satisfy the following properties:

(i) $D_X(x) \in [0, 1]$,
(ii) $D_X(x_1) \leq D_X(x_2)$, for $x_1 \leq x_2$,
(iii) $\lim_{x \to \infty} D_X(x) = 1$,
(iv) $\lim_{x \to -\infty} D_X(x) = 0$,
(v) $\lim_{\varepsilon^+ \to 0} D_X(x + \varepsilon) = D_X(x)$,

with the second property showing that $D_X(x)$ is a nondecreasing function and the fifth property pointing out its continuity on the right.

When the range of X is finite or countably infinite, then the random variable is *discrete* and a *discrete probability distribution* (known as *probability mass function*) $p_X(x)$ can be defined assigning a certain probability to each value in the range of X, that is, $\Pr(X = x_i) = p_X(x)$ for each sample value x_i. The probability mass function has to satisfy the following properties:

(i) $p_X(x_i) \in [0, 1]$, $i \in \mathbb{N}$,
(ii) $p_X(x) = 0$ for $x \neq x_i$,
(iii) $\sum_i p_X(x_i) = 1$.

Then, the cumulative distribution function $D_X(x)$ of a discrete random variable is given by

$$D_X(x) = \Pr(X \leq x) = \sum_{x_i \leq x} p_X(x_i).$$

In the case of an uncountably infinite range, X is a *continuous* random variable and, if it has a first derivative that is piecewise continuous and exists everywhere except possibly at a finite number of points, then a *probability density function* can be defined:

$$f_X(x) = \frac{d}{dx} D_X(x),$$

which can be integrated in order to find the probability. The probability density function has to satisfy the following properties:

(i) $f_X(x) \geq 0$,
(ii) $\int_{-\infty}^{\infty} f_X(x)\, dx = 1$,
(iii) $\Pr(a < X \leq b) = \int_a^b f_X(x)\, dx$.

Then, the cumulative distribution function $D_X(x)$ of a continuous random variable X is given by

$$D_X(x) = \Pr(X \leq x) = \int_{-\infty}^{x} f_X(t)\, dt.$$

Now, one can determine the *expectation value* (or *mean*) μ_X of a random variable X:

$$\mu_X = \sum_i x_i p_X(x_i), \quad i \in \mathbb{N}, X \text{ discrete},$$
$$\mu_X = \int_{-\infty}^{\infty} x f_X(x) \quad dx, X \text{ continuous}.$$

Based upon the above relations, the *n th moment* $E(X^n)$ *of a random variable* X can be introduced:

$$E(X^n) = \sum_i x_i^n p_X(x_i), \quad i \in \mathbb{N}, X \text{ discrete},$$
$$E(X^n) = \int_{-\infty}^{\infty} x^n f_X(x)\, dx, \quad X \text{ continuous}.$$

Clearly, the first ($n = 1$) moment of X is its expectation value μ_X.

Finally, the concepts of *variance* and *standard deviation* of a random variable X can be defined:

$$\sigma_X^2 = E[(X - \mu_X)^2] \geq 0,$$

which for a discrete random variable becomes

$$\sigma_X^2 = \sum_i (x_i - \mu_X)^2 p_X(x_i),$$

while for a continuous random variable, one obtains

$$\sigma_X^2 = \int_{-\infty}^{\infty} (x - \mu_X)^2 f_X(x)\, dx.$$

In fact, by expanding the expression for σ_X^2, one gets the useful formula for the variance of a random variable X:

$$\sigma_X^2 = E[X^2] - (E[X])^2 = E[X^2] - \mu_X^2.$$

Finally, the positive square root of σ_X^2 yields the *standard deviation*, σ_X, of a random variable X.

There are many important distributions for random variables, most notably the binomial distribution, the Poisson distribution, and the normal (Gaussian) distribution (see, e.g. [233]). At this point, one last remark concerning the so-called conditional distributions is deemed necessary. Following the definition of the *conditional probability* of an event A given event B:

Definition 6.1.3

$$\Pr(A|B) = \frac{\Pr(A \cap B)}{\Pr(B)}, \quad \Pr(B) > 0,$$

we can find the cumulative distribution function of a random variable X given the event B:

$$D_X(x|B) = \Pr(X \leq x|B) = \frac{\Pr((X \leq x) \cap B)}{\Pr(B)}.$$

Again, one can distinguish between a discrete and a continuous random variable. Thus, for a discrete random variable, the above equation yields for the *conditional probability mass function*

$$p_X(x_i|B) = \Pr(X = x_i|B) = \frac{\Pr((X = x_i) \cap B)}{\Pr(B)},$$

while in the case of a continuous random variable, we have for the *conditional probability density function*

$$f_X(x|B) = \frac{d}{dx} D_X(x|B).$$

6.2 Fuzzy Random Variables

In Section 6.1, we have reviewed classical random variables and their basic properties. In order to handle fuzzy data or observations, one can introduce fuzzy-valued random variables to grasp vagueness (see, e.g. [78] for a short review), in other words, randomness and vagueness are now allowed to appear simultaneously. In the literature, one can find different approaches to the notion of a fuzzy random variable, most notably the (mathematically equivalent) definitions introduced by Erich Peter Klement et al. [176] and Madan L. Puri, and Dan A. Ralescu [245], Huibert Kwakernaak [186, 187], and Rudolf Kruse and Klaus Dieter Meyer [183]. In what follows, we will adopt the Kruse–Meyer approach in which a fuzzy random variable is studied as a fuzzy perception/observation of a classical real-valued random variable. This approach is actually a combination of the other authors' considerations. However, we stress that we are not going to introduce the notions of expected value, variance, and distribution function for fuzzy random variables. The reader is referred to [183] for a detailed presentation.

First, let us see what is meant by perception/observation in this context. Assume a measurable space (Ω, A) with Ω denoting the set of all possible outcomes of a random experiment, A a σ-algebra of subsets of Ω[1] and $(\mathbb{R}, B)$ the Borel-measurable space (see footnote in page 241). Further, let the probability space (Ω, A, P) with P a set function defining a probability measure on the space (Ω, A). Suppose that the results of the random experiment are described by $u : \Omega \to \mathbb{R}$ which assigns a random value u to each random choice. Then, u is a random variable.

The perception/observation of the aforementioned random variable means that for each $\omega \in \Omega$, we can investigate whether $u(\omega) \in V_i$ for some i, where $V_i \subseteq \mathbb{R}$. Now, if we examine another mapping, say $X : \Omega \to B(\mathbb{R})$ with $X(\Omega) = V_i \iff V(\omega) \in V_i$, then we associate with each $\omega \in \Omega$ not a real number $u(\omega)$ as in the case of ordinary random variables, but a set V called *random set* (see, e.g. [210] for a detailed study of random sets).[2] The random variable u of which X is a perception is called an *original* of X. In general, given a random set X, the corresponding true original u is not known, we only have a possible set of originals. If no further information is available, then each random variable u with $u(\omega) \in X(\omega)$, for all $\omega \in \Omega$, is a possible candidate for being the original.

Definition 6.2.1 Suppose we have a random sample with fuzzy outcomes (*fuzzy random sample*) defined by the mapping $X : \Omega \to F(\mathbb{R})$ or $(X_1, \dots, X_n) : \Omega \to (F(\mathbb{R}))^n$. Then, this mapping is called a *fuzzy random variable* if and only if

1 Assume that X is a set. Then, a σ-algebra F is a nonempty collection of subsets of X such that the following hold:

(i) X is in F;
(ii) if A is in F, then so is the complement of A; and
(iii) if A_n is a sequence of elements of F, then the union of the elements A_n is in F.

2 A random set is a Borel-measurable function from Ω to the set of all nonempty and compact subsets of $\mathbb{R}$.

there exists a system $X_\alpha(\omega)$, for all $\omega \in \Omega$, and for all $\alpha \in [0, 1]$ of subsets of $\mathbb{R}$, such that

$$X_\alpha^{\mathrm{l}} : \Omega \to \mathbb{R}, \quad X_\alpha^{\mathrm{u}} : \Omega \to \mathbb{R}, \quad \forall \alpha \in (0, 1]$$

are real-valued random variables and $X_\alpha(\omega) = [X_\alpha^{\mathrm{l}}(\omega), X_\alpha^{\mathrm{u}}(\omega)]$, where l, u stand for "lower" and "upper," respectively.

In other words, a fuzzy random variable is a (fuzzy) perception of an unknown random variable $u : \Omega \to \mathbb{R}$, with u being a possible original of X.

Definition 6.2.2 The set of all possible originals is given by

$$\chi = \{u : \Omega \to \mathbb{R}\},$$

where u is a Borel-measurable mapping.

Now, an interesting theorem (see Ref. [183] for a proof) describing the behavior of fuzzy random variables is the following:

Theorem 6.2.1 *(i) If $n \in \mathbb{N}$, $(X_1, \ldots, X_n)$ is a fuzzy random vector, and $(x_1, \ldots, x_n) \in \mathbb{R}^n$, then the linear combination $\sum_{i=1}^n x_i X_i$ is a fuzzy random variable, (ii) if $k \in \mathbb{N}$ is an odd number and X is a fuzzy random variable, then X^k is a fuzzy random variable, (iii) if $k \in \mathbb{N}$ is an even number and $X : \Omega \to F(\mathbb{R})$ is such that $|X|$ is a fuzzy random variable, then X^k is a fuzzy random variable.*

Now, we come to the definition of the moment of a fuzzy random variable. Let the fuzzy random variable $X : \Omega \to F(\mathbb{R})$, and $k \in \mathbb{N}$. Then, the k*-th moment* of X with respect to the original χ is defined as

Definition 6.2.3

$$(EX)^k(q) = \bigvee \{\mu_X(u) \mid u \in \chi \text{ and } E|u^k| < \infty \text{ and } Eu^k = q\},$$

for all $q \in \mathbb{R}$ and where $\mu_X(u)$ is the so-called *acceptability degree*, that is, the degree for a number $q \in \mathbb{R}$ to be the *expected value* of X is the maximal value of $\mu_X(u)$ such that u is an original of X. Further, one can see by use of the extension principle, that EX^k is the image of the fuzzy subset $\mu_X : \chi \to [0, 1]$ if one considers the mapping $E^k : \chi \to \mathbb{R}, u \to Eu^k$ [183].

In practice, to compute the expected value for a series of results given by random experimental results that are in the class of fuzzy sets $K_n(\mathbb{R})$, $n \in \mathbb{N}$, one proceeds as follows:

Given a finite probability space $\Omega = \{\omega_i,\ i = 1, \dots, k\}$, the corresponding probabilities p_i, $i = 1, \dots, k$ for the ω_is, respectively, and the fuzzy random variable $X : \Omega \to K_n(\mathbb{R})$, then the expected value is calculated as

$$EX = \sum_{i=1}^{k} p_i X(\omega_i),$$

and the fuzzy number $E(X) \in F(\mathbb{R})$ is the *fuzzy expectation value* of X if $E(X_\alpha) = [EX_\alpha^{l}, EX_\alpha^{u}]$, for all $\alpha \in (0, 1]$.

Next, we consider the concept of *variance of a fuzzy random variable* with respect to χ, whereby we will use the notion of the moment of a fuzzy random variable.

Definition 6.2.4

$$\operatorname{var} X = \bigvee \{\mu_X(u) \mid u \in \chi \text{ and } E|u - Eu|^2 < \infty \text{ and } E(u - Eu)^2 = q\},$$

for all $q \in \mathbb{R}$.

Finally, we come to the definition of the *distribution function of a fuzzy random variable*. First, we need the notion of a *normal set representation* of a fuzzy set.

Definition 6.2.5 We say that a fuzzy set $S \in F(\mathbb{R})$ belongs to the class $Q(\mathbb{R})$ of all fuzzy sets with a *normal set representation* if and only if there exists a set representation $\{A_\alpha \mid \alpha \in (0, 1)\}$ of S such that the following hold:

(i) $\bigwedge A_\alpha > -\infty \Rightarrow \bigwedge A_\alpha \in A_\alpha$;
(ii) $\bigvee A_\alpha < \infty \Rightarrow \bigvee A_\alpha \in A_\alpha$;
(iii) $\bigvee A_\alpha = -\infty$ or $\bigvee A_\alpha = \infty \Rightarrow A_\alpha$ is convex for all $\alpha \in (0, 1)$.

With the help of this definition, we come to the notion of the distribution function:

Definition 6.2.6 The (one-dimensional) distribution function $(F_X(x))(p)$ of a fuzzy random variable $X : \Omega \to Q(\mathbb{R})$ is a mapping $F_X : \mathbb{R} \to F(\mathbb{R})$ such that

(i) $(F_X(x))(p) = \bigvee \{\mu_X(V) \mid V \in \overline{\chi} \text{ and } (P \otimes P')[V \leq x] = p\}$, for all $p \in [0, 1]$;
(ii) $(F_X(x))(p) = 0$, for all $p \in \mathbb{R} \setminus [0, 1]$,

where μ_X is the fuzzy set of all possible originals of the fuzzy random vector X, $\overline{\chi}$ is the class of all $A \otimes A' \to$ Borel-measurable random vectors with $A \otimes A'$ a product σ-algebra, and $P \otimes P'$ is a product probability measure. The latter is associated with the product space $(\Omega \otimes \Omega', A \otimes A', P \otimes P')$ that is the probability space from which (Ω, A, P) is the perception. For the notion of a multidimensional distribution function, the reader is referred to [183].

Finally, two fuzzy random variables X and Y are *identically distributed* if X_α^1, Y_α^1 and X_α^2, Y_α^2 are identically distributed for all $\alpha \in (0, 1]$, while X and Y are *independent* if each variable from the set

$$\{X_\alpha^1, X_\alpha^2 \mid \alpha \in (0, 1]\}$$

is independent from each variable from the set

$$\{Y_\alpha^1, Y_\alpha^2 \mid \alpha \in (0, 1]\}$$

(see Ref. [183]). Furthermore, $(X_1, \dots, X_n)$ is a *normal* (or *Gaussian*) *fuzzy random sample* of size n if all the X_i, $i = 1, \dots, n$, are *independent and identically distributed* (iid) normal fuzzy random variables, whereby X is a normal fuzzy random variable when $X = E(X) \oplus R$ with $\oplus$ denoting the extended operation of addition, and R is a normal (not fuzzy) random variable with zero mean and variance σ^2, so that $R \sim N(0, \sigma^2)$ [122].

6.3 Point Estimation

Classical statistical analysis is based on random variables, point estimations, statistical hypotheses, and so on. The first major question encountered in statistical inference concerns the *point estimation for one of more unknown parameters*. Just to give an idea, a point estimator estimates a parameter by giving a specific numerical value. Thus, for example, the best point estimate of the population mean μ is the calculated sample mean $\overline{x}$. So the general question becomes how can we choose an estimator on a sample of a fixed size taken on a random variable with a probability density function containing one or more unknown parameters, in order to have a best estimate of that parameter?

Let us formulate the classical problem as it is encountered in statistical inference theory and, in fact, best considered as a problem of decision theory: Let a random variable X with a probability density function $f_X(x; \xi_1, \xi_2, \dots, \xi_k)$, where ξ_i, $i = 1, \dots, k$ are unknown parameters. Then, given the value $(x_1, \dots, x_n)$ of a random sample $(X_1, \dots, X_n)$ of size n from the population $f(x; \theta_1, \dots, \theta_k)$, we ask, based on this value, for the estimation ("best guess") of the parameters θ_i. If the estimation of the parameters is given as a single value, then we speak of a *point estimation*, otherwise, we refer to an *interval estimation*. In this sense, an interval estimation of a parameter gives an estimation of the parameter as an interval or a range of values.

In the present section, we shall examine the point estimation. Let $\hat{\theta}_i$ be a point estimation of the parameter θ_i, $i = 1, \dots, n$. In fact, this estimation is but a *decision* d_i, so that $\hat{\theta}_i = d_i$. This decision depends on the parameter θ_i, and it is a function of the value $(x_1, \dots, x_n)$ of the random sample $(X_1, \dots, X_n)$, so $\hat{\theta}_i = d_i(x_1, \dots, x_n)$.

Further, $\hat{\theta}_i = d_i(x_1, \ldots, x_n)$ is a value of the function $\hat{\Theta}_i = d_i(X_1, \ldots, X_n)$. The latter is called the *estimator function* (or *decision function*) or simply *estimator* of the parameter θ_i and, indeed, the process of finding the point estimation of the parameters $\theta_1, \ldots, \theta_k$ amounts to finding the estimators $\hat{\theta}_1, \ldots, \hat{\theta}_k$ of these parameters. In fact, the finding of the estimator $\hat{\theta}_i$ depends on the properties of this function and, usually, the determination of these properties leads to the way of finding the estimator. In this sense, we can seek for various kinds of point estimators, such as the *sufficient*, the *unbiased*, the *consistent*, the *efficient*, or the *maximum likelihood estimator*. For more details see Ref. [161] or [191] where a more advanced approach to point estimators is provided.

In what follows, we choose to briefly present three, very commonly used, classical point estimators, the unbiased estimator, the consistent estimator, and the maximum likelihood estimator.

6.3.1 The Unbiased Estimator

Suppose we have a random sample $(X_1, \ldots, X_n)$ from a population $f(x; \theta)$ and the point estimator $\hat{\theta} = d(x_1, \ldots, x_n)$. Then, the following theorem holds

Theorem 6.3.1

$$E[(\hat{\theta} - \theta)^2] = \text{var}(\hat{\theta}) + [\theta - E(\hat{\theta})]^2$$

Proof:

$$\begin{aligned}
E[(\hat{\theta} - \theta)^2] &= E\{\{[\hat{\theta} - E(\hat{\theta})] - [\theta - E(\hat{\theta})]\}^2\} \\
&= E\{[\hat{\theta} - E(\hat{\theta})]^2 + [\theta - E(\hat{\theta})]^2 - 2[\hat{\theta} - E(\hat{\theta})][\theta - E(\hat{\theta})]\} \\
&= E\{[\hat{\theta} - E(\hat{\theta})]^2 + [\theta - E(\hat{\theta})]^2 - 2[E(\hat{\theta}) - E(\hat{\theta})][\theta - E(\hat{\theta})]\} \\
&= \text{var}(\hat{\theta}) + [\theta - E(\hat{\theta})]^2
\end{aligned}$$

The term $\theta - E(\hat{\theta})$ is called the *bias* of the estimator $\hat{\theta}$. □

Consequently, if it is possible to find an estimator $\hat{\theta}$ of the parameter θ that has very small bias $\theta - E(\hat{\theta})$ and $\text{var}(\hat{\theta})$, then the mean squared error $E[(\hat{\theta} - \theta)^2]$ will be correspondingly small. Naturally, one desires a zero bias, $\theta - E(\hat{\theta}) = 0$, or $E(\hat{\theta}) = \theta$. So we come to the following definition:

Definition 6.3.1 Suppose we have a random sample $(X_1, \ldots, X_n)$ from a population $f(x; \theta)$. Then, the point estimator $\hat{\theta} = d(x_1, \ldots, x_n)$ of the parameter θ is called *unbiased* when its expected value equals θ, that is, when $E(\hat{\theta}) = \theta$. Otherwise, $\hat{\theta}$ is a *biased* estimator.

6.3.2 The Consistent Estimator

Again, suppose that we have a random sample $(X_1, \dots, X_n)$ from a population $f(x;\theta)$ and let an estimator of θ be $\hat{\theta}_n = d_n(x_1, \dots, x_n)$, whereby the index n denotes the quantity of the sample. Intuitively speaking, a good estimator is one for which the so-called *risk function* $R(d_n;\theta)$ will decrease with increasing n. Let us briefly introduce the notion of the aforementioned risk function.

We know from decision theory that the estimators $\hat{\theta}_1, \dots, \hat{\theta}_k$ (i.e. the decision rules) of the parameters $\theta_1, \dots, \theta_k$ for the values $(x_1, \dots, x_n)$ of the random sample $(X_1, \dots, X_n)$ establish a mapping of the sample space to the decision space. The wrong choice of the estimators produces a loss or cost, expressing the difference between estimated and true values. This loss is quantified by the so-called *loss function* $L(\hat{\theta}_1, \dots, \hat{\theta}_k; \theta_1, \dots, \theta_k)$ and its expected value is the aforementioned risk function:

$$R(d_1, \dots, d_k; \theta_1, \dots, \theta_k) = E[L(\hat{\theta}_1, \dots, \hat{\theta}_k; \theta_1, \dots, \theta_k)]$$

Hence, good estimators are those which minimize the risk function.

So, suppose we have a sequence of estimators $\{\hat{\theta}_n\} = \hat{\theta}_1, \hat{\theta}_2, \hat{\theta}_3, \dots$ with $\hat{\theta}_1 = d_1(x_1), \hat{\theta}_2 = d_2(x_1, x_2), \hat{\theta}_3 = d_3(x_1, x_2, x_3), \dots$ for the parameter θ, generated by $\hat{\theta}_n = d_n(x_1, x_2, \dots, x_n)$ for $n = 1, 2, 3, \dots$. Then, we demand

$$\lim R(d_n;\theta) \xrightarrow[n\to\infty]{} 0$$

and if the risk function is the mean squared error $E[(\hat{\theta}_n - \theta)^2]$, then we have the following theorem:

Theorem 6.3.2

$$\lim E[(\hat{\theta}_n - \theta)^2] \xrightarrow[n\to\infty]{} 0.$$

Proof: From Theorem 6.3.1 we have $E[(\hat{\theta}_n - \theta)^2] = \mathrm{var}(\hat{\theta}_n) + [\theta - E(\hat{\theta}_n)]^2$. But $E(\hat{\theta}_n) = \theta$ and $\lim \mathrm{var}(\hat{\theta}_n) \xrightarrow[n\to\infty]{} 0$. Therefore, $\lim E[(\hat{\theta}_n - \theta)^2] \xrightarrow[n\to\infty]{} 0$. □

Now, having said all that, we finally come to the following definition:

Definition 6.3.2 Suppose that we have a random sample $(X_1, \dots, X_n)$ from a population $f(x;\theta)$. The estimator $\hat{\theta}_n = d_n(x_1, \dots, x_n)$ of the parameter θ is called *consistent* if it is unbiased and if it holds that $\lim \mathrm{var}(\hat{\theta}_n) \xrightarrow[n\to\infty]{} 0$ and $E(\hat{\theta}_n) = \theta$.

As an example, one can readily show that for a random sample $(X_1, \dots, X_n)$ from a population with the normal distribution $N(\mu, \sigma^2)$, $s^2 = \frac{1}{n-1}\sum_{i=1}^{n}(x_i - \overline{x})^2$ is a consistent estimator of the population variance σ^2.

6.3.3 The Maximum Likelihood Estimator

First, we shall present the definition of the *likelihood function*:

Definition 6.3.3 A likelihood function of n random variables $(X_1, \dots, X_n)$ is the joint distribution

$$L(\theta_1, \dots, \theta_k) = g(X_1, \dots, X_n; \theta_1, \dots, \theta_k), \quad k \geq 1$$

Apparently, when $X_1, \dots, X_n$ is a random sample from the population $f(x; \theta_1, \dots, \theta_k)$, then the likelihood function of this sample is

$$L(\theta_1, \dots, \theta_k) = g(X_1, \dots, X_n; \theta_1, \dots, \theta_k) = \prod_{i=1}^{n} f(x_i; \theta_1, \dots, \theta_k).$$

Now, we can define the maximum likelihood estimator as follows:

Definition 6.3.4 Let $X_1, \dots, X_n$ be a random sample from the population $f(x; \theta_1, \dots, \theta_k)$. The estimators $\hat{\Theta}_i = d_i(X_1, \dots, X_n)$ of the parameters θ_i, $i = 1, 2, \dots, k$ are the *maximum likelihood estimators* if their values $\hat{\theta}_i = d_i(x_1, \dots, x_n)$ maximize the likelihood function [$L(\theta_1, \dots, \theta_k)$ of the sample:

$$L(\hat{\theta}_1, \dots, \hat{\theta}_k) = \max L(\theta_1, \dots, \theta_k).$$

When the likelihood function contains only one parameter θ and is differentiable w.r.t. θ, then the maximum likelihood estimator $\hat{\theta}$ of θ is the solution of the equation

$$\frac{dL(\theta)}{d\theta} = 0.$$

In fact, often the equation

$$\frac{d \ln L(\theta)}{d\theta} = 0$$

is used in applications, since $L(\theta)$ and $\ln L(\theta)$ are maximized for the same values of θ.

In the case where the likelihood function L depends on more than one parameters $\hat{\theta}_1, \dots, \hat{\theta}_k$, then the maximum likelihood estimators $\hat{\theta}_1, \dots, \hat{\theta}_k$ are found as the solution of the system of equations

$$\frac{\partial L(\theta_1, \dots, \theta_k)}{\partial \theta_i} = 0, \quad i = 1, 2, \dots, k$$

or

$$\frac{\partial \ln L(\theta_1, \dots, \theta_k)}{\partial \theta_i} = 0, \quad i = 1, 2, \dots, k.$$

6.4 Fuzzy Point Estimation

We start with the problem of fuzzy point estimation that generalizes what has been presented in Section 6.3. First of all, we should point out that often fuzzy point estimation is considered as more basic than fuzzy interval estimation, since the latter can be determined by the point estimations of the lower and upper boundaries of the interval. In fact, if the point estimation is characterized by very low confidence, it can be relaxed to an interval estimation (for an application see, e.g. [7]).

Now, let us adopt the more formal approach to fuzzy point estimation and suppose we have a fuzzy random sample, that is, a fuzzy random vector, $(X_1, \dots, X_n) : \Omega \to (F(\mathbb{R}))^n$. A fuzzy parameter for our fuzzy point estimation is a perception of the unknown parameter we seek.

Definition 6.4.1 The *fuzzy parameter* $\theta_C(X_1, \dots, X_n)$ of a fuzzy random sample $(X_1, \dots, X_n) : \Omega \to (F(\mathbb{R}))^n$ with respect to a mapping assigning to each probability distribution function its parameter, is a fuzzy set

$$\begin{aligned}&\theta_C[X_1, \dots, X_n](m)\\&\quad = \bigvee \{\mu_{(X_1,\dots,X_n)}(V_1, \dots, V_n) \mid (V_1, \dots, V_n) \in \chi_C^n \text{ and } \theta_C(D_{V_1}) = m\},\end{aligned}$$

where $m \in \mathbb{R}$, $n \in \mathbb{N}$, $\theta_C : C \to \mathbb{R}$, C is the class of probability distribution functions, χ_C^n is the set of all possible originals, $\mu_{(X_1,\dots,X_n)}(V_1, \dots, V_n)$ is the acceptability that the random vector $(V_1, \dots, V_n)$ on $\Omega \times \Omega'$ is the original of $(X_1, \dots, X_n)$, and D_V is the probability distribution function of V from the class C.

So we come to the following definition of the fuzzy point estimation:

Definition 6.4.2 Suppose $(X_1, \dots, X_n)$ is a fuzzy random sample from the class of probability distribution functions C and $\theta_C : C \to \mathbb{R}$. Then, a *fuzzy point estimator* of the parameter θ_C is

$$G_n : \Omega \to E(\mathbb{R}), \quad \omega \to G_n[(X_1)_\omega, \dots, (X_n)_\omega],$$

where $E(\mathbb{R})$ is the class of all fuzzy subsets of $\mathbb{R}$.

As we have done with classical point estimation, we can distinguish between different kinds of fuzzy point estimators.

Definition 6.4.3 Suppose $G_n : \Omega \to F(\mathbb{R})$ is a fuzzy random vector and $\mu \in F(\mathbb{R})$. Then, G_n is an *unbiased fuzzy point estimator* if $E(G_n) = \mu$ holds.

Similarly, when we desire to obtain a parameter estimation by using a sequence of fuzzy random variables, then we have the following:

Definition 6.4.4 Suppose C is a class of distribution functions, $\theta_C : C \to \mathbb{R}$ and $G_n : [F(\mathbb{R})]^n \to F(\mathbb{R})$. Let S be the class of sequences of fuzzy random variables (X_k) such that the following holds for the *convex hulls* for every $(X_k) \in S$, $k \in \mathbb{N}$:

$$\mathrm{conv}(\theta_C[X_i]) = \mathrm{conv}(\theta_C[X_1, \dots, X_n]).$$

Then, (G_n) is a *consistent estimator* of θ_C with respect to S if for all sequences $(X_k) \in S$ one has the convergence

$$(G_n[X_1, \dots, X_n])_{n\in\mathbb{N}} \to \mathrm{conv}(\theta_C[X_1]).$$

Concerning the maximum likelihood estimator presented in Section 6.3.3 due to its simplicity and the consequent usefulness in statistical inference, its implementation in the realm of fuzzy data is fairly complex in most practical situations. Indeed, one "classical" approach is to consider a maximum likelihood estimation of a desired parameter as the crisp value that maximizes the probability of observing the fuzzy data (see, e.g. [59]). We shall not go into details, however, we point out that there have been more efficient ways to handle the aforementioned difficulty and the reader is referred, for instance, to the so-called "expectation–maximization algorithm" (see Ref. [95] and references therein).

A final, worth noticing, comment is deemed proper at this point. It concerns the interesting and different approach to the problem of fuzzy point estimation presented in [3] by introducing the methods of the *fuzzy uniformly minimum variance unbiased estimation* and the *fuzzy Bayesian estimation*, both based on the notions of the Yao-Wu signed distance and the L_2-metric. When the fuzzy random variables become crisp random variables, the aforementioned methods are reduced to the classical uniformly minimum variance unbiased estimation and the Bayesian estimation.

6.5 Interval Estimation

In order to determine an interval of plausible values for an unknown sample population parameter, the notion of *interval estimation* is used in classical statistics. In other words, the unknown parameters are estimated (notably, but not only, these parameters are the mean and the variance of the population) as an interval or as an entire range of numerical values within which the aforesaid parameter is estimated to lie. One of the most prevalent forms of interval estimation is the frequentist approach known as *confidence interval* (see, e.g. [215]) that was introduced by the Polish mathematician Jerzy Neyman [227] in 1937.

Let a random sample $(X_1, \dots, X_n)$ of size n with the value $(x_1, \dots, x_n)$. This value consists in measured data characterized, in general, by uncertainties expressed as errors that have been neglected in the previously presented process of finding a point estimator $\hat{\theta}$. So, in order to increase the reliability of the estimate, one ought to take into account these errors and give the point estimator $\hat{\theta}$ in some interval, say, $(\hat{\theta} - \varepsilon, \hat{\theta} + \varepsilon)$, $\varepsilon > 0$. Let us see how this can be realized in our case. Suppose that our random sample has a continuous distribution $f(x)$ with an unknown mean μ and a known variance σ^2. Now, it is a well-known fact from classical statistics that if X is a random variable with the distribution $f(x)$, then for the random variable defined as

$$Y = \frac{\overline{X} - \mu}{\sigma/\sqrt{n}},$$

where $\overline{X} = \frac{1}{n}\sum_{i=1}^{n} X_i$, we can always determine the distribution $g(y)$, where $Y = y(X)$. For large samples, $g(y)$ is the standard normal distribution (i.e. the Gaussian distribution with $\mu = 0$ and $\sigma^2 = 1$), as it is inferred by the Central Limit Theorem (see Ref. [161]).

Now, for two arbitrary values y_1 and y_2 of Y, we have

$$\Pr(y_1 < Y < y_2) = \int_{y_1}^{y_2} g(y)\,dy = 1 - \delta$$

or

$$\Pr\left(\overline{X} - \frac{\sigma}{\sqrt{n}}y_2 < \mu < \overline{X} - \frac{\sigma}{\sqrt{n}}y_1\right) = 1 - \delta.$$

Here, δ denotes the *confidence level* (also known as *significance level*).[3] The interval defined by the inequality

$$\overline{X} - \frac{\sigma}{\sqrt{n}}y_2 < \mu < \overline{X} - \frac{\sigma}{\sqrt{n}}y_1$$

is called *confidence interval* of the parameter μ with the probability $100(1 - \delta)\%$. The length of this interval is the *confidence length* or *width*:

$$\overline{x} - \frac{\sigma}{\sqrt{n}}y_1 - \left(\overline{x} - \frac{\sigma}{\sqrt{n}}y_2\right) = (y_2 - y_1)\frac{\sigma}{\sqrt{n}}.$$

The confidence length depends mainly on the size of the sample but also on its variability as well as on the confidence level. One always desires the confidence length to be the least possible. Obviously, this can be achieved by minimizing the difference $y_2 - y_1$. In fact, for a constant $1 - \delta$, it can be rather easily shown that the difference $y_2 - y_1$ gets its minimum value when $g(y_2) = g(y_1)$.

3 In the literature, it is often the case that the confidence level is designated by $1 - \delta$ when δ is the significance level. Furthermore, usually the desired confidence level is chosen prior to examining the measurements. Very often, the 95% confidence level is applied. However, attention should be paid to the following common misunderstanding: a 95% confidence level does ***not*** mean that 95% of the sample data lie within the confidence interval.

6.6 Interval Estimation for Fuzzy Data

The study of the notion of the fuzzy interval or confidence estimation has started mainly in the 1980s by Roger A. McCain [212] and begun to escalate after 2000 with a plethora of results (see, e.g. [169] for more information on the historical development and on various approaches to the problem). Here, we have chosen to present a rather practical method presented by Norberto Corral and Maria Ángeles Gil [79] for the construction of an interval estimation of an unknown parameter θ for a given sample fuzzy information. One has to stress that, although the method yields rather crude confidence intervals (the probability of the parameter being within the estimated interval may be larger than the confidence level), it has the great advantage of being always applicable, that is, for any membership function and any class of distribution functions.

Suppose that we have a random experiment X with probability space (Ω, A, P), where Ω denotes the set of all possible outcomes of a random experiment, A being a σ-algebra of subsets of Ω, and P defining a probability measure. Let θ be the parameter whose value lies in an interval of the experiment. Further, suppose that the set of all fuzzy observations defines our fuzzy information. Let us first recall some basic definitions:

Definition 6.6.1 A *fuzzy information* for a random experiment is a fuzzy event $\tilde{x}$ on Ω characterized by a Borel-measurable function $\mu_{\tilde{x}}$ that represents the grade of membership of x to $\tilde{x}$.

Definition 6.6.2 A fuzzy partition (orthogonal system) with fuzzy events on X, is called *fuzzy information system* S if $\sum_{\tilde{x}\in S} \mu_{\tilde{x}}(x) = 1, \forall x \in X$.

It is assumed that the increased sampling from X will not yield a precise observation but a *sample fuzzy information*:

Definition 6.6.3 A sample fuzzy information of size n from an experiment X is the algebraic product of n elements belonging to a sample information system S of the experiment. Indeed, the fuzzy information system on a random sample of size n of X is called a *fuzzy random sample* of size n from S.

Now, based on Definition 6.6.3, we can proceed to a formal definition of the confidence interval:

Definition 6.6.4 For a given fuzzy random sample of size n from S, denoted by $S^{(n)}$, and an interval $[\theta_{\mathrm{l}}(S^n), \theta_{\mathrm{u}}(S^n)]$, where the subscripts l and u stand for "lower" and "upper," respectively, we have

$$P_\theta\{\theta_l(S^n) \le \theta \le \theta_u(S^n)\} = \sum_{\substack{(\tilde{x}_1,\dots,\tilde{x}_n) \in S^{(n)} \\ \theta_l(\tilde{x}_1,\dots,\tilde{x}_n) \le \theta \le \theta_u(\tilde{x}_1,\dots,\tilde{x}_n)}} \int_X \mu_{(\tilde{x}_1,\dots,\tilde{x}_n)}(x_1,\dots,x_n) dP_\theta(x_1,\dots,x_n) \ge \delta,$$

as a δ-*confidence interval* for the parameter θ with $\delta \in [0,1]$ being the *confidence level*, and $\theta_l(S^n)$, $\theta_u(S^n)$ the *lower* and *upper confidence limits* of θ, respectively. A constant interval $[\theta_l(\tilde{x}_1,\dots,\tilde{x}_n), \theta_u(\tilde{x}_1,\dots,\tilde{x}_n)]$ is often also called a δ-confidence interval for the parameter θ.

Suppose that we have a sample fuzzy information $(\tilde{x}_1,\dots,\tilde{x}_n)$ with

$$s_{(\tilde{x}_1,\dots,\tilde{x}_n)} = \{(x_1,\dots,x_n) \in X^n \mid \mu_{(\tilde{x}_1,\dots,\tilde{x}_n)}(x_1,\dots,x_n) > 0\}$$

its support and n the size of the sample. Then, the following theorem provides a way to determine a δ-confidence interval for the parameter θ:

Theorem 6.6.1 *If* $\theta_l(\tilde{x}_1,\dots,\tilde{x}_n) = \bigwedge\{\Theta_l(s_{(\tilde{x}_1,\dots,\tilde{x}_n)})\}$ *and* $\theta_u(\tilde{x}_1,\dots,\tilde{x}_n) = \bigvee\{\Theta_u(s_{(\tilde{x}_1,\dots,\tilde{x}_n)})\}$, *with the δ-confidence interval* $[\Theta_l(X^{(n)}), \Theta_u(X^{(n)})]$ *for* θ *such that*

$$P_\theta\{\Theta_l(X^{(n)} \le \theta \le \Theta_u(X^{(n)})\} = \sum_{\substack{(x_1,\dots,x_n) \in X^n \\ \Theta_l(x_1,\dots,x_n) \le \theta \le \Theta_u(x_1,\dots,x_n)}} \int_X dP_\theta(x_1,\dots,x_n) \ge \delta,$$

then the interval $[\theta_l(\tilde{x}_1,\dots,\tilde{x}_n), \theta_u(\tilde{x}_1,\dots,\tilde{x}_n)]$ *determines a δ-confidence interval for the parameter θ.*

(For a proof of this theorem, see Ref. [79]).

6.7 Hypothesis Testing

Very often, samples of measurement data can be interpreted by a priori assuming a structure (i.e. a specific distribution) of the measurement results and then apply certain statistical tests to determine the probability of the initial assumption (hypothesis) being true or not. In other words, a statistical hypothesis is the assumption about a population parameter, and it is an important part of empirical evidence-based research (see, e.g. [192] for a general overview).

The procedure starts with the examination of a random sample of the population considered. First, it is assumed either that the sample data are the result of pure chance (*null hypothesis*, denoted as H_0) or they are sufficiently affected by some nonrandom influence (*alternative hypothesis*, denoted as H_1). Then, an appropriate statistical test is chosen in order to assess the truth of the null hypothesis. In the next step, one determines the probability that the given data would occur when H_0 is assumed. This probability is called the *p-value* (or *p-level*), and it is used to interpret the result obtained. The smaller the *p-value*, the stronger is the evidence against H_0. In other words, the p-value is a measure of how likely the data would be observed if H_0 were true. Finally, one compares the calculated p-value with the selected *significance* or *confidence level* δ (see Section 6.3). If $p \leq \delta$, then the observed influence is statistically significant, H_0 must be rejected, and H_1 holds. If $p > \delta$, then H_0 is true.

Now, concerning the magnitude of the p-value, one of two types of error may appear. When the p-value is small, there is a possibility that H_0 is true, but an unlikely event has been measured (*type I error* or *false positive*) and we have incorrectly rejected H_0, while if the p-value is large, there is a possibility that H_0 is false, but an unlikely event has been measured (*type II error* or *false negative*), and we have incorrectly accepted H_0. The probability of making a type I error, that is, the probability of rejecting H_0, given that it is true, is δ (i.e. the significance level), while the probability of making a type II error, that is, the probability of accepting H_0, given that H_1 is true, is denoted by β. A usual way out of this apparent impasse is provided by the demand for independent verification of the data.

In practice, in order to find when the null hypothesis can be rejected or not, the concept of the test function can be used. To this purpose, suppose that we have the null hypothesis $H(\theta_0) : \theta = \theta_0$, and let $A(\theta_0)$ be the acceptance set of $H(\theta_0)$ on a δ significance level. Then, with the set $G(x) = \{\theta \mid x \in A(\theta)\}$ we have

$$\theta \in G(x) \leftrightarrow x \in A(\theta),$$

so that

$$P_\theta\{\theta \in G(x)\} \geq 1 - \delta.$$

In other words, any set of acceptance on a δ significance level yields a *confidence set* $G(x)$ on the confidence level $1 - \delta$. The confidence set lets us conclude, for each θ_0, whether the null hypothesis $H(\theta_0)$ should be accepted or rejected on the δ significance level for the measured x. Indeed, we can define a *test function* $\Phi(x; \theta_0)$:

Definition 6.7.1

$$\Phi(x; \theta_0) = \begin{cases} 0, & \theta_0 \in G(x), \\ 1, & \theta_0 \notin G(x), \end{cases}$$

where $\Phi(x;\theta_0) = 0$ signifies the acceptance, while $\Phi(x;\theta_0) = 1$ signifies the rejection of the null hypothesis. Equivalently, by introducing the so-called *indicator function* $I_{G(x)}(\theta)$ of the set $G(x)$, one has

$$\Phi(x;\theta_0) = \begin{cases} 0, & I_{G(x)}(\theta_0) = 1, \\ 1, & I_{G(x)}(\theta_0) = 0. \end{cases}$$

The following example borrowed from [62] illustrates very clearly the use of the test function.

Example 6.7.1 Let $X_1, \ldots, X_n$ be iid from the normal distribution $N(\theta, 1)$ with the unknown mean θ. Then, one can deduce a confidence interval for θ at the level $1 - \delta$ of the form $G(X) = \left[\overline{X} - \frac{1}{\sqrt{n}} z_{1-\frac{\delta}{2}}, \overline{X} + \frac{1}{\sqrt{n}} z_{1-\frac{\delta}{2}}\right]$, with z_δ the δ-quantile of the standard normal distribution. Suppose that in a random sample of size $n = 25$ one observes $\overline{x} = 0.75$ and the null hypothesis $H_0 : \theta = 0.5$ is to be tested against the alternative hypothesis $H_1 : \theta \neq 0.5$ at the significance level $\delta = 0.05$. Here $G(x) = [0.358, 1.142]$, consequently the test function reads

$$\Phi(x;0.5) = 1 \begin{cases} 0, & I_{G(x)}(0.5) = 1, \\ 1, & I_{G(x)}(0.5) = 0. \end{cases}$$

Hence, on the basis of the observed value $\overline{x} = 0.75$, the null hypothesis H_0 is accepted at the significance level $\delta = 0.05$.

6.8 Fuzzy Hypothesis Testing

Starting in the 1980s with the work of María Rosa Casals et al. [59], a rather large amount of work has been published in the field of fuzzy statistics (see Ref. [282] for a review and references therein). More generally, statistical hypothesis testing in a fuzzy environment was taken up by Przemysław Grzegorzewski and Olgierd Hryniewicz [151]. The problem of fuzzy hypothesis testing has been linked to the notion of the p-value described in Section 6.7 by Glen Meeden and Siamak Noorbaloochi [214], who instead of determining a null hypothesis and an alternative hypothesis have given a reformulation of the problem "as the problem of estimating the membership function of the set of good or useful or interesting parameter points." Here, we shall present a different approach introduced by Jalal Chachi et al. [62], who alternatively, in the form of six steps, have given a constructive method to connect fuzzy hypothesis testing with confidence intervals.

Now, let us assume that both the hypothesis parameter θ and the confidence interval are fuzzy [61], that is, $\tilde{\theta} = \tilde{\theta}_0$ and $I = I(\tilde{\theta}_0)$ are, respectively, the fuzzy parameter value (according to what we have said in Section 6.4, $\tilde{\theta}$ is considered as a fuzzy perception of θ) and the degree of hypothesis acceptance that depends on $\tilde{\theta}_0$, then we have for the set of the parameter values for which the tested hypothesis is accepted

$$\{\tilde{\theta}_0 \in \tilde{I}(X) \mid I(\tilde{\theta}_0) > 0\},$$

while

$$\{\tilde{\theta}_0 \in \tilde{I}(X) \mid 1 - I(\tilde{\theta}_0) > 0\},$$

for the set of the parameter values for which the tested hypothesis is rejected, with $\tilde{I}(X)$ the fuzzy confidence interval. The hypothesis to be tested is the null hypothesis $H_0 : \tilde{\theta} = \tilde{\theta}_0$ against the alternative $H_1 : \tilde{\theta} \neq \tilde{\theta}_0$ for observational data with unknown fuzzy mean $\tilde{\theta}$ but known variance σ^2, so that $(X_1, \dots, X_n) \overset{\text{iid}}{\sim} N(\tilde{\theta}, \sigma^2)$. To that purpose, we must find the degrees of acceptability for H_0 and H_1 and the "Chachi–Taheri–Viertl algorithm" is codified as follows (see Ref. [62] and the nice numerical examples therein):

(i) Convert the hypothesis to be tested to a set of crisp problems on (for $\alpha \in [0,1]$) the fuzzy parameter. Then, for each α-level for the samples $X_\alpha^{\mathrm{l}} = (X_{1\alpha}^{\mathrm{l}}, \dots, X_{n\alpha}^{\mathrm{l}})$ and $X_\alpha^{\mathrm{u}} = (X_{1\alpha}^{\mathrm{u}}, \dots, X_{n\alpha}^{\mathrm{u}})$, one must solve at the confidence level δ the classical hypothesis testing problems

$$H_0 : \theta_\alpha^{\mathrm{l}} = \theta_{0\alpha}^{\mathrm{l}} \text{ vs. } H_1 : \theta_\alpha^{\mathrm{l}} \neq \theta_{0\alpha}^{\mathrm{l}}, \tag{6.1}$$

$$H_0 : \theta_\alpha^{\mathrm{u}} = \theta_{0\alpha}^{\mathrm{u}} \text{ vs. } H_1 : \theta_\alpha^{\mathrm{u}} \neq \theta_{0\alpha}^{\mathrm{u}}, \tag{6.2}$$

where the fuzzy parameters are $\tilde{\theta}_\alpha = [\theta_\alpha^{\mathrm{l}}, \theta_\alpha^{\mathrm{u}}]$, $\tilde{\theta}_{0\alpha} = [\theta_{0\alpha}^{\mathrm{l}}, \theta_{0\alpha}^{\mathrm{u}}]$.

(ii) Determine the $1 - \delta$ confidence intervals $[L_1(X_\alpha^{\mathrm{l}}), L_2(X_\alpha^{\mathrm{l}})]$ and $[U_1(X_\alpha^{\mathrm{u}}), U_2(X_\alpha^{\mathrm{u}})]$ for the crisp parameters $\theta_\alpha^{\mathrm{l}}$ and $\theta_\alpha^{\mathrm{u}}$, respectively, for each $\alpha \in (0, 1]$.

(iii) Test the hypotheses (6.1) and (6.2) through the examination of the $1 - \delta$ confidence intervals $[L_1(X_\alpha^{\mathrm{l}}), L_2(X_\alpha^{\mathrm{l}})]$ and $[U_1(X_\alpha^{\mathrm{u}}), U_2(X_\alpha^{\mathrm{u}})]$ to see whether they contain $\theta_{0\alpha}^{\mathrm{l}}$ and $\theta_{0\alpha}^{\mathrm{u}}$, respectively. Here, the corresponding test functions are

$$\Phi(X_\alpha^{\mathrm{l}}; \theta_{0\alpha}^{\mathrm{l}}) = \begin{cases} 0, & \theta_{0\alpha}^{\mathrm{l}} \in [L_1(X_\alpha^{\mathrm{l}}), L_2(X_\alpha^{\mathrm{l}})], \\ 1, & \theta_{0\alpha}^{\mathrm{l}} \notin [L_1(X_\alpha^{\mathrm{l}}), L_2(X_\alpha^{\mathrm{l}})], \end{cases}$$

and

$$\Phi(X_\alpha^{\mathrm{u}}; \theta_{0\alpha}^{\mathrm{u}}) = \begin{cases} 0, & \theta_{0\alpha}^{\mathrm{u}} \in [U_1(X_\alpha^{\mathrm{u}}), U_2(X_\alpha^{\mathrm{u}})], \\ 1, & \theta_{0\alpha}^{\mathrm{u}} \notin [U_1(X_\alpha^{\mathrm{u}}), U_2(X_\alpha^{\mathrm{u}})]. \end{cases}$$

(iv) Gather the results in the third case to get a fuzzy confidence interval in order to proceed to the construction of a fuzzy test on the basis of the membership degree of each fuzzy parameter $\tilde{\theta}$ in the fuzzy confidence interval. To this purpose, it is necessary to group the α-values for which the null hypotheses (6.1) and (6.2) are accepted or rejected. This grouping can be performed by making a graph of α vs. the $1 - \delta$ confidence intervals so that every confidence interval has α as its height, and then determine confidence bounds from their intersection (obviously, $\alpha = 1$ is the maximal height). Then, the membership function of the fuzzy parameter $\tilde{\theta}$ is compared with the confidence bound obtained. From this comparison, the α-values for which the null hypotheses are accepted or rejected are found.

(v) Construct the fuzzy set $\tilde{C} = \{(\tilde{\theta}, C(\tilde{\theta})) : \tilde{\theta} \in F(\Theta)\}$ by applying the method given in [61]. This set is a fuzzy confidence interval for the fuzzy parameter $\tilde{\theta}$.

(vi) Construct the fuzzy test function

$$\tilde{\Phi}(X;\tilde{\theta}_0)(r) = \begin{cases} C(\tilde{\theta}_0), & r = 0, \\ 1 - C(\tilde{\theta}_0), & r = 1, \end{cases}$$

for the fuzzy random sample $X = (X_1, \dots, X_n)$. Evidently, the above fuzzy test function $\tilde{\Phi}(X;\tilde{\theta}_0) : (F(\mathbb{R}))^n \to F\{0,1\}$ is described by a fuzzy set leading to the acceptance (with degree of acceptance $C(\tilde{\theta}_0)$) of the tested null hypothesis, or to its rejection with degree $1 - C(\tilde{\theta}_0)$.

At this point, we must stress that if $C(\tilde{\theta}_0)$ is between zero and one so that it is not absolutely clear what to do, then necessarily one has to make a subjective "fuzzy" decision on the acceptance or rejection of the null hypothesis. In such a case, as it is expected, the null hypothesis is accepted, the more the values of $C(\tilde{\theta}_0)$ tend to one, and rejected, the more they tend to zero. Naturally, the most difficult decision to be made is when the value of $C(\tilde{\theta}_0)$ is $\frac{1}{2}$.

6.9 Statistical Regression

Regression analysis is a statistical methodology for the estimation of the conditionally expected value of a dependent random variable y (the *response variable*) given one or more independent nonrandom variables x (the *predictor variables*) and one or more involved unknown parameters β to be estimated from the data, in other words, we have $E(y|x) = f(x, \beta)$. According to the linearity of the parameters β in the function $f(x, \beta)$, one can build linear or nonlinear regression models to fit the measured data. If one has n pairs of data points (x_i, y_i), $i = 1, \dots, n$ and $n < m$, where m is the length of the vector of the unknown parameters β, the regression model is underdetermined and β cannot be specified. If $n = m$ and the function $f(x, \beta)$ is assumed linear, then the *regression equation* $y = f(x, \beta)$ can be exactly solved, i.e. to find β one can solve a $n \times n$ quadratic system which has one unique solution provided the xs are linearly independent. Finally, if $n > m$, which is the main problem in regression analysis, one can estimate values for the βs that best fit the data. The method of least squares belongs to the latter case. Further, the performance of regression analysis depends on the chosen process for collecting and measuring the data and the assumptions made in this process (for a more detailed account of regression see, e.g. [106]).

As an example, let us consider the most simple *linear regression* model. It involves two-dimensional data points (given, say, in Cartesian coordinates) and contains one independent and one dependent variable. So let us assume that the model function is linear in β and of first order in x with the regression equation

$$y = \beta_0 + \beta_1 x + \varepsilon, \tag{6.3}$$

where the unknowns are $\beta_0, \beta_1, \varepsilon$ and ε is a random error term denoting the deviation from the true line. Usually, the distribution of the random errors is modeled as a normal distribution with zero mean. In order to estimate the parameters, we shall apply the *method of least squares* and $\hat{\beta}_0, \hat{\beta}_1$ will denote the estimated by the given data values of β_0, β_1. Then, the predicted value of y, denoted by $\hat{y}$, is obtained as

$$\hat{y} = \hat{\beta}_0 + \hat{\beta}_1 x, \tag{6.4}$$

that is, a straight-line fit. Now, based on the n pairs of data points, we can write (6.3) as

$$y_i = \beta_0 + \beta_1 x_i + \varepsilon_i, \quad i = 1, \dots, n \tag{6.5}$$

and we have

$$S = \sum_{i=1}^{n} \varepsilon_i^2 = \sum_{i=1}^{n} (y_i - \beta_0 - \beta_1 x_i)^2$$

The estimated values $\hat{\beta}_0$ and $\hat{\beta}_1$ generate the least possible value of S. For their calculation, we start with the derivatives

$$\frac{\partial S}{\partial \beta_0} = -2 \textstyle\sum_{i=1}^{n} (y_i - \beta_0 - \beta_1 x_i),$$
$$\frac{\partial S}{\partial \beta_1} = -2 \textstyle\sum_{i=1}^{n} x_i (y_i - \beta_0 - \beta_1 x_i),$$

from which it follows that

$$\sum_{i=1}^{n} (y_i - \hat{\beta}_0 - \hat{\beta}_1 x_i) = 0,$$

$$\sum_{i=1}^{n} x_i (y_i - \hat{\beta}_0 - \hat{\beta}_1 x_i) = 0,$$

respectively. From these equations, we get

$$\sum_{i=1}^{n} y_i - n\hat{\beta}_0 - \hat{\beta}_1 \sum_{i=1}^{n} x_i = 0,$$

$$\sum_{i=1}^{n} x_i y_i - \hat{\beta}_0 \sum_{i=1}^{n} x_i - \hat{\beta}_1 \sum_{i=1}^{n} x_i^2 = 0,$$

respectively. Hence,

$$\sum_{i=1}^{n} y_i = n\hat{\beta}_0 + \hat{\beta}_1 \sum_{i=1}^{n} x_i,$$

$$\sum_{i=1}^{n} x_i y_i = \hat{\beta}_0 \sum_{i=1}^{n} x_i + \hat{\beta}_1 \sum_{i=1}^{n} x_i^2.$$

From these equations, one readily obtains the estimated parameter $\hat{\beta}_1$:

$$\hat{\beta}_1 = \frac{\sum_{i=1}^{n} x_i y_i - \frac{1}{n}\left(\sum_{i=1}^{n} x_i \sum_{i=1}^{n} y_i\right)}{\sum_{i=1}^{n} x_i^2 - \frac{1}{n}\left(\sum_{i=1}^{n} x_i\right)^2}.$$

By denoting the mean as $\overline{x}$ and $\overline{y}$, the latter can be written in the more usual form

$$\hat{\beta}_1 = \frac{\sum_{i=1}^{n}(x_i - \overline{x})(y_i - \overline{y})}{\sum_{i=1}^{n}(x_i - \overline{x})^2},$$

while the other estimated parameter is found to be

$$\hat{\beta}_0 = \overline{y} - \hat{\beta}_1 \overline{x}.$$

Therefore, (6.4) becomes now

$$\hat{y} = \overline{y} + \hat{\beta}_1(x - \overline{x}).$$

Evidently, for $x = \overline{x}$ the last equation yields $\hat{y} = \overline{y}$, so $(\overline{x}, \overline{y})$ is a point of the fitted line.

The estimates of the errors ε_i are given by $\hat{\varepsilon}_i = y_i - \hat{y}_i$, thus, we have

$$\hat{\varepsilon}_i = (y_i - \overline{y}) - \hat{\beta}_1(x_i - \overline{x})$$

from which it follows that

$$\sum_{i=1}^{n} \hat{\varepsilon}_i = \sum_{i=1}^{n}(y_i - \overline{y}) - \hat{\beta}_1 \sum_{i=1}^{n}(x_i - \overline{x}) = 0.$$

However, one should point out that, in practical applications, the rounding of the measured data always results in a nonzero error.

6.10 Fuzzy Regression

Simply put, statistical regression as presented in Section 6.9 is based on crisp random errors, while fuzzy regression is based on fuzzy errors. The difference between these two kinds of uncertainty has led to the necessity of extending statistical regression for the case of a fuzzy environment. This extension started in 1982 with the pioneering work of Hideo Tanaka et al. [283], who applied the methodology of linear programming to develop a *fuzzy linear regression* model. In fact, Tanaka's approach constitutes one of two main approaches in fuzzy regression, namely the so-called *possibilistic regression analysis* that relies on the notion of possibility, and the approach known as *fuzzy least squares method* that aims at the minimalization of the errors between the observed and the estimated data.

Returning to Tanaka's approach, it must be stressed that it can be applied only to linear functions. However, it is very simple in its computational implementation, while the fuzzy least squares method has an advantage compared to Tanaka's approach on that it keeps the degree of fuzziness between observed and estimated results to a minimum (see, e.g. [103] and references therein for some critiques on Tanaka's model). This possibilistic regression model is based on the idea of fuzziness minimization through a minimization of the total support of the regression fuzzy coefficients [258], subject to the inclusion of all the observed data.

The basic form of the model is the general linear function

$$\tilde{Y} = \tilde{A}_0 + \tilde{A}_1 x_1 + \tilde{A}_2 x_2 + \cdots + \tilde{A}_n x_n,$$

with $\tilde{Y}$ the fuzzy response, $\tilde{A}_i$, $i = 1, \dots, n$ the fuzzy coefficients (or parameters), and $x_1, \dots, x_n$ the components of a nonfuzzy input vector. The $\tilde{A}_i$s are assumed to be triangular fuzzy numbers and the fuzzy coefficients are characterized by a membership function $\mu_A(\alpha)$. As an example, suppose that we have a fuzzy relationship between the variables (i.e. fuzzy-dependent variables), while the *observed data are crisp.* Further, the triangular fuzzy numbers are assumed symmetric. Then, the membership function of the jth coefficient can be defined as

$$\mu_{A_j}(a) = \max\left\{1 - \frac{a - a_j}{c_j}, 0\right\},$$

where a is the mean value of the fuzzy number $\tilde{A}_j$, a_j is the mode (center value of A_j), and c_j is the spread (width around the center value). The choice of symmetric triangular fuzzy numbers assures that the structure of the model depends only on the data involved in the determination of the upper and lower bounds, while other data points do not play any role. So we have [258]

$$\tilde{A}_j = \{a_j, c_j\}_L = \{\tilde{A}_j \mid a_j - c_j \leq \tilde{A}_j \leq a_j + c_j\}_L, \quad j = 0, 1, \dots, n$$

Thus, we obtain

$$\tilde{Y}_i = \tilde{A}_0 + \sum_{j=1}^{n} \tilde{A}_j x_{ij} = (a_0, c_0)_L + \sum_{j=1}^{n} (a_j, c_j)_L x_{ij}.$$

Now, if the support is just enough to contain all the sample's data points, then the confidence in an out-of-sample projection is limited unless one extends the support. To this purpose, one chooses a value $h < 1$ (called the *h-certain factor*) of $\mu_{A_j}(a)$ with the h-line cutting the two sides of the triangle graph of $\mu_{A_j}(a)$ at points with coordinates $(a_j - (1-h)c_j^L, a_j + (1-h)c_j^R)$ on the a-axis and L, R denoting "left" and "right" to a_j, respectively. The interval $[a_j - (1-h)c_j^L, a_j + (1-h)c_j^R]$ on the a-axis is the feasible data interval. This h-certain factor extends, by controlling the size of this data interval, the support of the membership function. In fact, the increase of h leads to the increase of c_j^L and c_j^R.

When the observed data are fuzzy, the application of the aforementioned h-certain factor is also possible. Assuming that the observed fuzzy result can be described by a symmetric triangular fuzzy number $\tilde{Y}_i = (y_i, e_i)$, where y_i is the mode (center value) and e_i is the spread, then the actual data points belong to the interval $[y_i - (1-h)e_i, y_i + (1-h)e_i]$. So the optimization of the model requires the minimization of the spread (width around the center value),

$$\min\left(c_0 + \sum_{j=1}^{n} c_j |x_{ij}|\right), \quad c_j \geq 0.$$

Then, the observed fuzzy data that are adjusted for the h-certain factor, are contained in the estimated fuzzy result [258]:

$$a_0 + \sum_{j=1}^{n} a_j x_{ij} + (1-h)\left(c_0 + \sum_{j=1}^{n} c_j |x_{ij}|\right) > y_i + (1-h)e_i,$$

$$a_0 + \sum_{j=1}^{n} a_j x_{ij} - (1-h)\left(c_0 + \sum_{j=1}^{n} c_j |x_{ij}|\right) < y_i - (1-h)e_i,$$

$$c_j \geq 0, \quad i = 0, 1, \ldots, m, \quad j = 0, 1, \ldots, n.$$

Thus, the increase of the h-certain factor extends the confidence interval and, consequently, the probability for out-of-sample values to be covered by the model.

Let us now examine the *fuzzy least-squares regression*. Remembering (6.5) from Section 6.9 one has, in a fuzzy environment,

$$\tilde{Y}_i = \beta_0 + \beta_1 \tilde{X}_i + \tilde{\varepsilon}_i, \quad i = 1, \ldots, m$$
$$\Longleftrightarrow \tilde{\varepsilon}_i = \tilde{Y}_i - \beta_0 - \beta_1 \tilde{X}_i$$

and one has to optimize

$$\min \sum_{i=1}^{n} (\tilde{Y}_i - \beta_0 - \beta_1 \tilde{X}_i)^2.$$

The most commonly used approach for this is the *method of distance measures* that was introduced by Phil M. Diamond [98]. Namely, by defining a measure of the distance d between two triangular fuzzy numbers $\text{tfn}(r_1, s_1, t_1)$, $\text{tfn}(r_2, s_2, t_2)$:

$$d(\text{tfn}(r_1, s_1, t_1), \text{tfn}(r_2, s_2, t_2))^2$$
$$= (r_1 - r_2)^2 + [(r_1 - s_1) - (r_2 - s_2)]^2 + [(r_1 + t_1) - (r_2 + t_2)]^2,$$

with the model written as

$$\tilde{Y}_i = \tilde{\beta}_0 + \tilde{\beta}_1 \tilde{X}_i + \tilde{\varepsilon}_i, \quad i = 1, \dots, m \tag{6.6}$$

one has to optimize

$$\min_{A,B} \sum_{i=1}^{m} d(\tilde{A} + \tilde{B}x_i, \tilde{Y}_i)^2.$$

Assuming $\tilde{B} > 0$, it follows for the distance from (6.6)

$$d(\tilde{A} + \tilde{B}x_i, \tilde{Y}_i)^2 = (a + bx_i - y_i)^2 + (a + bx_i - c_A^L - c_B^L x_i - y_i + c_{Y_i}^L)^2$$
$$+ (a + bx_i + c_A^R + c_B^R x_i - y_i + c_{Y_i}^R)^2,$$

while a similar expression can be derived for $\tilde{B} < 0$.

A 6×6 system of equations yields the parameters of $\tilde{A}$, $\tilde{B}$ (see, e.g. [86] and references therein for an implementation of the above algorithm).

Exercises

The following exercises are inspired by examples presented in [41].

6.1 Let the fuzzy random vector $X_1, \dots, X_n$ be independent and identically distributed (iid) from the normal distribution $N(\theta, 1)$, where θ is the unknown mean. Suppose there is a random sample of size $n = 50$, and by performing an experiment, we observe that $\overline{x} = 0.65$. Test the null hypothesis $H_0 : \theta = 0.5$ against the alternative hypothesis $H_1 : \theta \neq 0.5$ at the significance level $\delta = 0.06$.

6.2 Suppose we have a random variable X with the Poisson probability mass function. The probability for $X = x$ is given by $R(x) = \frac{a^x e^{-a}}{x!}, x \in \mathbb{N}_0, a \in \mathbb{R}^+$. The fuzzy Poisson probability mass function $\overline{P}(x)$ is obtained when a is replaced by the positive fuzzy number $\overline{a}$. Let $x = 10$ and $\overline{a} = \text{tfn}(5, 7, 9)$. Find the α-cut of the fuzzy probability $\overline{P}(8)[\alpha]$, $\alpha \in [0, 1]$.

6.3 Let the random variable X with the normal probability density function $N(\mu, 500)$ and a random sample $X_1, \dots, X_n$ from $N(\mu, 500)$ with sample size

$n = 70$ and a mean equal to 50. Find the fuzzy estimator $\overline{\mu}$ as a triangular fuzzy number.

6.4 Let the random variable X with the normal probability density function $N(\mu, \sigma^2)$ and a random sample $X_1, \ldots, X_n$ from $N(\mu, \sigma^2)$ with sample size $n = 10$. Find an unbiased fuzzy estimator $\overline{\sigma}^2$ for the variance.

6.5 An experiment is performed and the following ten crisp data pairs (x, y) are measured:

i	x	y
1	20	27
2	24	44
3	22	38
4	18	30
5	8	21
6	4	26
7	32	38
8	14	30
9	30	40
10	11	19

Conclude whether there is any additional information missing in order to find the fuzzy coefficients $\tilde{A}_i$, $i = 1, \ldots, 10$, of the fuzzy linear regression model for this data set. Then, by assuming that the necessary missing information is known, determine the fuzzy coefficients $\tilde{A}_i$.

7

Fuzzy Logics

Logic is about the process of reaching an answer by thinking about known facts. And logic in general can be divided into *formal logic*, *informal logic*, *symbolic logic*, and *mathematical logic*. We know that in mathematical reasoning, a statement is either *true* (e.g. "3 is an odd number") or *false* (e.g. "4 is greater than 5"), since nothing else is meaningful. However, outside mathematics, there are statements that are neither true or false. For example, the statement "Brazil will win the next world cup in football" is such a statement. Put simply, fuzzy logic is a mathematical logic to reason about statements that are true, false, or something in between. The discussion that follows is based mainly on [157, 217, 218].

7.1 Mathematical Logic

The word "logic" derives from the Greek word λόγος (logos), which means reason. In fact, logic is yet another ancient Greek invention. In particular, as was noted in Section 2.1, Aristotle is credited with the invention of logic. Logic was born when he begun to study the structure of arguments. Aristotle called *syllogisms* all these structures. An example of a syllogism is the following: John goes to the movies on Wednesday; today is Wednesday; therefore, John will go to the movies. This deduction is an instance of the general *inference rule*:

> Assume that "A" and "if A then B" are true. Then, "B" is also true.

Aristotle invented logic, but it took more than two millennia to have a mathematical description of logic and its rules (see Ref. [207] for a comprehensive presentation of the history of logic).

In (formal) logic, we are concerned about statements, their properties, and proofs. In addition, we do care about *form* and not about *matter* or "content," if you prefer this expression. In essence this means that in formal logic, we are

A Modern Introduction to Fuzzy Mathematics, First Edition.
Apostolos Syropoulos and Theophanes Grammenos.

concerned about rules as the inference rule above, but we are not concerned about the nature of A and B. For instance, consider the following syllogism:

- If you have a valid password, then you can log on to your Facebook account.
- You have a valid password.

Therefore,

- You can log on to your Facebook account.

This syllogism and the previous one about John's habits have the same form, but of course, logic is not concerned about one's password or John's habits. When logic is using mathematical techniques or when formal logic is employed in mathematical reasoning, then it may be called mathematical logic.

In logic, we can combine statements or sentences, as they are usually called, in various ways to form more complicated sentences. Negation is the simplest logical operation and if A is a sentence, then its negation will be written as $\neg A$. The *truth table* for negation (i.e. its "multiplication table") follows:

A	$\neg A$
1	**0**
0	**1**

Thus, if A is true, then $\neg A$ is false, and when A is false, then $\neg A$ is true. Recall that the symbols **1** and **0** denote the truth values true and false, respectively.

Given two sentences, A and B, their *conjunction* is the sentence A and B, which is written as $A \wedge B$. The truth table of this operation follows:

A	B	$A \wedge B$
1	**1**	**1**
0	**1**	**0**
1	**0**	**0**
0	**0**	**0**

Thus, $A \wedge B$ is true if and only if both A and B are true.

The *disjunction* of two sentences A and B is the expression A or B, which means "A or B or both." The disjunction of two sentences is written as $A \vee B$ and its truth table follows:

A	B	$A \vee B$
1	**1**	**1**
0	**1**	**1**
1	**0**	**1**
0	**0**	**0**

Thus, $A \vee B$ is false if and only if both A and B are false.

The *conditional* or *implication* is an expression of the form "if A, then B," written as $A \Rightarrow B$. Typically, if A, the *antecedent*, is true and B, the *consequent*, is false, then the whole expression is false. However, in all other cases, it is really not clear what is the truth value of the whole expression. For instance, the following sentences, borrowed from [217], should clearly demonstrate this point:

(i) If $1 + 1 = 2$, then Paris is the capital of France.
(ii) If $1 + 1 \neq 2$, then Paris is the capital of France.
(iii) If $1 + 1 \neq 2$, then Rome is the capital of France.

Since we have agreed that $A \Rightarrow B$ is false only when $A = \mathbf{1}$ and $B = \mathbf{0}$, we have made the convention that in all other cases, the expression shall be true:

A	B	$A \Rightarrow B$
1	**1**	**1**
0	**1**	**1**
1	**0**	**0**
0	**0**	**1**

The expressions $((\neg A) \vee B)$ and $(A \Rightarrow B)$ are *logically equivalent* since they receive the same truth value under every assignment of truth values to their statement letters, as the following truth table shows:

A	B	$\neg A \vee B$
1	**1**	$\neg\mathbf{1} \vee \mathbf{1} = \mathbf{0} \vee \mathbf{1} = \mathbf{1}$
0	**1**	$\neg\mathbf{0} \vee \mathbf{1} = \mathbf{1} \vee \mathbf{1} = \mathbf{1}$
1	**0**	$\neg\mathbf{1} \vee \mathbf{0} = \mathbf{0} \vee \mathbf{0} = \mathbf{0}$
0	**0**	$\neg\mathbf{0} \vee \mathbf{0} = \mathbf{1} \vee \mathbf{0} = \mathbf{1}$

The expression $\mathscr{E} \equiv \mathscr{F}$ means that $\mathscr{E}$ and $\mathscr{F}$ are logically equivalent.

The linguistic expression "A if and only if B" is denoted by the mathematical expression $A \Leftrightarrow B$. This expression is called a *biconditional*, and it is true only when A and B have the same truth value:

A	B	$A \Leftrightarrow B$
1	**1**	**1**
0	**1**	**0**
1	**0**	**0**
0	**0**	**1**

The operators $\neg$, $\wedge$, $\vee$, $\Rightarrow$, and $\Leftrightarrow$ will be called *propositional connectives*. The letters A, B, C,...will be called *statement letters*. A *statement form* will be formed by following these rules:

(i) All statement letters with or without numerical subscripts are statement forms.

(ii) If $\mathscr{B}$ and $\mathscr{C}$ are statement forms, then so are $(\neg\mathscr{B})$, $(\mathscr{B} \vee \mathscr{C})$, $(\mathscr{B} \wedge \mathscr{C})$, $(\mathscr{B} \Rightarrow \mathscr{C})$, and $(\mathscr{B} \Leftrightarrow \mathscr{C})$.

(iii) $\mathscr{C}$ is a statement form if and only if there is a finite sequence $\mathscr{B}_1, \ldots, \mathscr{B}_n$, $n \geq 1$, such that $\mathscr{B}_n = \mathscr{C}$ and, if $1 \leq i \leq n$, then $\mathscr{B}_i$ is either a statement letter or a negation, conjunction, disjunction, conditional, or biconditional constructed from previous expressions in the sequence.

Example 7.1.1 The following expressions are statement forms: D, $(\neg S_1)$, $(B_4 \vee (\neg A_2))$, and $(C_2 \Rightarrow (A \wedge B))$.

To simplify things, we may omit the outer pair of parentheses of a statement form. Also, we assume the following decreasing order of operator precedence: $\neg, \wedge, \vee, \Rightarrow, \Leftrightarrow$. Thus, $((\neg A) \vee B)$ can be written as $\neg A \vee B$. What we presented so far is called *propositional logic.*

Consider a discrete set Ω of N elements $x \in \Omega$. Also, consider the powerset of Ω. Then, the subsets $A, B, \ldots$ can also be considered as logical statements. Thus, set operations can be viewed as logical connectives. In particular,

Set operation	Logical connective
Union ($\cup$)	Disjunction ($\vee$)
Intersection ($\cap$)	Conjunction ($\wedge$)
Complement (^{C})	Negation ($\neg$)
Ω	**1** (true)
$\varnothing$	**0** (false)

Based on this “equivalence,” we can reformulate De Morgan’s laws as follows:

$$\neg(A \vee B) \equiv \neg A \wedge \neg B$$

$$\neg(A \wedge B) \equiv \neg A \vee \neg B$$

Similarly, the law of the excluded middle and the law of contradiction can be expressed as follows:

$$A \vee \neg A \equiv \mathbf{1}$$

$$A \wedge \neg A \equiv \mathbf{0}$$

The law of the excluded middle is a *tautology* (i.e. a statement form that is always true, no matter what the truth values of its statement letters may be), while the law of contradiction is *contradictory* (i.e. a statement form that is always false, no matter what the truth values of its statement letters may be).

Propositional logic, in particular, and any logic, in general, can be used to answer questions about specific statement forms (e.g. whether it is a tautology or a contradiction). However, more complex things are not possible using truth tables. More specifically, we need to define a formal theory $\mathscr{F}$ that consists of

(i) A countable set of symbols that are the symbols of $\mathscr{F}$. A finite sequence of symbols is called an *expression* of $\mathscr{F}$.
(ii) A subset of the set of expressions of $\mathscr{F}$ called the *well-formed formulas* of $\mathscr{F}$.
(iii) A set of well-formed formulas called the set of *axioms* of $\mathscr{F}$. When we can decide whether a well-formed formula is an axiom, then $\mathscr{F}$ is an *axiomatic* theory.
(iv) There is a finite set $R_1, \dots, R_n$ of relations among well-formed formulas called *rules of inference*. The important thing is that we can *effectively* decide whether a set of well-formed formulas are in relation R_i to a well-formed formula B. If this is true, then we say that B is a *direct consequence* of the given well-formed formulas.

Informally, this methodology was introduced by Euclid (Εὐκλείδης). For the propositional logic, the following tautologies are the axioms of the formal theory:

$$A \Rightarrow (B \Rightarrow A) \tag{7.1}$$

$$(A \Rightarrow (B \Rightarrow C)) \Rightarrow ((A \Rightarrow B) \Rightarrow (A \Rightarrow C)) \tag{7.2}$$

$$(\neg A \Rightarrow \neg B) \Rightarrow (B \Rightarrow A) \tag{7.3}$$

When a sentence A is *provable*, we write $\vdash A$. In propositional logic, we use *modus ponens* as the inference rule. This rule specifies that if $\vdash A$ and $\vdash A \Rightarrow B$, then infer $\vdash B$. Another common proof rule is the *adjunction rule*: if $\vdash A$ and $\vdash B$, then infer $\vdash A \wedge B$. The opposite of this rule is the *simplification rule*. This rule has two forms: if $\vdash A \wedge B$, then $\vdash A$ or if $\vdash A \wedge B$, then $\vdash B$. Instead of writing the steps of proof as a list, one can write a proof tree. For instance, the following is an example of a trivial proof tree:

$$\frac{\vdash A \qquad \vdash A \Rightarrow B}{\vdash B}$$

More formally, we can say that a proof is a sequence $A_1, \dots, A_n$ of sentences such that each A_i is either an axiom of propositional logic or it follows from some preceding $A_j, A_k, j, k < i$, by modus ponens. Such systems are known as *Hilbert systems*. A more advanced way to prove theorems in a logic is to use *Gentzen systems*. However, we will not discuss these systems, and the interested reader should consult a more specialized text (e.g. see Ref. [218]).

Propositional logic cannot be used to express certain kinds of deductions. For example, consider the following deduction:

All human beings are rational.
Some animals are human beings.
Therefore, some animals are rational.

The correctness of this deduction depends on the meaning of various connectives and on the meaning of expressions such as "any," "all," "some." To allow the expression of such statements, we introduce *predicates* that express attributes, properties, etc., and quantifiers. Thus, $P(x)$ is a predicate that asserts that x has the property P, while $Q(x, y)$ is a predicate that asserts that x and y have a property in common (e.g. it may express the fact that x and y are brothers). The number of *terms* (i.e. the symbols in a predicate) equals to the arity of the predicate. Moreover, the sentence "there is an x with the property P" is written as $(\exists x)P(x)$, whereas the sentence "all z have the property Q" is written as $(\forall z)Q(z)$. The symbols "$(\forall x)$" and "$(\exists x)$" are called a *universal quantifier* and an *existential quantifier*, respectively. For instance, the previous deduction becomes

$$\frac{\begin{array}{l}(\forall x)\big(H(x) \Rightarrow R(x)\big)\\ (\exists x)\big(A(x) \wedge H(x)\big)\end{array}}{(\exists x)\big(A(x) \wedge R(x)\big)}$$

Here H, R, and A designate the properties of being human, rational, and an animal, respectively. Also, to simplify expressions, we drop the parentheses whenever this is possible.

One interesting question is whether there are other quantifiers. Andrzej Mostowski [225] gave an affirmative answer to this question by introducing a generalization of quantifiers. Here is an outline of what he meant by this. Assume that I is an arbitrary set and $I^* = I \times I \times \cdots \times I$ is its infinite Cartesian product. A mapping $F : I^* \to \{\mathbf{0}, \mathbf{1}\}$ is called a *propositional function* on I provided that it satisfies the following condition: there is a finite set K of integers such that if

$$x = (x_1, x_2, \ldots) \in I^*, \quad y = (y_1, y_2, \ldots) \in I^*, \quad \text{and} \quad x_i = y_i \text{ for } i \in K,$$

then $F(x) = F(y)$. Clearly, the value of F depends on a finite number of arguments. Assume that $\phi : I \to I'$ is an injective mapping, where I' is not necessarily different from I. Then, $\phi(x)$, where $x = (x_1, x_2, \ldots) \in I^*$, is the tuple $(\phi(x_1), \phi(x_2), \ldots)$. Also, if F is a propositional function on I, then F_ϕ is the propositional function on I' such that $F_\phi(\phi(x)) = F(x)$. A quantifier limited to I is a function $\mathbf{Q}$ that assigns one of the elements $\mathbf{1}$ and $\mathbf{0}$ to each propositional function F on I that takes one argument only and satisfies the *invariance* condition

$$\mathbf{Q}(F) = \mathbf{Q}(F_\phi)$$

for each F and each permutation ϕ of I.

Suppose that (m_ξ, n_ξ) is the finite or transfinite sequence of all pairs of cardinal numbers that satisfy the equation $m_\xi + n_\xi = \text{card}(I)$. For every function T that assigns one of the truth values to each pair (m_ξ, n_ξ), we write:

$$\mathbf{Q}_T(F) = T(\text{card}(\{x|F(x) = \mathbf{1}\}), \text{card}(\{x|F(x) = \mathbf{0}\})).$$

It can be shown that for each $\mathbf{Q}_T$, which is a quantifier limited to I, there is a T such that $\mathbf{Q}_T = \mathbf{Q}$. The quantifier determined by $T^*(m_\xi, n_\xi) = \neg T(n_\xi, m_\xi)$ is the dual of $\mathbf{Q}_T$ and is denoted by $\mathbf{Q}_T^*$. Finally, an *unlimited quantifier* or just quantifier is a function that assigns a quantifier $\mathbf{Q}_I$ limited to I to each set I and satisfies the equation $\mathbf{Q}_I(F) = \mathbf{Q}_{I'}(F_\phi)$ for each propositional function F on I with one argument and for each 1–1 mapping from I to I'.

Example 7.1.2 Suppose that $\{T(m_\xi, n_\xi) = \mathbf{1}\} \equiv \{m_\xi \neq 0\}$, then $\mathbf{Q}_T$ is the existential quantifier $\exists$ limited to I. The dual of $\mathbf{Q}_T$ is the general quantifier $\forall$ limited to I.

7.2 Many-Valued Logics

It seems that Aristotle not only invented logic but also he invented multivalued logics, as he was the first who recognized that there are propositions that cannot be classified as either true or false. In particular, in chapter IX of his treatise *De Interpretatione* (*Περὶ Ἑρμηνείας*, On Interpretation), which is part of his Organon, he ponders about *future contingents* and their truth values. He concludes that [12]:

> ὥστε δῆλον ὅτι οὐκ ἀνάγκη πάσης καταφάσεως καὶ ἀποφάσεως τῶν ἀντικειμένων τὴν μὲν ἀληθῆ τὴν δὲν ψευδῆ εἶναι• οὐ γὰρ ὥσπερ ἐπὶ τῶν ὄντων οὕτως ἔχει καὶ ἐπὶ τῶν μὴ ὄντων, δυνατῶν δὲ εἶναι ἢ μὴ εἶναι, ἀλλ' ὥσπερ εἴρηται.[1]

Unfortunately, the idea that there are more than two truth values was not particularly appealing until the 1920s when Jan Łukasiewicz introduced a three-valued logic [198]. Besides $\mathbf{1}$ and $\mathbf{0}$, this logic includes the truth value $\mathbf{2}$ (later he used the symbol $\mathbf{1/2}$ to denote this value), which may be interpreted as "possibility." Later on, in his *Philosophical Remarks on Many-Valued Systems of Propositional Logic* [199], he presented the truth tables for negation and implication for the

1 Translation: *It is clear then that it is not necessary for every affirmation or negation taken from among opposite propositions that the one be true, the other false. For what is nonexistent but has the potentiality of being or not being does not behave after the fashion of what is existent, but in the manner just explained.*

three-valued logic:

A	$\neg A$
1	**0**
1/2	**1/2**
0	**1**

A	B	$A \Rightarrow B$
0	**0**	**1**
1/2	**0**	**1/2**
1	**0**	**0**
0	**1/2**	**1**
1/2	**1/2**	**1**
1	**1/2**	**1/2**
0	**1**	**1**
1/2	**1**	**1**
1	**1**	**1**

In the same paper, he discussed what happens when the truth values are "certain numbers of the interval [0, 1]":

$$A \Rightarrow B = 1 \qquad \text{for } A \leq B$$

$$A \Rightarrow B = 1 - A + B \qquad \text{for } A > B$$

$$\neg A = 1 - A.$$

Emil Leon Post [242] described another logic with m truth values $t_1, t_2, \dots, t_m$. The following are the truth tables for negation and disjunction:

A	$\neg A$
t_1	t_2
t_2	t_3
…	…
t_m	t_1

A	B	$A \vee B$	
t_1	t_1	t_1	
…	…	…	
t_{i_1}	t_{j_1}	t_{i_1}	$i_1 \leq j_1$
…	…	…	
t_{i_2}	t_{j_2}	t_{j_2}	$i_2 \geq j_2$
…	…	…	
t_m	t_n	t_m	

The generalized negation permutes the truth values cyclically, while the generalized disjunction yields from the two truth values the one with smaller subscript. Note that Post did not provide the truth table for conjunction, mainly because he was not really interested in many-valued logics.

Stephen Cole Kleene [175] presented yet another logic with three truth values: true, false, and undefined. In the case of negation: $\neg \mathbf{1} = \mathbf{0}$, $\neg \mathbf{U} = \mathbf{U}$, and $\neg \mathbf{0} = \mathbf{1}$.

The truth tables for the other logical operations are given as follows:

A	B	$A \vee B$	$A \wedge B$	$A \Rightarrow B$	$A \Leftrightarrow B$
1	1	1	1	1	1
1	U	1	U	U	U
1	0	1	0	0	0
0	1	1	0	1	0
0	U	U	0	1	U
0	0	0	0	1	1
U	1	1	U	1	U
U	U	U	U	U	U
U	0	U	0	U	U

In a very short note, Gödel [155] introduced a many-valued logic to show that intuitionistic propositional logic cannot be viewed as a many-valued logic. In this logic, the truth values are elements of the set $\{\mathbf{1}, 2, \ldots, n\}$, where $\mathbf{1}$ is a designated element. The various logical connectives are defined as follows:

$$A \vee B = \min(A, B) \qquad A \wedge B = \max(A, B)$$

$$A \Rightarrow B = \begin{cases} 1, & \text{for } A \geq B \\ B, & \text{for } A < B \end{cases} \qquad \neg A = \begin{cases} n, & \text{for } A \neq n \\ \mathbf{1}, & \text{for } A = n \end{cases}$$

Dmitri Anotolevich Bochvar [32] introduced a three-valued logic, where every statement is either meaningless (denoted by $-\mathbf{1}$), true, or false. In this logic, if some statement A is meaningless, then the statements "A is false" and "A is true" are meaningful and false, respectively. In addition, he considered two kinds of statements: internal and external. The following table shows the difference between the two kinds.

Internal statements	External statements
"A"	"A is true"
"not-A"	"A is false"
"A and B"	"A is true and B is true"
"A or B"	"A is true or B is true"
"if A, then B"	"if A is true, then B is true"
	"A is meaningless"

In order to see the difference between the two kinds of statements, assume that A is a meaningless statement (i.e. its truth value is -1). Then, the statement "not-A" is also meaningless, but the statement "A is false" is not meaningless but false. Also, if A is a meaningless statement, then any possible combination of A with any statement B, regardless of its truth value, by means of the various internal connectives, will yield another meaningless statement. However, no combination of external connectives will yield a meaningless statement. The following tables are the truth tables for internal denial ("not-A," denoted by $\sim A$) and internal conjunction (denoted by $A \sqcap B$):

A	$\sim A$
1	0
0	1
-1	-1

A	B	$A \sqcap B$
1	1	1
1	0	0
0	1	0
0	0	0
1	-1	-1
-1	1	-1
0	-1	-1
-1	0	-1
-1	-1	-1

The statements "A is true" and "A is false" are denoted by $\vdash A$ and $\neg A$, respectively. Their truth tables follow:

A	$\vdash A$
1	1
0	0
-1	0

A	$\neg A$
1	0
0	1
-1	0

The operations "A or B" (denoted by $A \bigsqcup B$), "if A, then B" (denoted by $A \sqsupset B$), and the internal equivalence operation (denoted by $A \sqsupset\sqsubset B$) are defined as follows:

$$A \sqcup B \stackrel{\text{def}}{=} \sim(\sim A \sqcap \sim B)$$

$$A \sqsupset B \stackrel{\text{def}}{=} \sim(A \sqcap \sim B)$$

$$A \sqsupset\sqsubset B \stackrel{\text{def}}{=} (A \sqsupset B) \sqcap (B \sqsupset A)$$

The various external logical connectives are defined as follows:

$$A \wedge B \stackrel{\text{def}}{=} \vdash A \sqcap \vdash B$$

$$A \vee B \stackrel{\text{def}}{=} \vdash A \sqcup \vdash B$$

$$A \Rightarrow B \stackrel{\text{def}}{=} \vdash A \sqsupset \vdash B$$

$$A \Leftrightarrow B \stackrel{\text{def}}{=} (A \Rightarrow) \sqcap (B \Rightarrow A)$$

$$A \equiv B \stackrel{\text{def}}{=} (A \Leftrightarrow B) \sqcap (\sim A \Leftrightarrow \sim B)$$

$$\downarrow A \stackrel{\text{def}}{=} \backsim(\vdash A \sqcup \neg A)$$

$$\bar{(A)} \stackrel{\text{def}}{=} \backsim \vdash A$$

The expression $A \wedge B$ is read as "A is true and B is true"; the expression $A \vee B$ is read as "A is true or B is true"; the expression $A \Rightarrow B$ is read as "if A is true, then B is true"; the expression $A \Leftrightarrow B$ is read as "A is of the same strength as B"; and the expression $A \equiv B$ is read as "A is equivalent to B." The truth tables for the remaining two connectives follow:

A	$\downarrow A$
1	0
0	0
−1	1

A	$\bar{A}$
1	0
0	1
−1	1

So far we have described logics that have three different truth values. However, there are logics with more truth values, and there are logics with an infinite number of truth values. Łukasiewicz and Alfred Tarski [200] discussed these ideas in detail. In particular, they described many-valued logics that are now known as Łukasiewicz logics. Each of these finite logics is denoted by $\mathscr{L}_n$, where n is the number of truth values. For a logic $\mathscr{L}_n$, its truth values follow:

$$\frac{0}{n-1}, \frac{1}{n-1}, \ldots, \frac{n-2}{n-1}, \frac{n-1}{n-1}.$$

Thus, when $n = 3$, the truth values are 0, 1/2, and 1. When the set of truth values is infinite (i.e. when its cardinality is equal to $\aleph_0$, where $\aleph_0$ denotes the cardinality of $\mathbb{N}$, that is, the set of positive integers including zero), then the logic is denoted by $\mathscr{L}_{\aleph_0}$ and its truth values are all fractions k/l, for $0 \leq k < l$, and the truth-value 1. For all these logics, given two truth values A and B, negation and implication are defined as follows:

$$\neg A = 1 - A \quad \text{and} \quad A \Rightarrow B = \min(1, 1 - A + B).$$

In the case of the infinite-valued logic, it was proposed that the following sentences

(Ł1) $A \Rightarrow (B \Rightarrow A)$
(Ł2) $(A \Rightarrow B) \Rightarrow ((B \Rightarrow C) \Rightarrow (A \Rightarrow C))$
(Ł3) $((A \Rightarrow B) \Rightarrow B) \Rightarrow ((B \Rightarrow A) \Rightarrow A)$
(Ł4) $(A \Rightarrow B) \Rightarrow ((B \Rightarrow A) \Rightarrow (B \Rightarrow A))$
(Ł5) $(\neg A \Rightarrow \neg B) \Rightarrow (B \Rightarrow A)$

form a set of axioms for $\mathscr{L}_{\aleph_0}$. Later on, it was shown that axiom (Ł4) is deducible from the remaining axioms.

In the 1950s Robert McNaughton [213] described an infinite-valued logic whose truth values belong to the unit interval, that is, any truth value $x \in [0, 1]$. In this logic 1 denotes *complete truth* and 0 denotes *complete falsity*. The logic has two basic logical connectives: negation and implication. Conjunction and disjunction are defined in terms of negation and implication. These basic logic connectives are defined as follows:

$$\neg A = 1 - A \quad \text{and} \quad A \Rightarrow B = \min(1 - A + B, 1).$$

The other two operations are defined as follows:

$$A \vee B = \max(A, B) \quad \text{and} \quad A \wedge B = \min(A, B).$$

Let us see how these two connectives are defined in terms of the basic connectives:

$$A \vee B = (A \Rightarrow B) \Rightarrow B$$
$$A \wedge B = \neg(\neg A \vee \neg B)$$

7.3 On Fuzzy Logics

When Zadeh introduced fuzzy sets by extending the notion of sets, he also introduced fuzzy logic. However, it was Goguen [143] who actually introduced a fuzzy logic, in the sense of a logical system. Zadeh himself was more interested in applications of fuzzy sets and not so much in the logical aspect of fuzzy set theory. After all, Zadeh was an engineer and not a mathematician or a computer scientist. Since the term "fuzzy logic" was used for different things, Zadeh proposed the use of the term *fuzzy logic in the narrow sense* to mean fuzzy logic, and the term *fuzzy logic in the wide sense* to mean the application of fuzzy sets to everything else. Zadeh himself explained the difference between the two terms as follows[2]:

> Many of the misconceptions about fuzzy logic stem from differing interpretations of what "fuzzy logic" means. In the narrow sense, fuzzy logic is a logical system that focuses on modes of reasoning that are approximate rather than exact. In this sense, fuzzy logic, or FLn for short, is an extension of classical multivalued logical systems, but with an agenda that is quite different in spirit and in substance.
> In the wide sense, which is in predominant use today, fuzzy logic, or FLw for short, is almost synonymous with the theory of fuzzy sets. The agenda of FLw is much broader than that of FLn, and logical reasoning–in its traditional sense–is an important but not a major part of FLw. In this perspective, FLn may be viewed as one of the branches of FLw. What is important to note is that, today, most of the practical applications of fuzzy logic involve FLw and not FLn.

---

2 Since we did not have access to Zadeh's original paper, we present the quotation as it appeared in [26, p. 43].

The use of the word "wide" was not a good idea. So, many authors proposed the replacement of this word with the word "broad", mainly because it is more natural in this case (see Ref. [26]). Thus, this chapter is about fuzzy logic in the narrow sense and the rest of the book about fuzzy logic in the *broad* sense. And of course, fuzzy logic in the broad sense is not really logic...

In fuzzy logic, the truth values are the elements of the set [0, 1]. Thus, one can say that fuzzy logic is an infinite-valued logic since the cardinality of the unit interval is $2^{\aleph_0}$, which is clearly greater than $\aleph_0$. It is not difficult to convince one about the need to have more than two truth values, nevertheless, one can easily abolish the idea of an infinite number of truth values as absurd. Petr Hájek [157] proposes the use of simple questions like *Do you like Haydn?* to demonstrate the usefulness of infinite-valued logics. To answer such a question, one is offered a number of possible answers like: absolutely yes, more or less, rather not, absolutely not. But there is no reason to restrict the number of answers to 4 or 20 as one might not be happy with any of these answers. A better approach is to allow one to choose the required values by using a slider that can go from 0 (absolutely no) to 1 (absolutely yes). Since we cannot have such a slider at our disposal, we should be free to just choose an element of [0, 1] as our truth value.

We have talked about fuzzy logics and not fuzzy logic because one can use any t-norm together with an implication connective to define all logical connectives. But how can we define such a logical connective? A good idea is to use the exponential element and use it to define the new logical connective. Indeed, the following result shows how this can be done.

Lemma 7.3.1 *Assume that* $*$ *is a continuous t-norm. Then, there exists a unique operation* $a \Rightarrow b$ *that satisfies the condition* $(a * c) \leq y$ *if and only if* $c \leq (a \Rightarrow b)$*, for all* $a, b, c \in [0, 1]$*. This simply means that* $a \Rightarrow b = \max\{c|a * c \leq b\}$*, which means that* $\Rightarrow$ *and* $*$ *form an adjoint pair.*

Proof: For each $a, b \in [0, 1]$ assume that

$$(a \Rightarrow b) = \bigvee\{c|a * c \leq b\}.$$

Also, assume that for a fixed c, $f(a) = a * c$, the function f is continuous and non-decreasing and so it commutes with sups. Thus,

$$a * (a \Rightarrow b) = a * \bigvee\{c|a * c \leq b\} = \{a * c|a * c \leq b\} \leq b.$$

Hence $(a \Rightarrow b) = \vee\{c|a * c \leq b\}$. Uniqueness is obvious. □

The operation $a \Rightarrow b$ is called the *residuum* of the t-norm. The next step is to give a concrete definition of this operation for specific t-norms. The result that follows answers the question: What are the residua of the Gödel, Łukasiewicz, and the product t-norms?

Theorem 7.3.1 *The following operations are residua of the three t-norms that define the Łukasiewicz, Gödel, and product logics. For $a \leq b$, $a \Rightarrow b = 1$. For $a > b$,*

(i) Łukasiewicz implication: $a \Rightarrow b = 1 - a + b$.
(ii) Gödel implication: $a \Rightarrow b = b$.
(iii) Goguen implication: $a \Rightarrow b = b/a$ (residuum of product t-norm).

Proof: Suppose that $a > b$. Then,

(i) $a * c = b$ if and only if $a + c - 1 = b$ if and only if $c = 1 - a + b$; thus

$$1 - a + b = \max\{c | a * c \leq b\}.$$

(ii) $a * c = b$ if and only if $\min(a, c) = b$ if and only if $c = b$.
(iii) $a \cdot c = b$ if and only if $c = b/a$ (since $a > 0$).

□

7.4 Hájek's Basic Many-Valued Logic

We can use any t-norm and its residuum to define a propositional logic. First, the t-norm will be used as a (strong) conjunction &, its residuum ⇒ as the truth function of the implication →, and the truth constant **0** for 0 (false). The table that follows shows the basic logical connectives and how they are defined:

$$A \wedge B \stackrel{\text{def}}{=} A \,\&\, (A \rightarrow B)$$
$$A \vee B \stackrel{\text{def}}{=} ((A \rightarrow B) \rightarrow B) \wedge ((B \rightarrow A) \rightarrow A)$$
$$\neg A \stackrel{\text{def}}{=} A \rightarrow \mathbf{0}$$
$$A \equiv B \stackrel{\text{def}}{=} (A \rightarrow B) \,\&\, (B \rightarrow A)$$

These operations specify a propositional calculus that we call $PC(*)$, where $*$ is a t-norm.

Each statement letter A is assigned a numerical value $T(A) \in [0, 1]$ that is interpreted as its *truth degree*. Also, this extends to the evaluation of all statement forms as follows:

$$T(\mathbf{0}) = 0$$
$$T(A \rightarrow B) = (T(A) \Rightarrow T(B))$$
$$T(A \,\&\, B) = (T(A) * T(B))$$

The evaluation rule for implication was exactly the reason why we had to introduce a new symbol for implication. In order to present the evaluation rules for conjunction and disjunction, we need the following result:

Lemma 7.4.1 *For each continuous t-norm $*$, the following identities are true in* $[0, 1]$*:*

(i) $\min(A, B) = A * (A \Rightarrow B)$,
(ii) $\max(A, B) = \min\{((A \Rightarrow B) \Rightarrow B), ((B \Rightarrow A) \Rightarrow A)\}$.

Proof:

(i) If $A \leq B$, then $A \Rightarrow B = 1$ and $A * (A \Rightarrow B) = A$. If $A > B$, assume that $f(C) = C * A$ and it is continuous on $[0, 1]$ and $f(1) = A$. Then, for some C with $0 \leq C \leq 1$, $f(C) = B$. For the maximal C satisfying $B = C * A$ we get $C = A \Rightarrow B$. Thus $B = A * (A \Rightarrow B)$.
(ii) Suppose that $A \leq B$. Then, $(A \Rightarrow B) = 1$ and $(A \Rightarrow B) \Rightarrow B = (1 \Rightarrow B) = B$. Moreover, $B \leq (B \Rightarrow A) \Rightarrow A$ because $B * (B \Rightarrow A) \leq A$ *by residuation). Thus, $\min\{((A \Rightarrow B) \Rightarrow B), ((B \Rightarrow A) \Rightarrow A)\} = B$. A similar argument holds when $B \leq A$.

□

The following result is a direct consequence of the previous lemma:

Lemma 7.4.2 *For any statement forms A and B*

$$T(A \wedge B) = \min(T(A), T(B))$$
$$T(A \vee B) = \max(T(A), T(B))$$

Definition 7.4.1 A statement form A is an 1-tautology of $PC(*)$, if $T(A) = 1$ for each evaluation T.

Naturally, the set of 1-tautologies depends on the choice of the t-norm, and clearly, different t-norms may have different sets of 1-tautologies. We have presented the basic ingredients of the logic, and we can proceed with the presentation of the axioms of basic logic, BL.

Definition 7.4.2 The following statement forms are the axioms of BL:

(A1) $(A \to B) \to ((B \to C) \to (A \to C))$
(A2) $(A \mathbin{\&} B) \to A$
(A3) $(A \mathbin{\&} B) \to (B \mathbin{\&} A)$

(A4) $(A \& (A \to B)) \to (B \& (B \to A))$
(A5) $(A \to (B \to C)) \to ((A \& B) \to C)$
(A6) $((A \& B) \to C) \to (A \to (B \to C))$
(A7) $(A \to (B \to C)) \to (((B \to A) \to C) \to C)$
(A8) $\mathbf{0} \to A$

Note that the only inference rule is modus ponens. Also, all axioms of BL are 1-tautologies in each $PC(*)$.

Lemma 7.4.3 *BL proves the following properties:*

(i) $A \to (B \to A)$
(ii) $(A \to (B \to C)) \to (B \to (A \to C))$
(iii) $A \to A$
(iv) $(A \& (A \to B)) \to B$
(v) $(A \to B) \to ((A \& C) \to)B \& C))$
(vi) $(A \wedge B) \to A$, $(A \wedge B) \to B$, $(A \& B) \to (A \wedge B)$
(vii) $(A \wedge B) \to (B \wedge A)$
(viii) $((A_1 \to B_1) \& (A_2 \to B_2)) \to ((A_l \& A_2) \to (B_1 \& B_2))$

Proof:

(i) By (A2) and (A5) and by using modus ponens, we get the result. The corresponding proof tree will look like the following ones:

$$\frac{\vdash (A \& B) \to A \qquad \vdash ((A \& B) \to A) \to (A \to (B \to A))}{\vdash A \to (B \to A)}$$

(ii) By (A1), $\vdash ((B \& A) \to (A \& B)) \to ((A \& B) \to C) \to (B \& A) \to C)$ thus $\vdash ((A \& B) \to C) \to ((B \& A) \to C)$. Applying (A5) two times, we get the following provable chain of implications:

$$\vdash (A \to (B \to C)) \to ((A \& B) \to C) \to ((B \& A) \to C) \to (B \to (A \to C)).$$

(iii) Applying (ii) to (i), we get $\vdash B \to (A \to A)$ [i.e. in (ii) replace C with A and apply modus ponens]. Then, let B be any axiom and apply modus ponens.

(iv) BL $\vdash (A \to B) \to (A \to B)$ by (iii), hence
BL $\vdash A \to ((A \to B) \to B)$ by (ii) and
BL $\vdash (A \& (A \to B)) \to B$ by (A5).

(v) BL $\vdash (A \& (A \to B)) \to B$ and
BL $\vdash B \to (C \to (B \& C))$, thus
BL $\vdash (A \& (A \to B)) \to (C \to (B \& C))$ by (A1). Thus,
BL $\vdash A \to ((A \to B) \to (C \to (B \& C)))$,

$\mathrm{BL} \vdash A \to (C \to ((A \to B) \to (B \mathbin{\&} C)))$,
$\mathrm{BL} \vdash (A \mathbin{\&} C) \to ((A \to B) \to (B \mathbin{\&} C))$,
$\mathrm{BL} \vdash (A \to B) \to ((A \mathbin{\&} C) \to (B \mathbin{\&} C))$.

(vi) $\mathrm{BL} \vdash (A \wedge B) \to A$ by the definition of $\wedge$ and (A2);
$\mathrm{BL} \vdash (A \wedge B) \to B$ is just (iv).
$\mathrm{BL} \vdash (A \mathbin{\&} B) \to (A \mathbin{\&} (A \to B))$ follows from (A5) and (v).

(vii) This is just (A4).

(viii) $\mathrm{BL} \vdash (A_1 \to B_1) \to ((A_1 \mathbin{\&} A_2) \to (B_1 \mathbin{\&} A_2))$,
$\mathrm{BL} \vdash ((A_1 \to B_1) \mathbin{\&} (A_2 \to B_2)) \to ((A_1 \mathbin{\&} A_2) \to (B_1 \mathbin{\&} B_2)) \mathbin{\&} (A_2 \to B_2)$,
$\mathrm{BL} \vdash (((A_1 \mathbin{\&} A_2) \to (B_1 \mathbin{\&} B_2)) \mathbin{\&} (A_2 \to B_2)) \to$,
$(((A_1 \mathbin{\&} A_2) \to (B_1 \mathbin{\&} A_2)) \mathbin{\&} ((B_1 \mathbin{\&} A_2) \to (B_1 \mathbin{\&} B_2)))$,
$\mathrm{BL} \vdash (((A_1 \mathbin{\&} A_2) \to (B_1 \mathbin{\&} A_2)) \mathbin{\&} ((B_1 \mathbin{\&} A_2) \to (B_1 \mathbin{\&} B_2))) \to$,
$((A_1 \mathbin{\&} A_2) \to (B_1 \mathbin{\&} B_2))$.
Therefore $\mathrm{BL} \vdash (A_1 \to B_1) \to ((A_2 \to B_2) \to ((A_1 \mathbin{\&} A_2) \to (B_1 \mathbin{\&} B_2)))$. □

7.5 Łukasiewicz Fuzzy Logic

Łukasiewicz fuzzy logic is actually the $\mathscr{L}_{\aleph_0}$ logic. If we start with axioms of basic logic and then add the following statement form as axiom:

$$\neg\neg A \to A,$$

then we get a system that we will call Ł. It can be shown that this system is equivalent to $\mathscr{L}_{\aleph_0}$. This equivalence is based on the following lemmas:

Lemma 7.5.1 *Ł proves (Ł1), (Ł2), (Ł3), and (Ł4).*

Lemma 7.5.2 *$\mathscr{L}_{\aleph_0}$ proves the axioms (A1)-(A7) of BL as well as the axiom $\neg\neg A \to A$.*

We omit the proofs since they are based on many intermediate results that we have not presented here.

In the systems we have briefly described so far, the proofs are about sentences whose truth value is **1**. However, this is particularly strange since we are interested in partial truth. Thus, is it reasonable to ask if it is possible to prove *partially* true conclusions from *partially* true premises? If we extend the language $\mathscr{L}_{\aleph_0}$ by adding truth constants $\mathbf{r}$ for each $r \in \mathbb{Q} \cap [0, 1]$, then this means that our calculus will have symbols like $\overline{0.7}$ and $\overline{0.47}$ as truth constants in addition to **0** and **1**.[3] The result

3 Note that we write **0** and **1** instead of $\overline{0}$ and $\overline{1}$ to avoid changing the notation used so far.

is a new language that is called *Rational Pavelka Logic*, RPL. This new language has the two connectives ¬ and →. Negation and conjunction are defined as follows:

$$\neg A \stackrel{\text{def}}{\equiv} A \to \mathbf{0}$$
$$A \,\&\, B \stackrel{\text{def}}{\equiv} \neg(A \to \neg B).$$

In addition, the language has the connectives ∧, ∨, and $\underline{\vee}$. The following equations summarize the evaluation rules of all sentences of RPL:

$$T(A \wedge B) = T(A \,\&\, (A \to B))$$
$$T(A \vee B) = T((A \to B) \to B)$$
$$T(A \,\underline{\vee}\, B) = T(\neg A \to B)$$
$$T(\,\overline{r}\,) = r$$

The axioms of RPL are the axioms of $\mathscr{L}_{\aleph_0}$ plus the axioms that follow, whereas the deduction rule is modus ponens:

$$(\,\overline{r} \to \overline{s}\,) \equiv \overline{r \Rightarrow s}$$
$$\neg\overline{r} \equiv \overline{1 - r}$$

Example 7.5.1 We have that $\overline{0.7} \to \overline{0.9} \equiv \mathbf{1}$, $\overline{0.9} \to \overline{0.7} \equiv \overline{0.8}$, and $\neg\overline{0.9} \equiv \overline{0.1}$.

A *graded sentence* is a pair (A, r), where A is a sentence and $r \in \mathbb{Q} \cap [0, 1]$. Alternatively, one can write $(\overline{r} \to A)$.

Lemma 7.5.3 *The following is a derived deduction rule in RPL:*

$$\frac{(A, r) \qquad (A \to B, s)}{(B, r * s)}$$

That is, whenever a theory T (i.e. a set of sentences) proves both (A, r) and $(A \to B, s)$, then T proves $(B, r * s)$.

Proof: If $T \vdash \overline{r} \to A$ and $T \vdash \overline{s} \to (A \to B)$, then $T \vdash (\overline{r} \,\&\, \overline{s}) \to (A \,\&\, (A \to B))$, therefore $T \vdash \overline{r * s} \to B$. □

7.6 Product Fuzzy Logic

We will now briefly present the propositional calculus $PC(*_\Pi)$, where $*_\Pi$ is the product t-norm, and the corresponding *product logic* (Π). The implication used in the calculus is the Goguen implication (see Theorem 7.6.1), while the negation is the Gödel negation:

$$\neg x = \begin{cases} \mathbf{1}, & \text{if } x = \mathbf{0}; \\ \mathbf{0}, & \text{if } x > \mathbf{0}. \end{cases}$$

Definition 7.6.1 The axioms of product logic are defined in terms of the new connective $\odot$ that is called *product conjunction*. It holds that $\overline{r} \odot \overline{s} = \overline{r \cdot s}$. The axioms of the product logic are those of the basic logic plus

(Π1) $\neg\neg C \to ((A \odot C \to B \odot C) \to (A \to B))$,
(Π2) $B \wedge \neg B \to \mathbf{0}$.

Lemma 7.6.1 *The product logic proves the following sentences:*

(i) $\neg(A \odot B) \to \neg(A \wedge B)$
(ii) $(A \to \neg A) \to \neg A$
(iii) $\neg A \wedge \neg\neg A$

Proof:

(i) The following are equivalent forms of the sentences to be proved:
$((A \odot B) \to \mathbf{0}) \to ((A \wedge B) \to \mathbf{0})$, $((A \to (B \to \mathbf{0})) \odot (A \wedge B)) \to \mathbf{0}$,
$((A \to \neg B) \odot (A \wedge B)) \to \mathbf{0}$.
Now, the following chains of implications are provable:
$((A \to \neg B) \odot (A \wedge B)) \to ((A \to \neg B) \odot A) \to \neg B$,
$((A \to \neg B) \odot (A \wedge B)) \to ((A \to \neg B) \odot B) \to B$,
$((A \to \neg B) \odot (A \wedge B)) \to (B \to \neg B) \to \mathbf{0}$.
(ii) We have $\neg(A \odot A) \to \neg A$ by (i), therefore $(A \odot A \to \mathbf{0}) \to (A \to \mathbf{0})$,
$(A \to (A \to \mathbf{0})) \to (A \to \mathbf{0})$, thus $(A \to \neg A) \to \neg A$.
(iii) The following implications are provable:
$(\neg A \to \neg\neg A) \to \neg\neg A$ by (ii), $\neg\neg A \to (\neg A \vee \neg\neg A)$ by BL, thus
$(\neg A \to \neg\neg A) \to (\neg A \vee \neg\neg A)$. On the other hand,
$\neg\neg A \to \neg A) \to (\neg\neg A \to \neg\neg\neg A)$ by BL, thus $(\neg\neg A \to \neg\neg\neg A) \to \neg\neg\neg A$ by (ii),
$\neg\neg\neg A \to \neg A$ by BL, thus $(\neg\neg A \to \neg A) \to \neg A, (\neg\neg A \to \neg A) \to (\neg A \vee \neg\neg A)$. We get $(\neg A \vee \neg\neg A)$ by applying axiom (A6) to $\neg A$, $\neg\neg A$, and $\neg A \vee \neg\neg A$. □

Lemma 7.6.2 *The axiom (Π2) can be equivalently replaced by each of the following sentences:*

$$\neg(A \odot A) \to \neg A, \quad (A \to \neg A) \to \neg A, \quad \neg A \vee \neg\neg A.$$

Proof: We have just shown that all three sentences are provable in product logic. Now we will prove that each of them, together with BL+(Π1), proves (Π2).

(i) Let us take $(A \rightarrow \neg A) \rightarrow \neg A$ with BL+(Π1). Then, we have the following chain of provable implications:

$$(A \rightarrow \neg A) \rightarrow (A \odot (A \rightarrow \neg A)) \rightarrow (A \odot \neg A) \rightarrow \mathbf{0}.$$

(ii) Now we take $\neg(A \odot A) \rightarrow \neg A$; we get $(A \odot A \rightarrow \mathbf{0}) \rightarrow (A \rightarrow \mathbf{0})$, and hence $(A \rightarrow (A \rightarrow \mathbf{0})) \rightarrow A \rightarrow \mathbf{0}$, which is $(A \rightarrow \neg A) \rightarrow \neg A$. This was (i).

(iii) The last step is to take $\neg A \vee \neg\neg A$ and we see that the following are provable: $\neg\neg A \rightarrow (((A \odot A) \rightarrow (A \odot \mathbf{0})) \rightarrow (A \rightarrow \mathbf{0}))$ [axiom (Π1], $\neg A \rightarrow (D \rightarrow (A \rightarrow \mathbf{0}))$, thus $(\neg\neg A \vee \neg A) \rightarrow ((A \odot A \rightarrow \mathbf{0}) \rightarrow (A \rightarrow \mathbf{0}))$ (note that $\mathbf{0}$ is equivalent to $A \odot \mathbf{0}$ in BL), hence we get $(A \odot A \rightarrow \mathbf{0}) \rightarrow (A \rightarrow \mathbf{0})$, that is, $\neg(A \odot A) \rightarrow \neg A$; this was (ii).

□

The direct analog of RPL over product logic is not possible, and this happens because of the discontinuity of Goguen implication. It is possible to overcome this difficulty, but we will not discuss the technicalities involved, so we will not present how this can be done.

7.7 Gödel Fuzzy Logic

Gödel logic, G, is another logic generated by a continuous t-norm. In this case, the strong conjunction operator is the min operator. The set of axioms of G includes all axioms of BL plus the following one:

(G) $\quad A \rightarrow (A \mathbin{\&} A).$

This axiom together with $(A \mathbin{\&} A) \rightarrow A$, which is a special case of (A2), prove that $A \mathbin{\&} A \equiv A$ (i.e. the connective & is *idempotent*).

Lemma 7.7.1 *G proves that* $(A \mathbin{\&} B) \equiv (A \wedge B)$.

Proof: Obviously, $\mathrm{BL} \vdash (A \mathbin{\&} B) \rightarrow (A \wedge B)$. Also, $\mathrm{BL} \vdash (A \wedge B) \rightarrow A$, $\mathrm{BL} \vdash (A \wedge B) \rightarrow B$, therefore $\mathrm{BL} \vdash ((A \wedge B) \mathbin{\&} (A \wedge B)) \rightarrow (A \mathbin{\&} B)$ and $\mathrm{G} \vdash (A \wedge B) \rightarrow ((A \wedge B) \mathbin{\&} (A \wedge B))$. □

In other words, this lemma shows that one of the two conjunction connectives is redundant. Therefore, we should simplify G by having only one conjunction connective. Let us call this new simplified system G′. This new logical system has the connectives $\wedge$, $\rightarrow$, and $\mathbf{0}$ (nullary connective) and its set of axioms includes (A1)–(A3), (A5)–(A7) of BL, where & is replaced by $\wedge$, and the axiom

$$\text{(G4)} \qquad A \rightarrow (A \wedge A)$$

In addition, there are three more connectives that are defined as follows:

$$\neg A \stackrel{\text{def}}{\equiv} A \rightarrow \mathbf{0}$$

$$A \equiv B \stackrel{\text{def}}{\equiv} (A \rightarrow B) \wedge (B \rightarrow A)$$

$$A \vee B \stackrel{\text{def}}{\equiv} ((A \rightarrow B) \rightarrow B) \wedge ((B \rightarrow A) \rightarrow A)$$

It is necessary to show that the introduction of $\wedge$ brings nothing new and the following result states exactly this.

Lemma 7.7.2 G′ *proves* $(A \wedge (A \rightarrow B)) \rightarrow (B \wedge (B \rightarrow A))$ *and* $(A \wedge B) \equiv A \wedge (A \rightarrow B)$.

Proof: It is easy to show that if a sentence Φ is provable in BL using the connectives $\rightarrow$, &, and $\mathbf{0}$, then the sentence Φ', which is Φ after replacing each occurrence of & with $\wedge$, is provable in G′. Next, we check that (A4) was not used in the proofs of Lemma 7.4.3, and we get $\mathrm{G} \vdash (A \wedge (A \rightarrow B)) \rightarrow B$ [by (iv)], $\mathrm{G} \vdash (A \wedge (A \rightarrow B)) \rightarrow (B \rightarrow A)$ [by (i), (A2), and (A1)], and so $(A \wedge (A \rightarrow B) \rightarrow (B \wedge (B \rightarrow A))$ by (vii), using axiom (G4) for the sentence $A \wedge (A \rightarrow B)$, which proves the first part.

The second sentence can be proved by first observing that $\mathrm{G}' \vdash B \rightarrow (A \rightarrow B)$, $\mathrm{G}' \vdash (A \wedge B) \rightarrow (A \wedge (A \rightarrow B))$. Going the other way around, $\mathrm{G}' \vdash (A \wedge (A \rightarrow B)) \rightarrow B$, thus G′ proves $[A \wedge (A \rightarrow B)] \rightarrow [A \wedge A \wedge (A \rightarrow B)] \rightarrow [A \wedge B]$. □

Corollary 7.7.1 *G and G′ are equivalent in the sense that* $\mathrm{G} \vdash \Phi$ *if and only if* $\mathrm{G}' \vdash \Phi'$, *where* Φ' *is obtained from* Φ *by replacing all & symbols with* $\wedge$.

Thus, G and G′ are indistinguishable!

Gödel logic is related to intuitionist logic. The connection between the two logical systems will be outlined in what follows.

Definition 7.7.1 The intuitionist logic I has the connectives $\rightarrow$, $\wedge$, $\vee$, and $\neg$ plus the following axioms:

(I1) $(A \rightarrow B) \rightarrow ((B \rightarrow C) \rightarrow (A \rightarrow C))$
(I2) $A \rightarrow (A \vee B)$
(I3) $B \rightarrow (A \vee B)$
(I4) $(A \rightarrow C) \rightarrow ((B \rightarrow C) \rightarrow ((A \vee B) \rightarrow C)))$
(I5) $(A \wedge B) \rightarrow A$
(I6) $(A \wedge B) \rightarrow B$

(I7) $(C \to A) \to ((C \to B) \to (C \to (A \wedge B)))$
(I8) $(A \to (B \to C)) \to ((A \wedge B) \to C)$
(I9) $((A \wedge B) \to C) \to (A \to (B \to C))$
(I10) $(A \wedge \neg A) \to B$
(I11) $(A \to (B \wedge \neg B)) \to \neg B$

Lemma 7.7.3 *G proves all axioms of I.*

Proof: It can be verified that BL proves (I1)–(I7) and (I10), (I11) as they stand, and (I8), (I9) after replacing $\wedge$ with &. □

One can prove the converse by extending I with (A6). Note that in I, the connective $\vee$ cannot be defined from the other connectives.

7.8 First-Order Fuzzy Logics

For all three fuzzy logics presented so far (i.e. Łukasiewicz, Product, and Gödel logic), there is a corresponding first-order language (i.e. a corresponding predicate calculus). Beneath each of these logics lies a predicate language that consists of a nonempty set of predicates and a (possibly empty) set of object constants (i.e. something that we presume or hypothesize to exist in the world). Predicates like $P(t_1, t_2, \dots, t_n)$ are atomic sentences or formulas. If Φ and Ψ are formulas, then $\Phi \to \Psi$, $\Psi \,\&\, \Psi$, $(\forall x)\Psi$, $(\exists x)\Phi$, $\mathbf{0}$, and $\mathbf{1}$ are formulas. The following are logical axioms on quantifiers:

(i) $(\forall x)\Phi(x) \to \Phi(t)$ (t is substitutable for x in $\Phi(x)$);
(ii) $\Phi(t) \to (\exists x)\Phi(x)$ (t is substitutable for x in $\Phi(x)$);
(iii) $\forall x)(\Psi \to \Phi) \to (\Psi \to (\forall x)\Phi)$ (x not free in Ψ);
(iv) $\forall x)(\Phi \to \Psi) \to ((\exists x)\Phi \to \Psi)$ (x not free in Ψ); and
(v) $\forall x)(\Phi \vee \Psi) \to ((\forall x)\Phi \vee \Psi)$ (x not free in Ψ).

In the rest of this section, we discuss some logical calculi $\forall$C that are derived from C by adding the logical axioms on quantifiers to its axioms.

Let us start our very brief discussion with the Gödel predicate logic $\forall$G [18]. A Gödel set is a closed set $D \subseteq [0, 1]$ that contains 0 and 1. Given a Gödel set V, then an *interpretation* $\mathfrak{I} = (D, v_{\mathfrak{I}})$ into V for $\forall$G consists of

(i) a nonempty *domain* D;
(ii) for each n-ary predicate symbol, a function that maps elements of D^n to V;
(iii) for each n-ary function symbol, a function that maps elements of D^n to D;
(iv) for each variable, an element of D.

For atomic formulas:

$$v_{\mathfrak{I}}(p(t_1, \dots, t_n)) = v_{\mathfrak{I}}(p)(v_{\mathfrak{I}}(t_1), \dots, v_{\mathfrak{I}}(t_n)).$$

For a variable x, $v_{\mathfrak{I}}(x) = a$, where $a \in D$. For all formulas:

$$\begin{aligned}
v_{\mathfrak{I}}(\bot) &= 0 \\
v_{\mathfrak{I}}(A \wedge B) &= \min(v_{\mathfrak{I}}(A), v_{\mathfrak{I}}(B)) \\
v_{\mathfrak{I}}(A \vee B) &= \max(v_{\mathfrak{I}}(A), v_{\mathfrak{I}}(B)) \\
v_{\mathfrak{I}}(A \to B) &= \begin{cases} 1, & \text{if } v_{\mathfrak{I}}(A) \le v_{\mathfrak{I}}(B) \\ v_{\mathfrak{I}}(B), & \text{otherwise} \end{cases} \\
v_{\mathfrak{I}}(\neg A) &= \begin{cases} 1, & \text{if } v_{\mathfrak{I}}(A) > 0 \\ 0, & \text{if } v_{\mathfrak{A}} = 0, \end{cases} \\
v_{\mathfrak{I}}(\forall x A(x)) &= \bigwedge \{v_{\mathfrak{I}}(A(a)) | a \in D\} \\
v_{\mathfrak{I}}(\exists x A(x)) &= \bigvee \{v_{\mathfrak{I}}(A(a)) | a \in D\}
\end{aligned}$$

If $v_{\mathfrak{I}}(A) = 1$, we say that $\mathfrak{I}$ satisfies A, and write $\mathfrak{I} \models A$. Also, if $v_{\mathfrak{I}}(A) = 1$ for every V-interpretation $\mathfrak{I}$, we say A is valid in $\forall G_V$ and write $\forall G_V \models A$. If Γ is a set of sentences, we define

$$v_{\mathfrak{I}}(\Gamma) = \bigwedge \{v_{\mathfrak{I}}(A) | A \in \Gamma\}.$$

Given a possibly infinite set of formulas Γ, we say that Γ *entails* A, $\Gamma \models_V A$, if and only if for all $\mathfrak{I}$ into V, $v_{\mathfrak{I}}(\Gamma) \le v_{\mathfrak{I}}(A)$.

Definition 7.8.1 For a Gödel set V, we define the first-order Gödel logic $\forall G_V$ as the set of all pairs (Γ, A) such that $\Gamma \models_V A$.

Similar things can be said about the Łukasiewicz predicate logic ∀Ł [17]. Thus, we will not repeat what has been already discussed. Instead, we just present the valuations in what follows:

$$\begin{aligned}
v_{\mathfrak{I}}(\bot) &= 0 \\
v_{\mathfrak{I}}(A \to B) &= \min(1, 1 - v_{\mathfrak{I}}(A) + v_{\mathfrak{I}}(B)) \\
v_{\mathfrak{I}}(\forall x A(x)) &= \bigwedge \{v_{\mathfrak{I}}(A(a)) | a \in D\} \\
v_{\mathfrak{I}}(\exists x A(x)) &= \bigvee \{v_{\mathfrak{I}}(A(a)) | a \in D\}
\end{aligned}$$

In general, one uses modus ponens and the *adjunction rule*

$$\frac{\vdash A \qquad \vdash B}{\vdash A \wedge B}$$

plus the axioms in order to deliver a proof.

7.9 Fuzzy Quantifiers

Zadeh [315] was the first who introduced fuzzy quantifiers. Naturally, many people worked on this subject, but it was Helmut Thiele who introduced t-norm and t-conorm fuzzy quantifiers (e.g. see Ref. [285]). This section is based on Ingo Glöckner's monograph [141].

The definition given in Section 7.1 is not very "useful." A better formulation follows:

Definition 7.9.1 A generalized Boolean quantifier on a base set $E \neq \emptyset$ is a mapping $Q : \mathscr{P}(E)^n \to \{\mathbf{0}, \mathbf{1}\}$, where $n \in \mathbb{N}$ is the arity of Q.

Example 7.9.1 The usual quantifiers can be expressed as

$$\forall(X) = 1 \Leftrightarrow X = E \quad \text{and} \quad \exists(X) = 1 \Leftrightarrow X \neq \emptyset$$

for all $Y \in \mathscr{P}(E)$.

Example 7.9.2 Assume that $E \neq \emptyset$ is an arbitrary base set. Then, we define

$$\begin{aligned} \mathbf{all}_E(X_1, X_2) &= 1 \Leftrightarrow X_1 \subseteq X_2 \\ \mathbf{some}_E(X_1, X_2) &= 1 \Leftrightarrow X_1 \cap X_2 \neq \emptyset \end{aligned}$$

for all $X_1, X_2 \in \mathscr{P}(E)$. Suppose that

$$\begin{aligned} E &= \{\text{Anakin, Obi-Wan, Padmé}\} \\ \mathbf{men} &= \{\text{Anakin, Obi-Wan}\} \in \mathscr{P}(E) \\ \mathbf{married} &= \{\text{Anakin, Padmé}\} \end{aligned}$$

are a set of persons, a set of men in E, and a set of the married persons in E, respectively. Then, the previous definition can be used as follows:

$$\begin{aligned} \mathbf{some}(\mathbf{men}, \mathbf{married}) &= \mathbf{some}(\{\text{Anakin, Obi-Wan}\}, \{\text{Anakin, Padmé}\}) \\ &= 1, \end{aligned}$$

but

$$\mathbf{all}(\mathbf{men}, \mathbf{married}) = \mathbf{all}(\{\text{Anakin, Obi-Wan}\}, \{\text{Padmé, Anakin}\}) = 0.$$

The definition of Boolean quantifiers is used as a basis to define fuzzy quantifiers:

Definition 7.9.2 An n-ary fuzzy quantifier on a base set $E \neq \emptyset$ is a mapping $\tilde{Q} : \mathscr{F}(E)^n \to [0, 1]$.

From this definition, it should be clear that $\tilde{Q} : \mathscr{F}(E)^n \to [0,1]$ assigns to an n-tuple of fuzzy sets $A_1, \dots, A_n \in \mathscr{F}(E)$ an interpretation $\tilde{Q}(A_1, \dots, A_n) \in [0,1]$.

Example 7.9.3 Let us see how one can define a fuzzy-∀ and a fuzzy-∃. These quantifiers would be mappings $\tilde{\forall}, \tilde{\exists} : \mathscr{F}(E) \to [0,1]$ defined as follows:

$$\tilde{\forall}(X) = \bigwedge \{X(e) | e \in E\}$$
$$\tilde{\exists}(X) = \bigvee \{X(e) | e \in E\}$$

for all $X \in \mathscr{F}(E)$. The fuzzy equivalent of the operators **all** and **some** are defined as follows:

$$\widetilde{\mathbf{all}}(X_1, X_2) = \bigwedge \{\max(1 - X_1(e), X_2(e)) | e \in E\}$$
$$\widetilde{\mathbf{some}}(X_1, X_2) = \bigvee \{\min(1 - X_1(e), X_2(e)) | e \in E\}$$

for all $X_1, X_2 \in \mathscr{F}(E)$.

7.10 Approximate Reasoning

Without trying to exaggerate, we could say that pure mathematics is the only field where we reason in precise terms. In all other areas, we reason in approximate terms. Since fuzzy set theory is about vagueness, one could use it to reason in all these cases where we reason in approximate terms. Zadeh [310] used his fuzzy restrictions to propose concrete tools that would allow one to mechanically reason approximately. However, we have presented a number of fuzzy logics and, obviously, it would be better to reformulate these tools in these new languages. Thus, we first present Zadeh's original ideas and then we will present their reformulation in these languages of logic.

We introduced fuzzy restrictions in Section 5.1. The following is a summary of what is a fuzzy restriction. Assume that $U_1, \dots, U_n$ is a collection of universes and that $(u_1, \dots, u_n) \in U_1 \times \cdots \times U_n$. Then, by a fuzzy restriction on $(u_1, \dots, u_n)$, written as $R(u_1, \dots, u_n)$, we mean a fuzzy relation in $U_1 \times \cdots \times U_n$ that defines the compatibility with $R(u_1, \dots, u_n)$ of values that are assigned to $(u_1, \dots, u_n)$. If p is a proposition of the form

$$p \stackrel{\text{def}}{=} (u_1, \dots, u_n) \text{ is } F,$$

where F is a fuzzy relation in $U_1 \times \cdots \times U_n$, then the previous "equation" may be interpreted as the assignment equation

$$R(u_1, \dots, u_n) = F.$$

Thus, for the propositions

$$p \stackrel{\text{def}}{=} u_1 \text{ is small} \quad \text{and} \quad q \stackrel{\text{def}}{=} (u_1, u_2) \text{ are approximately equal}$$

the relational assignment equations are

$$R(u_1) = \text{small} \quad \text{and} \quad R(u_1, u_2) = \text{approximately equal.}$$

Given these two propositions/equations, what can be said about u_2? To answer this question, we use the *compositional rule of inference*:

$$\frac{x_1 \text{ is } P_1 \qquad x_1 \text{ and } x_2 \text{ are } P_2}{x_2 \text{ is } P_1 \circ P_2}$$

Now assume that

$$U_1 = U_2 = \{1, 2, 3, 4\}, \quad \text{small} = 1/1 + 0.6/2 + 0.2/3 + 0/4 \quad \text{and}$$

$$\text{approximately equal} = \begin{array}{c} \begin{matrix} 1 & 2 & 3 & 4 \end{matrix} \\ \begin{bmatrix} 1 & 0.5 & 0 & 0 \\ 0.5 & 1 & 0.5 & 0 \\ 0 & 0.5 & 1 & 0.5 \\ 0 & 0 & 0.5 & 1 \end{bmatrix} \end{array} \begin{matrix} 1 \\ 2 \\ 3 \\ 4 \end{matrix}$$

We need to compute the composition small∘approximately equal. This can be expressed as the max–min composition of the relation matrices of small and approximately equal. Thus,

$$\text{small} \circ \text{approximately equal} = \begin{bmatrix} 1 & 0.6 & 0.2 & 0 \end{bmatrix} \circ \begin{bmatrix} 1 & 0.5 & 0 & 0 \\ 0.5 & 1 & 0.5 & 0 \\ 0 & 0.5 & 1 & 0.5 \\ 0 & 0 & 0.5 & 1 \end{bmatrix}$$

$$= \begin{bmatrix} 1 & 0.6 & 0.5 & 0.2 \end{bmatrix}$$

Therefore, the fuzzy restriction on u_2 is given by

$$u_2 = 1/1 + 0.6/2 + 0.5/3 + 0.2/4$$

A rough *linguistic approximation* of this fuzzy restriction is

$$\text{LA}(1/1 + 0.6/2 + 0.5/3 + 0.2/4) \cong \text{more or less small},$$

where LA is the operation of linguistic approximation and

$$\text{more or less } A = \int_U \sqrt{A(u)}/u.$$

If you attempt to calculate the values, you will notice that you do not get the expected value, but something close to it. This is why Zadeh called it approximate reasoning! From the premises p and q, we conclude that

u is more or less small.

Along those lines, Zadeh defined a *compositional modus ponens* as follows:

$$\frac{x \text{ is } P \qquad \text{If } x \text{ is } Q \text{ then } y \text{ is } S}{y \text{ is } P \circ \left(\text{cyl}\,\bar{Q} \oplus \text{cyl}\,S\right)}$$

where $A \oplus B = \int_U \min[1, A(u) + B(u)]/u$.

The previous inference rules can be reformulated so as to make it possible to have them as ingredients of predicate calculus. In order to achieve this, first we need to introduce *variates*. A variate consists of a symbol X, which is its *name*, and a nonempty set D, which is its *domain*. For instance, height (in centimeters) with domain the set of integers ≤ 250 is a variate. Roughly, we can take the domains to be collections of different things (e.g. plants and animals) that will allow us to interpret a predicate language. In addition, fixed fuzzy subsets of a domain will interpret some unary predicates. Furthermore, the name of a variate will play the role of an object constant, which is interpreted as the actual value of the variate. Thus, the expression "X is A" is a formula $A(X)$. For example, the typical rule "IF X is A THEN Y is B" may be interpreted as $A(X) \to B(Y)$. The compositional rule of inference is expressed as follows:

From "X is A" and "(X, Y) is R" infer "Y is B" if for all $v \in D_Y$, $r_B(v) = \bigvee_{u \in D_X} (r_A(u) * r_R(u, v))$.

Here $*$ is a continuous t-norm, and r_A, r_B, and r_R are fuzzy relations corresponding to the predicates A, B, and R. This condition means that the formula

$$\text{(Comp)} \qquad (\forall y)(B(y) \equiv (\exists x)(A(x) \mathbin{\&} R(x, y))$$

is 1-true in any structure $\mathbf{D}$ of domains of variates (i.e. the valuation of this formula is equal to 1).

Corollary 7.10.2 *For the formula Comp it holds that*

$$BL\forall \vdash (Comp \mathbin{\&} A(X) \mathbin{\&} \Phi(X, Y)) \to B(Y).$$

It can be shown that compositional modus ponens is actually a special case of the compositional inference rule.

Theorem 7.10.1 *Assume that $Comp_{MP}$ is the formula*

$$(\forall y)(B^*(y) \equiv (\exists x)(A^*(x) \mathbin{\&} (A(x) \to B(y)))).$$

Then, BL∀ proves

$$(Comp_{MP} \,\&\, A^*(X) \,\&\, (A(X) \rightarrow B(Y))) \rightarrow B^*(Y),$$

where A and A^* are predicates.

This can be visualized as follows:

$$\frac{Comp_{MP} \qquad A^*(X) \qquad A(X) \rightarrow B(Y)}{B^*(Y)}$$

In what follows, we will suppose that the predicates A and A^* are of the same *sort* (i.e. they are of the same *type* in programming languages parlance, that is, they are of the same kind) and B and B^* are also of the same sort. The following result holds of the generalized simplification rule:

Theorem 7.10.2 *Assume that $Comp_{SR}$ is*

$$(\forall y)(B^*(y) \equiv (\exists x)(A^*(x) \,\&\, A(x) \,\&\, B(y))).$$

Then BL∀ proves

$$Comp_{SR} \,\&\, A^*(x) \,\&\, A(x) \,\&\, B(y) \rightarrow B^*(y).$$

As in the previous case, this can be visualized as follows:

$$\frac{Comp_{SR} \qquad A^*(X) \qquad A(X) \,\&\, B(Y)}{B^*(Y)}$$

These rules are primarily used in fuzzy expert systems.

7.11 Application: Fuzzy Expert Systems

Generally speaking, an expert system is a computer program that uses tools and methods employed in artificial intelligence to solve problems that typically require human expertise, hence the term "expert system." The definition that follows, borrowed from [234], describes the functionality of an expert system.

> An expert system is a program that relies on a body of knowledge to perform a somewhat difficult task usually performed only by a human expert. The principal power of an expert system is derived from the knowledge the system embodies rather than from search algorithms and specific reasoning methods. An expert system successfully deals with problems for which clear algorithmic solutions do not exist.

In expert systems, knowledge is represented with IF–THEN rules. For example, these are two simplified rules that describe what should be done when a car encounters traffic lights.

IF	the “traffic light” is “green”
THEN	push pedal and go
IF	the “traffic light” is “red”
THEN	push brake and stop

Fuzzy expert systems [150, 261] represent knowledge with fuzzy IF–THEN rules. Typically, fuzzy rules contain linguistic variables, and the following rules are a typical example of fuzzy rules:

IF	player is a very good passer and ball handler and not primarily a shooter
THEN	player is a point guard
IF	player is not a great ball handler and a good perimeter shooter
THEN	player is a shooting guard

In this example, we have used linguistic *hedges* (i.e. words that are used to soften what we say or write). Obviously, terms such as “good” and “great” are modeled with fuzzy sets. But what is the effect of hedges? Zadeh [308] gave an interpretation of hedges in the framework of fuzzy set theory. In particular, in this interpretation, hedges behave like operators that are applied to fuzzy sets and modify them. If H is a hedge and A is a membership function, then A_H will denote A after applying H to it. Here are some hedges, and how they modify a fuzzy set:

very $A_{\text{very}}(x) = (A(x))^2$.
more or less (morl) $A_{\text{morl}}(x) = \sqrt{A(x)}$.
slightly $A_{\text{slightly}}(x) = \sqrt[3]{A(x)}$.
extremely $A_{\text{extremely}}(x) = (A(x))^3$.
somewhat $A_{\text{somewhat}}(x) = \min\{A_{\text{morl}}(x), 1 - A_{\text{slightly}}(x)\}$.

The modifier $(A(x))^p$, where $p = -\infty$, could correspond to *exactly*, since all membership degrees become equal to 1.

The fuzzy inference engine is the core of a fuzzy expert system. This module performs fuzzy deductions using fuzzy logic. Traditionally, fuzzy inference engines use approximate deductions (i.e. deductions using the inference rules presented in the previous section). Fuzzy inference is a multistep process. Given some input, the output is produced by the following steps:

Step 1. Define fuzzy sets
Step 2. Relate observations to fuzzy sets
Step 3. Define fuzzy rules

Step 4. Evaluate each case for all fuzzy rules
Step 5. Combine information from rules
Step 6. Defuzzify results

The most interesting steps of this process are the fuzzification step, the rule evaluation step, and the defuzzification step. In what follows, we will briefly explain these steps.

7.11.1 Fuzzification

A process that makes a crisp quantity fuzzy is called fuzzification. In reality, many of the quantities that we think to be crisp are not actually crisp because they carry considerable uncertainty. And if this uncertainty has its roots to imprecision, ambiguity, or vagueness, then the quantity is possibly fuzzy and can be represented by a membership function.

In Section 1.3, we noted that hardware is actually vague since it operates within a specific range of accuracy. Thus, we can assume that the readings on a digital voltmeter are actually vague readings (after all, there are no infinitely accurate devices). Figure 7.1 shows the reading of such a device and, assuming a $\pm 1\%$ "error" range, how one can create a fuzzy set from such readings.

7.11.2 Evaluation of Rules

Evaluation rules are actually fuzzy implications. Thus, the rule IF A THEN B is actually a verbose way to say $A \to B$. The most important problem as regards the

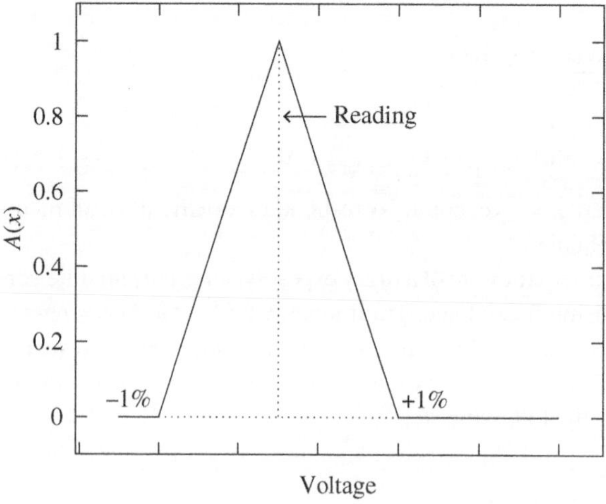

Figure 7.1 Membership function of readings of a digital voltmeter.

evaluation of rules is how to choose the proper implication. In our opinion this is a bit silly since implication operators are used to build logics and logics have many properties (e.g., completeness results, truth values, etc.). Thus, the real question is not which implication the operator chooses but which logic to choose. Naturally, our choice should depend on the properties of a specific logic. These properties have been discussed thoroughly by Hájek [157].

7.11.3 Defuzzification

The term "defuzzification" refers to a process that makes crisp a fuzzy set. The most obvious method to transform a fuzzy set into a crisp one is to get an α-cut of it. Another method that is very frequently used, is the center of gravity or centroid of area method. In the discrete case, we have a fuzzy set A and n elements x_i and from these, we compute the defuzzified value $\overline{x}$:

$$\overline{x} = \frac{\sum_{i=1}^{n} x_i \cdot A(x_i)}{\sum_{i=1}^{n} A(x_i)}.$$

For continuous membership functions, $\overline{x}$ is computed by the following formula:

$$\overline{x} = \frac{\int x_i \cdot A(x_i)\, dx}{\int A(x_i)\, dx}.$$

A very important usage of fuzzy expert systems is in *medical diagnosis*. Since exactness is something one encounters only in abstract worlds, a medical professional is typically forced to use vague, imprecise, and ambiguous information in his/her practice. Naturally, a medical expert system should be able to handle such kind of information as well. In other words, medical information is incomplete, inaccurate, and inconsistent [2]. In particular, typically, medical data are divided into four categories. Below we present these categories and give an outline of what may make data inaccurate, vague, etc.

Medical history. Typically, a patient gives his/her own medical history. Of course, what the patient says is subjective and may say things that are not important, while he/she may forget to mention things that are important (e.g. diseases he/she or members of their family got).

Physical examination. It is possible that a physician in a physical examination of a patient may overlook important indications, fail to carry out a complete examination, or may make some mistakes. Thus, a physical examination may not produce valid medical data.

Results of laboratory tests. Although the results of laboratory tests are objective data, however, measurement errors, organizational problems (e.g. mislabeling samples), or improper patient behavior (e.g. eating fruits late at night may affect an examination of blood glucose level the next morning) may introduce some sort of inaccuracy in medical data.

Results of histological, ultrasonic, PET scans, MRI scans, etc. These results depend on correct interpretation by medical or other staff. In many cases, uncertainty is part of the evaluation process.

The medical expert system CADIAG-2 [2] was an attempt to create a fuzzy expert system to assist physicians in medical diagnosis. The symptom S_i takes values in $[0,1] \cup \{\nu\}$. The value indicates the degree to which a patient exhibits symptom S_i. If the value of S_i is equal to ν, this means the symptom has not yet been examined. The system built a fuzzy relation $R_{PS} : \Pi \times \Sigma \to [0,1] \cup \{\nu\}$, where Π is a set of patients and Σ a set of symptoms. In order to deliver results, the fuzzy logical operations have been extended. The value ν is assumed to be less than zero, so the min and max operations are adjusted accordingly. The fuzzy expert system used rules of the following form

$$\text{IF } (\textit{antecedent}) \text{ THEN } (\textit{consequent}) \text{ WITH } (o, c)$$

where o is the frequency of occurrence and c the strength of confirmation.

A different approach to fuzzy medical diagnosis is based on *intermediate diagnostic units* [162] (IDU). This approach considers a universe of diseases $D = \{D_1, \dots, D_m\}$ and a universe of symptoms $S = \{S_1, \dots, S_n\}$. Each patient corresponds to a fuzzy subset of S and each symptom S_i belongs to this fuzzy subset to a degree that corresponds to the degree the patient has this symptom. A model of a disease consists of a lattice structure composed of IDUs that are ordered by symptoms. The IDUs are formed from accumulated medical knowledge. There are many levels of IDUs. First-level IDUs consist of symptoms that have common characteristics or criteria of diagnostic interest, whereas IDUs of higher levels consist of symptoms from IDUs from lower levels. Also, there are three types of IDUs and each type is associated with a formula that can be used to compute its membership degree. The associative IDUs consist of symptoms that must be present in order to make a diagnosis. For the nonassociative IDUs only one symptom, the most significant, is enough to make a diagnosis. Finally, the excluding IDUs include symptoms that can reduce the feasibility of a diagnosis when their intensity increases.

In order to derive a diagnosis using this model, one has to calculate the membership degree of each IDU from the symptoms of each patient. The estimated membership degree depends on a parameter r whose value is initially set by an expert. This parameter may be adjusted later on. For an associative IDU, the membership degree is computed by the following formula:

$$d = \frac{1}{n} \sum_{i=1}^{n} 1 - \sqrt{|r_i - s_i| k_i},$$

where r is a fuzzy relation on IDUs $\times$ S and r_i is the degree to which the specific IDU is related to the ith symptom, s is again a fuzzy relation on patients $\times$ S and s_i is the symptom intensity for symptom i for a specific patient, while $k_i = 0$ if $s_i \geq r_i$ and $k_i = 1$ otherwise.

Nowadays, most people do not use standalone applications. Instead, it is customary to use some Web interface to some application or a real Web application. This is particularly useful in rural areas of underdeveloped countries that lack good medical facilities. A Web application that employs an alternative approach to computer-aided medical diagnosis has been described in [84]. This application employs generalized fuzzy numbers, because it is extremely useful to be able to model the personal confidence level to a physician. Thus, the system has rules whose antecedent part uses generalized fuzzy numbers to quantify linguistic variables. On the other hand, the variables in the consequent part are quantified by "intuitionistic" fuzzy sets. The membership degree describes the degree of association of the disease to the patient, while the nonmembership degree describes the degree of nonassociation of the disease to the patient. To understand how all these things mix together, consider the following fuzzy decision rule:

> If temperature is high with confidence level (CL) 0.7, headache is medium with CL 0.8, stomach pain is very low with CL 0.8, cough is very high with CL 0.5, and chest pain is low with CL 0.9, then the possibility of viral fever is $(0.9, 0.0)$, for malaria is $(0.5, 0.3)$, for typhoid is $(0.4, 0.6)$, for stomach problem is $(0.3, 0.7)$, for chest problem is $(0.1, 0.8)$.

7.12* A Logic of Vagueness

Fuzzy set theory, in general, and fuzzy logic, in particular, are not universally accepted. Today, more than 50 years since Zadeh published his ideas on fuzzy sets and logic, there are scholars who do not think that fuzzy logic is a logic! In addition, many others do not buy the idea that vagueness is a property of our world. In fact, as it was already mentioned, some confuse fuzzy sets with probability theory and the first author [275] has presented a number of arguments that refute this wrong idea. As far as it regards fuzzy logic, the French logician Jean-Yves Girard has put fuzzy logic under the "umbrella" of *Broccoli logics* [138] (i.e. new logics that are based on classical logic but have new connectives, new rules, etc.). However, Girard [140, p. 506] did not stop there and had to offer the following critique on fuzzy logic:

> [s]peaking of fuzzy "logic", what is wrong is not the idea of going beyond the boolean truth values $\{0, 1\}$, it is the fact of confining one to the static, dead, domain of truth.

Clearly, this statement is a critique against all many-valued logics. But why is it problematic to confine a logic to a "static, dead, domain of truth" and what does this really mean?

Girard [136] is the inventor of linear logic, a logic that has found many applications in computer science. In linear logic, propositions are not static as they are in classical logic. Obviously, since fuzzy logic is actually an extension of classical logic, propositions are static too. Thus, once a proposition gets a truth value, it will never change. And this is exactly what static means. But why is this bad? Let us give a simple example to make this clear. Assume that a certain moment a summit definitely belongs to a cloud. As it is natural, a few minutes later, the same summit may not definitely belong to the cloud. Of course, one could say that the point here is that time is not taken into account, nevertheless, we are not really interested in time, but instead, we want to emphasize that truth values are dynamic entities. In linear logic, a proposition is viewed as some kind of resource that can obviously be exhausted. The following example borrowed from Girard [137] should make things clear. Think of modus ponens:

if A and $A \Rightarrow B$, then B.

When applying this rule, A is still valid. It is like saying that A stands for "to spend €5" and B stands for "to get a pack of cigarettes." Then if $A \Rightarrow B$ means that one spends €5 to get a pack of cigarettes, then an application of modus ponens if A holds means that one will spend €5, he/she will get the pack of cigarettes, and in the end, he/she will still have €5. Of course, it does not really matter if A and B are fuzzy predicates since this problem is still there. Of course, there are mechanisms that allow one to repeatedly use a proposition in proofs, but this is not the norm. The next question is how can one modify fuzzy logic to accommodate these ideas? Of course, this is not an easy task, but we can get inspiration from computer science.

The *proofs-as-processes* slogan by Samson Abramsky [1] was about the description of proofs in linear logic as expressions in some process calculus like Robin Milner's π-calculus. In the framework of theoretical computer science, processes are perfect entities that either stop or work ad infinitum. However, in reality, processes may be similar to some degree, perform tasks to a certain degree, etc. This simply means that one could use processes and their interaction as a starting point and, by going back to logic, one could possibly define some vague logic that would have dynamic truth values. After all, as Girard [139] has pointed out: "Instead of teaching logic to nature, it is more reasonable to learn from her." Thus, it is possible to take all these under consideration and produce a logic of vagueness.

Exercises

7.1 Write down the truth tables of the following expressions: $((A \Rightarrow B) \wedge (\neg B))$, $((A \Rightarrow (B \Rightarrow C)) \Leftrightarrow (A \Rightarrow B))$, and $(A \vee (\neg A))$.

7.2 Show that $(A \Rightarrow B)$ is logically equivalent to $(\neg(A \wedge (\neg B)))$.

7.3 Prove in BL the following:
   a) $A \rightarrow (B \rightarrow (A \,\&\, B))$;
   b) $(A \rightarrow B) \rightarrow ((A \,\&\, C) \rightarrow (B \,\&\, C))$; and
   c) $((A \rightarrow B) \wedge (A \rightarrow C)) \rightarrow (A \rightarrow (B \wedge C))$.

7.4 Show that $\mathbf{0}$ is equivalent to $A \odot \mathbf{0}$ in BL.

7.5 Assume that $U = \{1, 2, 3, 4\}$, $P = 0/1 + 0.6/2 + 1/3 + 0.5/4$, $S = 0/1 + 1/2 + 0.6/3 + 0.2/4$. Compute $\mathrm{cyl}\overline{P} \oplus \mathrm{cyl}S$.

7.6 If A is a fuzzy set, explain how the hedge "very very" will modify it.

8

Fuzzy Computation

Computation is about calculating or enumerating mechanically. Typically, the word mechanically means that one builds a device and sets it in motion in order to perform the desired calculation or enumeration. Many and different real or conceptual devices capable of performing computations have been proposed. Most of them operate in a crisp environment and in a crisp manner. However, there are some devices that profit from the use of vagueness in their overall operation. These devices and the related theory are described in this chapter. The material presented in this chapter is based on [220, 274].

8.1 Automata, Grammars, and Machines

A *finite automaton* can be seen as a machine equipped with scanning head that can read the contents of sequence of cells, while the head can move in only one direction. At any moment, the machine is in a state. Initially, the scanning head is positioned on the leftmost cell, while a number of symbols are printed on consecutive cells starting with the leftmost cell. Also, the machine enters a default initial state. Each machine is associated with a number of *transition rules*. A transition rule has the general form $q_i \stackrel{s}{\longrightarrow} q_j$, where q_i and q_j are states and s is a symbol. The meaning of this rule is that if the automaton is in state q_i, it will enter state q_j only if the next symbol is s. When the machine starts, it reads the first symbol and if there is a transition rule that includes this symbol and the current state, then the scanning head moves to the right and the machine enters a new state. If the machine enters the final state, then it accepts the input and terminates. The machine may suspend without completion when no transition rule applies. The alphabet of the machine consists of the symbols that the scanning head can

A Modern Introduction to Fuzzy Mathematics, First Edition.
Apostolos Syropoulos and Theophanes Grammenos.

recognize. The following figure shows a finite automaton with two states whose alphabet is $\{a, b\}$.

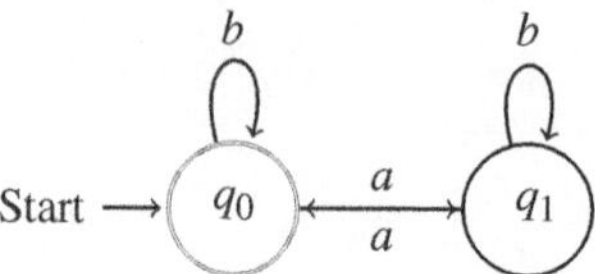

The symbols in the circles are the states, the symbol in the double circle is the accepting state, and the symbols over the arcs are the symbols that the automaton consumes. Thus, this is a compact way to write the various transition rules. This automaton determines if an input sequence of symbols over its alphabet contains an even number of a's. For example, if the input is the sequence "*abba*", then the following table shows what has to be done in order to have this sequence accepted by this automaton.

Current state	Unread symbols	Transition rule
q_0	*abba*	$q_0 \xrightarrow{a} q_1$
q_1	*bba*	$q_1 \xrightarrow{b} q_1$
q_1	*ba*	$q_1 \xrightarrow{b} q_1$
q_1	*a*	$q_1 \xrightarrow{a} q_0$
q_0		Input accepted!

More generally, automata are able to examine whether character sequences or just *strings* belong to some formal language.

Definition 8.1.1 Assume that Σ is an arbitrary set of symbols, which is called *alphabet*, and that ϵ denotes the *empty string*. Then,

$$\Sigma^* = \{\epsilon\} \cup \Sigma \cup \Sigma \times \Sigma \cup \Sigma \times \Sigma \times \Sigma \cup \cdots .$$

is the set of all finite strings over Σ. A *formal language L*, or just language L, is a subset of Σ^*, that is, $L \subseteq \Sigma^*$.

Typically, a language is defined by a grammar:

Definition 8.1.2 A grammar is defined to be a quadruple $G = (V_N, V_T, S, \Phi)$ where V_T and V_N are disjoint sets of terminal and nonterminal (syntactic class) symbols, respectively; S, a distinguished element of V_N, is called the starting symbol. Φ is a finite nonempty relation from $(V_T \cup V_N)^* V_N (V_T \cup V_N)^*$ to $(V_T \cup V_N)^*$. In general, an element (α, β) is written as $\alpha \to \beta$ and is called a production or rewriting rule [287].

Grammars are classified as follows:

Unrestricted grammars. There are no restrictions on the form of the production rules.

Context-sensitive grammars. The relation Φ contains only productions of the form $\alpha \rightarrow \beta$, where $|\alpha| \leq |\beta|$, and in general, $|\gamma|$ is the length of the string γ.

Context-free grammars. The relation Φ contains only productions of the form $\alpha \rightarrow \beta$, where $|\alpha| = 1$ and $\alpha \in V_N$.

Regular grammars. The relation Φ contains only productions of the form $\alpha \rightarrow \beta$, where $|\alpha| \leq |\beta|$, $\alpha \in V_N$, and β has the form aB or a, where $a \in V_T$ and $B \in V_N$.

Syntactically complex languages can be defined by means of grammars. To each class of languages, there is a class of automata (machines) that *accept* (i.e., they can answer the decision problem "$s \in L$?," where s is a string and L is a language) this class of languages, which are generated by the respective grammars. In particular, *finite automata* accept languages generated by regular grammars, *push-down automata* accept languages generated by context-free grammars, *linear bounded automata* accept languages generated by context-sensitive grammars, and *Turing machines* accept *recursive* languages, that is, a subclass of the class of languages generated by unrestricted grammars.

A Turing machine is a conceptual computing device consisting of an *infinite tape*, a *controlling device*, and a *scanning head* (see Figure 8.1). The tape is divided into an infinite number of cells. The scanning head can read and write symbols in each cell. The symbols are elements of a set $\Sigma = \{S_1, \ldots, S_n\}$, $n \geq 1$, which is called the *alphabet*. Usually, there is an additional symbol, ␣, called the *blank* symbol,

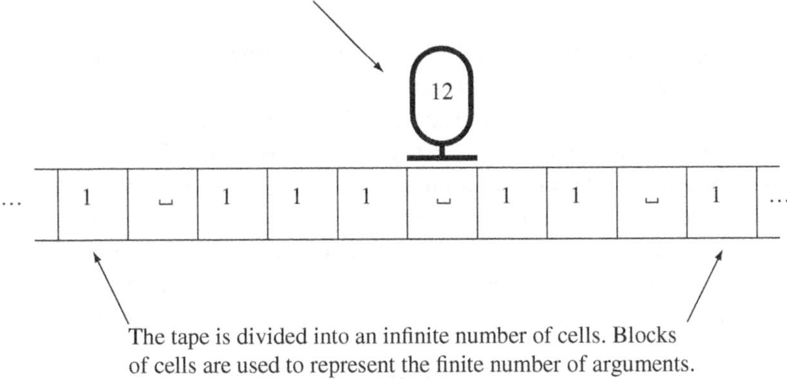

The Turing machine's scanning head moves back and forth along the tape. The number that the scanning head displays is its current state, which changes as it proceeds.
12
... 1 ␣ 1 1 1 ␣ 1 1 ␣ 1 ...
The tape is divided into an infinite number of cells. Blocks of cells are used to represent the finite number of arguments.

Figure 8.1 A typical Turing machine.

and when this symbol is written on a cell by the scanning head, the effect of this operation is the erasure of the symbol that was printed on this particular cell. At any moment, the machine is in a *state* q_i, which is a member of a finite set $Q = \{q_0, q_1, \dots, q_r\}, r \geq 0$. The controlling device is actually a look-up table that is used to determine what the machine has to do next at any given moment. In particular, the action a machine has to take depends on its current state and the symbol that is printed on the cell the scanning head has just finished scanning. If no action has been specified for a particular combination of state and symbol, the machine halts. Usually, the control device is specified by a finite set of *quadruples*, which are special cases of expressions.

Definition 8.1.3 An *expression* is a string of symbols chosen from the list q_0, q_1,...; $\llcorner$, S_1, ...; R, L.

A quadruple can have one of the following forms:

$$q_i S_j S_k q_l \tag{8.1}$$

$$q_i S_j L q_l \tag{8.2}$$

$$q_i S_j R q_l \tag{8.3}$$

Note that $L, R \notin \Sigma$ and $S_j, S_k \in \Sigma \cup \{\llcorner\}$. The quadruple (8.1) specifies that if the machine is in state q_i and the cell that the scanning head scans contains the symbol S_j, then the scanning head replaces S_j by S_k and the machine enters state q_l. The quadruples (8.2) and (8.3) specify that if the machine is in state q_i and the cell that the scanning head scans contains the symbol S_j, then the scanning head moves to the cell to the left of the current cell, or to the cell to the right of the current cell, respectively, and the machine enters the state q_l. Sometimes the following quadruple is also considered:

$$q_i S_j q_k q_l. \tag{8.4}$$

This quadruple is particularly useful if we want to construct a Turing machine that will compute *relatively computable functions* . These quadruples provide a Turing machine with a means of communicating with an external agency that can give correct answers to questions about a set $A \subset \mathbb{N}$. In particular, when a machine is in state q_i and the cell that the scanning head scans contains the symbol S_j, then the machine can be thought of as asking the question, "Is $n \in A$?" Here n is the number of S_1's that are printed on the tape. If the answer is "yes," then the machine enters state q_k; otherwise it enters state q_l. Turing machines equipped with such an external agency are called *oracle machines* , and the external agency is called an *oracle*.

Turing machines are used to compute the value of functions $f(n_1, \dots, n_m)$ that take values in $\mathbb{N}^m$. Each argument $n_i \in \mathbb{N}$, is represented on the tape by preprinting the symbol S_1 on $n_i + 1$ consecutive cells. Typically, such a block of cells is denoted by $\overline{n_1}$. Argument representations are separated by a blank cell (i.e., a cell on which the symbol ␣ is printed), while all other cells are empty (i.e., the symbol S_0 has been preprinted on each cell). It is customary to represent such a block of cell with the expression

$$\overline{(n_1, n_2, \dots, n_m)} = \overline{n_1}\_\overline{n_2}\_\cdots\_\overline{n_m}.$$

If α is an expression, then $\langle\alpha\rangle$ will denote the number of S_1 contained in α. In addition,

$$\langle\overline{m-1}\rangle = m \quad \text{and} \quad \langle\alpha\beta\rangle = \langle\alpha\rangle + \langle\beta\rangle.$$

It is also customary to use the symbol 1 for S_1. Thus, the sequence 3, 4, 2 will be represented by the following three blocks of 1's:

1111␣11111␣111

The machine starts at state q_0 and the scanning head is placed atop the leftmost 1 of a sequence of n blocks of 1's. If the machine has reached a situation in which none or more than one quadruple is applicable, the machine halts. Once the machine has terminated, the result of the computation is equal to the number of cells on which the symbol S_1 is printed.

Although the description presented so far is quite formal for our own taste, still fuzzy versions of the Turing machine are extensions of the "standard" formal definition that is given below.

Definition 8.1.4 A Turing machine $\mathscr{M}$ is an octuple $(Q, \Sigma, \Gamma, \delta, \_, \triangleright, q_0, H)$, where

- Q is a finite set of states;
- Σ is the input alphabet;
- Γ is an alphabet, called *working alphabet*, where $\Sigma \subseteq \Gamma$;
- $\_ \in \Gamma$ is the blank symbol;
- $\triangleright \in \Gamma$ is the *left end symbol*;
- $q_0 \in Q$ is the initial state;
- $H \subseteq Q$ is the set of halting or accepting and rejecting states;
- δ, the *transition* function, is a function from $(Q \setminus H) \times \Gamma$ to $Q \times \Gamma \times \{L, R, N\}$ such that $\mathscr{M}$ may perform an instruction $(q', Y, D) \in Q \times \Gamma \times \{L, R, N\}$, if $\mathscr{M}$ is in state q, the scanning head has just read the symbol X, and $\delta(q, X) = (q', Y, D)$. Depending on the value of D, the machine will move to the left (L), to the right (R), or it will stay still when $D = N$ and the scanning head will overwrite X with Y. Also, $\delta(q, \triangleright) = (q, \triangleright, R)$, which means that whenever the scanning head has read the symbol $\triangleright$, it immediately moves to the right.

Note that here we have defined a machine that has a unidirected tape and not a bidirectional tape. One can get the bidirectional version by eliminating all references to the $\rhd$ symbol.

A *configuration* of $\mathscr{M}$ is an element from

$$C(\mathscr{M}) = \{\rhd\}\Gamma^* Q\Gamma^+ \cup Q\{\rhd\}\Gamma^*,$$

where AB denotes strings that are formed by concatenating a string that belongs to A with a string that belongs to B. If $w_1 qaw_2$ is a configuration, then $w_1 \in \{\rhd\}\Gamma^*$, $w_2 \in \Gamma^*$, $a \in \Gamma$, and $q \in Q$. Moreover, this configuration means that a machine $\mathscr{M}$ is in state q, the content of the tape is $\rhd\ w_1 aw_2 \sqcup\sqcup\sqcup \ldots$, and the scanning head sits atop the $(|w_1| + 1)$th cell, where $|s|$ is the length of string s. The *initial* configuration is $q_0 \rhd x$, where x is the input fed to the machine. A configuration whose state component is in H is called a *halted* configuration.

A *step* is a relation $\vdash_{\mathscr{M}}$ on the set of configurations defined as follows:

(i) $\cdots x_{i-1} q x_i x_{i+1} \cdots \vdash_{\mathscr{M}} \cdots x_{i-1} q' y x_{i+1} \cdots$, if $\delta(q, x_i) = (q', y, N)$;

(ii) $\cdots x_{i-1} q x_i x_{i+1} \cdots \vdash_{\mathscr{M}} \cdots x_{i-2} q' x_{i-1} y x_{i+1} \cdots$, if $\delta(q, x_i) = (q', y, L)$;

(iii) $\cdots x_{i-1} q x_i x_{i+1} \cdots \vdash_{\mathscr{M}} \cdots x_{i-1} y q' x_{i+1} \cdots$, if $\delta(q, x_i) = (q', y, R)$; and

(iv) $\cdots x_{n-1} q x_n \vdash_{\mathscr{M}} \cdots x_{n-1} y q' \sqcup$, if $\delta(q, x_n) = (q', y, R)$.

A *computation* by $\mathscr{M}$ is a sequence of configurations $C_0, C_1, \ldots, C_n$, for some $n \geq 0$ such that

$$C_0 \vdash_{\mathscr{M}} C_1 \vdash_{\mathscr{M}} C_2 \vdash_{\mathscr{M}} \cdots \vdash_{\mathscr{M}} C_n.$$

A computation from C_0 to C_n can be written compactly as $C_0 \vdash^*_{\mathscr{M}} C_n$.

A computation by $\mathscr{M}$ on input x is a series of actions that start at configuration $C_0 = q_0 \rhd x$ and is either *infinite* (i.e. nonterminating) or stops at configuration $w_1 q w_2$, where $q \in H$. Assume that $H = \{q_a, q_r\}$, where q_a is the *accepting* state and q_r is the *rejecting* state. Then, a computation is called *accepting* if it finishes in the configuration $w_1 q_a w_2$ and *rejecting* if it finishes in the configuration $w_1 q_r w_2$. Furthermore, if a computation on input x is accepting or rejecting, we say that the corresponding machine *accepts* or *rejects* x, respectively. More generally, the words accepted by $\mathscr{M}$ form a formal language $L(\mathscr{M})$.

8.2 Fuzzy Languages and Grammars

Suppose that Σ is an alphabet. Then, a *fuzzy formal language* (or just fuzzy language) is a fuzzy subset of Σ^*. If we have a set of nonterminal symbols, V_N, and a set of terminal symbols, V_T, such that $V_T \cap V_N = \emptyset$, then a fuzzy language is a fuzzy subset of V_T^*.

Assume that λ_1 and λ_2 are two fuzzy languages over V_T. Then, the *union* of λ_1 and λ_2 is the fuzzy language denoted by $\lambda_1 \cup \lambda_2$ and defined by

$$(\lambda_1 \cup \lambda_2)(x) = \lambda_1(x) \vee \lambda_2(x), \quad \text{for all } x \in V_T^*.$$

Similarly, the *intersection* of λ_1 and λ_2 is the fuzzy language denoted by $\lambda_1 \cap \lambda_2$ and defined by

$$(\lambda_1 \cap \lambda_2)(x) = \lambda_1(x) \wedge \lambda_2(x), \quad \text{for all } x \in V_T^*.$$

The *concatenation* of λ_1 and λ_2 is the fuzzy language denoted by $\lambda_1\lambda_2$ and defined by

$$(\lambda_1\lambda_2)(x) = \bigvee \{\lambda_1(u) \wedge \lambda_2(v) | x = uv,\ u, v \in V_T^*\}.$$

Since the operators $\wedge$ and $\vee$ are distributive, concatenation is associative. Note that $\lambda^2(x) = (\lambda\lambda)(x)$, $\lambda^3(x) = (\lambda\lambda\lambda)(x)$, $\lambda^4(x) = (\lambda\lambda\lambda\lambda)(x)$, etc.

Suppose that λ is a fuzzy language in V_T. Then, the fuzzy subset λ^∞ of V_T^* defined by

$$\lambda^\infty(x) = \bigvee \{\lambda^n(x) | n = 0, 1, 2, \ldots\} \quad \text{for all } x \in V_T^*$$

is called the *Kleene closure* of λ.

A fuzzy grammar includes a set of rules for generating the elements of a fuzzy language. More specifically, a fuzzy grammar G is a quadruple (V_N, V_T, S, Φ) where Φ is a set of fuzzy production rules and everything else is as in Definition 8.2.2. The elements of Φ are expressions of the form

$$\alpha \xrightarrow{\rho} \beta, \quad \rho > 0, \tag{8.5}$$

where $\alpha, \beta \in (V_T \cup V_N)^*$ and ρ is the plausibility degree that α can generate β. Also, we write $\rho_{\alpha,\beta}$ to designate that this is the plausibility degree of rule of the form $\alpha \to \beta$. Given the *rewriting rule* (8.5) and two arbitrary strings s and t in $(V_T \cup V_N)^*$, then we have

$$s\alpha t \xrightarrow{\rho} s\beta t$$

and $s\beta t$ is said to be *directly derivable* from $s\alpha t$. Suppose that $r_1, r_2, \ldots, r_n \in (V_T \cup V_N)^*$ and

$$r_1 \xrightarrow{\rho_2} r_2, \ldots, r_{n-1} \xrightarrow{\rho_n} r_n,$$

where $\rho_2, \ldots, \rho_n > 0$, then r_n is *derivable* from r_1 in G. This is usually written as

$$r_1 \underset{G}{\Rightarrow} r_m \quad \text{or simply} \quad r_1 \Rightarrow r_n.$$

The following expression

$$r_1 \xrightarrow{\rho_2} r_2 \to \cdots \to r_{n-1} \xrightarrow{\rho_n} r_n$$

is called a *derivation chain* from r_1 to r_n.

A fuzzy grammar G generates a fuzzy language $L(G)$. Given a string x consisting of terminal symbols, we say that it is in $L(G)$ if and only if x is derivable from S. The membership degree of x in $L(G)$ is given by

$$L(G)(x) = \bigvee (\rho_{S,r_1} \wedge \rho_{r_1,r_2} \wedge \cdots \wedge \rho_{r_n,x}), \tag{8.6}$$

where the least upper bound is taken over all derivation chains from S to x. This means that (8.6) defines $L(G)$ as a fuzzy subset of $(V_{\mathrm{T}} \cup V_{\mathrm{N}})^*$. Also, if $L(G_1)$ and $L(G_2)$ are equal fuzzy sets, then the grammars G_1 and G_2 are *equivalent*. Naturally, it is important to know whether we can compute $L(G)(x)$ using Eq. (8.6):

Definition 8.2.1 A fuzzy grammar G is *recursive* if there is an *algorithm* that can be used to compute $L(G)(x)$ using Eq. (8.6).

Similar to crisp grammars, fuzzy grammars are classified as follows:

Fuzzy unrestricted grammars. Production rules are of the form $\alpha \xrightarrow{\rho} \beta$, $\rho > 0$, where $\alpha, \beta \in (V_{\mathrm{T}} \cup V_{\mathrm{N}})^*$.

Fuzzy context-sensitive grammars. Production rules are of the form $\alpha_1 A \alpha_2 \xrightarrow{\rho} \alpha_1 \beta \alpha_2$, $\rho > 0$, where $\alpha_1, \alpha_2, \beta \in (V_{\mathrm{T}} \cup V_{\mathrm{N}})^*$, $A \in V_{\mathrm{N}}$, and $\beta \neq \epsilon$. The production rule $S \to \epsilon$ is also allowed.

Fuzzy context-free grammars. Production rules are of the form $A \xrightarrow{\rho} \beta, A \in V_{\mathrm{N}}$, $\beta \in (V_{\mathrm{T}} \cup V_{\mathrm{N}})^*$, $\beta \neq \epsilon$, and $S \to \epsilon$.

Fuzzy regular grammars. Productions are of the form $A \xrightarrow{\rho} \beta B$ or $A \xrightarrow{\rho} \beta, \rho > 0$, where $\alpha \in V_{\mathrm{T}}$ and $A, B \in V_{\mathrm{N}}$. In addition, the rule $S \to \epsilon$ is allowed.

The following result states that fuzzy context-sensitive and hence also fuzzy context-free and fuzzy regular grammars are recursive.

Theorem 8.2.1 *If* $G = (V_{\mathrm{N}}, V_{\mathrm{T}}, S, \Phi)$ *is a fuzzy context-sensitive grammar, then* G *is recursive.*

Proof: The first thing we need to show is that, for any kind of grammar, the least upper bound in (8.6) may be taken over a subset of the set of all derivation chains from S to x, namely, the subset of chains in which no r_i, $i = 1, \ldots, n$, occurs more than once.

Assume that in a derivation chain C,

$$C = S \xrightarrow{\rho_1} r_1 \xrightarrow{\rho_1} r_2 \cdots \xrightarrow{\rho_n} r_n \xrightarrow{\rho_{n+1}} x,$$

r_i is the same as $r_j, j > i$. Suppose that C' is the chain resulting from replacing the subchain

$$r_i \xrightarrow{\rho_{i+1}} \cdots \xrightarrow{\rho_j} r_j \xrightarrow{\rho_{j+1}} r_{j+1}$$

in C by $r_i \xrightarrow{\rho_{j+1}} r_{j+1}$. Obviously, if C is a derivation chain from S to x, then C' is also a derivation chain. But

$$\bigwedge\{\rho_1, \dots, \rho_i, \rho_{i+1}, \dots, \rho_{j+1}, \dots, \rho_{n+1}\} \leq \bigwedge\{\rho_1, \dots, \rho_i, \rho_{j+1}, \dots, \rho_{n+1}\}$$

and so C may be deleted without affecting the least upper bound in (8.6). This means that we can replace the definition (8.6) for $L(G)(x)$ by

$$L(G)(x) = \bigvee \bigwedge\{\rho_{S,r_1}, \rho_{r_1,r_2}, \dots, \rho_{r_n,x}\}, \tag{8.7}$$

where the least upper bound is taken over all loop-free derivation chains from S to x.

We proceed by showing that for fuzzy context-sensitive grammars, the set over which the least upper bound is taken in (8.7) can be derivation chains of bounded length l_0, where l_0 depends on $|x|$ (i.e. the length of x) and the number of symbols in $V_T \cup V_N$.

If G is fuzzy context-sensitive, then because the production rules in Φ are not contracting, it follows that

$$|r + j| \geq |r_i| \quad \text{if } j > i. \tag{8.8}$$

Suppose that card$(V_T \cup V_N) = k$. Then, since there are at most k' distinct strings in $(V_T \cup V_N)^*$ having length equal to l, and since the derivation chain is loop-free, it follows from (8.8) that the total length of the chain is bounded by

$$l_0 = 1 + k + \cdots + k^{|x|}.$$

Now let us describe a method that can be used to generate all finite derivation chains from S to x having length less than or equal to l_0. We start with S and, using Φ, generate the set Q_1 of all strings in $(V_T \cup V_N)^*$ having length less than or equal to $|x|$ that are derivable from S in one step. Then, we construct Q_2, that is, the set of all strings in $(V_T \cup V_N)^*$ having length less than or equal to $|x|$ that are derivable from S in two steps. This can be done by noting that Q_2 is identical with the set of all strings in $(V_T \cup V_N)^*$ having length less than or equal to $|x|$ that are directly derivable from strings in Q_1. Next, we construct the sets $Q_3, Q_4, \dots, Q_k$ by following the same process. This process should stop the moment $k = l_0$ or when $Q_k = \emptyset$. Since each Q_i, $i = 1, \dots, k$ is a finite set, we have managed to find, in a finite number of steps, all loop-free derivation chains from S to x having length less than or equal to l_0 and so to compute $L(G)(x)$ from (8.6). Clearly, this makes the whole process an *algorithm*, which may not be an optimal one, that can be used to compute $L(G)(x)$. Here G is recursive. □

8.3 Fuzzy Automata

In general, a fuzzy automaton, or more formally, a *fuzzy finite state automaton* is, a triple $M = (Q, \Sigma, A)$, where Q is the set of states, X is the set of input symbols and A is a fuzzy subset of $Q \times \Sigma \times Q$, that is, $A : Q \times \Sigma \times Q \to [0, 1]$. Both Q and Σ are finite nonempty sets and Σ^* is the set of all finite words of elements of Σ.

Definition 8.3.1 Suppose that $M = (Q, \Sigma, A)$ is a fuzzy finite state automaton. Then, $A^* : Q \times \Sigma^* \times Q \to [0, 1]$ is defined as follows:

$$A^*(q, \epsilon, r) = \begin{cases} 1, & \text{if } q = r \\ 0, & \text{if } q \neq r \end{cases}$$

and

$$A^*(q, \sigma a, r) = \bigvee \{A^*(q, \sigma, s) \wedge A(s, a, r) | s \in Q\}, \quad \forall \sigma \in \Sigma^*, a \in \Sigma.$$

Definition 8.3.2 Assume that $M_1 = (Q_1, \Sigma_1, A_1)$ and $M_2 = (Q_2, \Sigma_2, A_2)$ are two fuzzy finite state automata. Then, a pair (α, β) of mappings, $\alpha : Q_1 \to Q_2$ and $\beta : \Sigma_1 \to \Sigma_2$ is called a *homomorphism* written $(\alpha, \beta) : M_1 \to M_2$ if $A_1(q, x, p) \leq A_2(\alpha(q), \beta(x), \alpha(p))$ for all $p, q \in Q_1$ and all $x \in \Sigma_1$.

The pair (α, β) is called a *strong homomorphism* if

$$A_2(\alpha(q), \beta(x), \alpha(p)) = \bigvee \{A_1(q, x, t) | t \in Q_1 \text{ and } \alpha(t) = \alpha(p)\}$$

for all $q, p \in Q_1$ and all $x \in \Sigma_1$.

Suppose that $M_1 = (Q_1, \Sigma_1, A_1)$ and $M_2 = (Q_2, \Sigma_2, A_2)$ are two fuzzy finite state automata. Also, suppose that $\overline{\Sigma}$ is a finite set and $f : \overline{\Sigma} \to \Sigma_1 \times \Sigma_2$ is a function. Assume that π_1 and π_2 are the projection maps of $\Sigma_1 \times \Sigma_2$ onto Σ_i, $i = 1, 2$. Then, we define $A_f : (Q_1 \times Q_2) \times \overline{\Sigma} \times (Q_1 \times Q_2) \to [0, 1]$ as follows:

$$A_f((q_1, q_2), \sigma, (p_1, p_2)) = A_1 \times A_2((q_1, q_2), (\pi_1(f(\sigma)), \pi_2(f(\sigma))),$$

for all $(q_1, q_2), (p_1, p_2) \in Q_1 \times Q_2$ and for all $\sigma \in \overline{\Sigma}$. Then, $(Q_1 \times Q_2, \overline{\Sigma}, A_f)$ is called the *general direct product* of M_1 and M_2 and it is usually denoted by $M_1 * M_2$.

Example 8.3.1 Suppose that $M_1 = (Q_1, \Sigma_1, A_1)$ and $M_2 = (Q_2, \Sigma_2, A_2)$ are two fuzzy finite state automata, where $Q_1 = \{q_1, q_2\}$, $\Sigma_1 = \{a\}$, $Q_2 = \{r_1, r_2\}$, $\Sigma_2 = \{a, b\}$, and A_1 and A_2 are defined as follows:

$$\begin{array}{lll} A_1(q_1, a, q_1) = 0 & A_1(q_2, a, q_2) = 0 & A_1(q_1, a, q_2) = 0.3, \\ A_1(q_2, a, q_1) = 0.4 & A_2(r_1, b, r_1) = 0.5 & A_2(r_1, b, r_2) = 0, \\ A_2(r_2, a, r_1) = 0 & A_2(r_2, a, r_2) = 0.6 & A_2(r_1, a, r_1) = 0, \\ A_2(r_2, b, r_1) = 0.9 & A_2(r_1, a, r_2) = 0.2 & A_2(r_2, b, r_2) = 0, \end{array}$$

Then,

$$A_1 \times A_2((q_1, r_1), (a, b), (q_2, r_2)) = 0.3,$$
$$A_1 \times A_2((q_2, r_1), (a, b), (q_1, r_1)) = 0.4,$$
$$A_1 \times A_2((q_1, r_1), (a, a), (q_2, r_2)) = 0.2,$$
$$A_1 \times A_2((q_2, r_2), (a, b), (q_1, r_1)) = 0.4,$$
$$A_1 \times A_2((q_2, r_1), (a, a), (q_1, r_2)) = 0.2,$$
$$A_1 \times A_2((q_1, r_2), (a, b), (q_2, r_1)) = 0.3,$$
$$A_1 \times A_2((q_1, r_2), (a, a), (q_2, r_2)) = 0.3,$$
$$A_1 \times A_2((q_2, r_2), (a, a), (q_1, r_2)) = 0.4.$$

Assume that $M_1 = (Q_1, \Sigma_1, A_1)$ and $M_2 = (Q_2, \Sigma_2, A_2)$ are two fuzzy finite state automata. Also, assume that $\omega : Q_2 \times \Sigma_2 \to \Sigma_1$ is a function and $Q = Q_1 \times Q_2$. Define

$$A^{\omega} : Q \times \Sigma_2 \times Q \to [0, 1]$$

as follows: for all $((q_1, q_2), b, (r_1, r_2)) \in Q \times \Sigma_2\ Q$,

$$A^{\omega}((q_1, q_2), b, (r_1, r_2)) = A_1(q_1, \omega(q_2, b), r_1) \wedge A_2(q_2, b, r_2).$$

Then, $M = (Q, \Sigma_2, A^{\omega})$ is a fuzzy finite state automaton, it is called the *cascade product* of M_1 and M_2 and we write $M = M_1 \omega M_2$.

Example 8.3.2 Suppose that $M_1 = (Q_1, \Sigma_1, A_1)$ and $M_2 = (Q_2, \Sigma_2, A_2)$ are two fuzzy finite state automata, where $Q_1 = \{q_1, q_2\}$, $\Sigma_1 = \{a, b\}$, $Q_2 = \{r_1, r_2\}$, $\Sigma_2 = \{a\}$, and A_1 and A_2 are defined as follows:

$$A_1(q_1, a, q_1) = 0 \qquad A_1(q_2, b, q_1) = 0.4 \qquad A_1(q_1, a, q_2) = 0.1,$$
$$A_1(q_2, b, q_2) = 0 \qquad A_1(q_1, b, q_1) = 0.2 \qquad A_2(r_1, a, r_1) = 0,$$
$$A_1(q_1, b, q_2) = 0 \qquad A_2(r_1, a, r_2) = 0.6 \qquad A_2(r_2, a, r_1) = 0.7,$$
$$A_1(q_2, a, q_2) = 0.3 \qquad A_2(r_2, a, r_2) = 0.$$

Suppose that $\omega : Q_2 \times \Sigma_2 \to \Sigma_1$ is defined as follows:

$$\omega(r_1, a) = a \quad \text{and} \quad \omega(r_2, a) = b.$$

Also, suppose that $Q = Q_1 \times Q_2$. Then, $A^{\omega} : Q \times \Sigma_2 \times Q \to [0, 1]$ is such that

$$A^{\omega}((q_1, r_1), a, (q_2, r_2)) = A_1(q_1, \omega(r_1, a), q_2) \wedge A_2(r_1, a, r_2) = 0.1$$
$$A^{\omega}((q_2, r_2), a, (q_1, r_1)) = A_1(q_2, \omega(r_2, a), q_1) \wedge A_2(r_2, a, r_1) = 0.4$$
$$A^{\omega}((q_1, r_2), a, (q_1, r_1)) = A_1(q_1, \omega(r_2, a), q_1) \wedge A_2(r_2, a, r_1) = 0.2$$
$$A^{\omega}((q_2, r_1), a, (q_2, r_2)) = A_1(q_2, \omega(r_1, a), q_2) \wedge A_2(r_1, a, r_2) = 0.3$$

and A^{ω} is 0 elsewhere.

As before, let $M_1 = (Q_1, \Sigma_1, A_1)$ and $M_2 = (Q_2, \Sigma_2, A_2)$ be two fuzzy finite state automata. Also, let $f : Q_2 \to \Sigma_1$ be a function. In addition, let

$$A^0 : Q \times (\Sigma_1^{Q_2} \times \Sigma_2) \to [0, 1]$$

be a function such that for all $((q_1, q_2), (f, b), (r_1, r_2)) \in Q \times (\Sigma_1^{Q_2} \times \Sigma_2) \times Q$,

$$A^0((q_1, q_2), (f, b), (r_1, r_2)) = A_1(q_1, f(q_2), r_1) \wedge A_2(q_2, b, r_2).$$

Then, $M = (Q, \Sigma_1^{Q_2} \times \Sigma_2, A^0)$ is a fuzzy finite state automaton. $M = M_1 \circ M_2$ is called the *wreath product* of M_1 and M_2.

Example 8.3.3 Suppose that M_1 and M_2 are the two fuzzy finite state automata of the previous example. Suppose that $\Sigma_1^{Q_2} = \{f_1, f_2, f_3, f_4\}$, where $f_1(r_1) = f_1(r_2) = a, f_2(r_1) = a, f_2(r_2) = b, f_3(r_1) = b, f_3(r_2) = a$, and $f_4(r_1) = f_4(r_2) = b$. Then,

$$A^0((q_1, r_1), (f_1, a), (q_2, r_2)) = A_1(q_1, f_1(r_1), q_2) \wedge A_2(r_1, a, r_2) = 0.1,$$
$$A^0((q_2, r_1), (f_1, a), (q_2, r_2)) = A_1(q_2, f_1(r_1), q_2) \wedge A_2(r_1, a, r_2) = 0.3,$$
$$A^0((q_1, r_2), (f_1, a), (q_2, r_1)) = A_1(q_1, f_1(r_2), q_2) \wedge A_2(r_2, a, r_1) = 0.1,$$
$$A^0((q_2, r_2), (f_1, a), (q_2, r_1)) = A_1(q_2, f_1(r_2), q_2) \wedge A_2(r_2, a, r_1) = 0.3,$$
$$A^0((q_1, r_1), (f_2, a), (q_2, r_2)) = A_1(q_1, f_2(r_1), q_2) \wedge A_2(r_1, a, r_2) = 0.1,$$
$$A^0((q_2, r_1), (f_2, a), (q_2, r_2)) = A_1(q_2, f_2(r_1), q_2) \wedge A_2(r_1, a, r_2) = 0.3,$$
$$A^0((q_1, r_2), (f_2, a), (q_1, r_1)) = A_1(q_1, f_2(r_2), q_1) \wedge A_2(r_2, a, r_1) = 0.2,$$
$$A^0((q_2, r_2), (f_2, a), (q_1, r_1)) = A_1(q_2, f_2(r_2), q_1) \wedge A_2(r_2, a, r_1) = 0.4,$$
$$A^0((q_1, r_1), (f_3, a), (q_1, r_2)) = A_1(q_1, f_3(r_1), q_1) \wedge A_2(r_1, a, r_2) = 0.2,$$
$$A^0((q_2, r_1), (f_3, a), (q_1, r_2)) = A_1(q_2, f_3(r_1), q_1) \wedge A_2(r_1, a, r_2) = 0.4,$$
$$A^0((q_1, r_2), (f_3, a), (q_2, r_1)) = A_1(q_1, f_3(r_2), q_2) \wedge A_2(r_2, a, r_1) = 0.1,$$
$$A^0((q_2, r_2), (f_3, a), (q_2, r_1)) = A_1(q_2, f_3(r_2), q_2) \wedge A_2(r_2, a, r_1) = 0.3,$$
$$A^0((q_1, r_1), (f_4, a), (q_1, r_2)) = A_1(q_1, f_4(r_1), q_1) \wedge A_2(r_1, a, r_2) = 0.2,$$
$$A^0((q_2, r_1), (f_4, a), (q_1, r_2)) = A_1(q_2, f_4(r_1), q_1) \wedge A_2(r_1, a, r_2) = 0.4$$
$$A^0((q_1, r_2), (f_4, a), (q_1, r_1)) = A_1(q_1, f_4(r_2), q_1) \wedge A_2(r_2, a, r_1) = 0.2,$$
$$A^0((q_2, r_2), (f_4, a), (q_1, r_1)) = A_1(q_2, f_4(r_2), q_1) \wedge A_2(r_2, a, r_1) = 0.4.$$

Definition 8.3.3 Assume that $M = (Q, \Sigma, A)$ is a fuzzy finite state automaton and that $q, r \in Q$. Then, r is called an *immediate successor* of q if there is an $a \in \Sigma$ such that $A(q, a, r) > 0$. Also, r is called a *successor* of q if there is a $\sigma \in \Sigma^*$ such that $A^*(q, \sigma, r) > 0$.

Proposition 8.3.1 *Assume that $M = (Q, \Sigma, A)$ is a fuzzy finite state automaton and that $q, r \in Q$. Then, the following are true:*

(i) q is a successor of q; and
(ii) if r is a successor of q and s is a successor of r, then s is a successor of q.

Definition 8.3.4 Assume that $M = (Q, \Sigma, A)$ is a fuzzy finite state automaton and that $q \in Q$. Then, the set of all successors of q are denoted by $S(q)$.

Definition 8.3.5 Assume that $M = (Q, \Sigma, A)$ is a fuzzy finite state automaton and that $T \subseteq Q$. Then, the set of all successors of T, which is denoted by $S_Q(T)$, is the set

$$S_Q(T) = \bigcup \{S(q) | q \in T\}.$$

Definition 8.3.6 Assume that $M = (Q, \Sigma, A)$ is a fuzzy finite state automaton and that $T \subseteq Q$. Also, suppose that ν is a fuzzy subset of $T \times \Sigma \times T$ and $N = (T, \Sigma, \nu)$. Then, the fuzzy finite state automaton N is called a *subautomaton* of M if

(i) $A|_{T \times \Sigma \times T} = \nu$ (i.e. ν is a restriction of A), and
(ii) $S_Q(T) \subseteq T$.

Subsystems and strong subsystems are special cases of subautomata:

Definition 8.3.7 Assume that $M = (Q, \Sigma, A)$ is a fuzzy finite state automaton and that δ is a fuzzy subset of Q. Then, (Q, δ, Σ, A) is called a *subsystem* of M if for all $q, r \in Q$ and for all $a \in \Sigma$,

$$\delta(q) \geq \delta(r) \geq A(r, a, q).$$

Theorem 8.3.1 *Assume that $M = (Q, \Sigma, A)$ is a fuzzy finite state automaton and that δ is a fuzzy subset of Q. Then, (Q, δ, Σ, A) is a subsystem of M if and only if for all $r, q \in Q$ and for all $\sigma \in \Sigma^*$,*

$$\delta(q) \geq \delta(r) \wedge A^*(r, \sigma, q).$$

Proof: Let (Q, δ, Σ, A) be a subsystem. Also, assume that $q, r \in Q$ and $\sigma \in \Sigma^*$. The result is proved by induction on the length of σ, that is, $|\sigma| = n$. If $n = 0$, then $\sigma = \epsilon$. If $r = q$, then $\delta(q) \wedge A^*(q, \epsilon, q) = \delta(q)$. If $q \neq r$, then $\delta(r) \wedge A^*(r, \epsilon, q) = 0 \leq \delta(q)$. Thus, the result is true when $n = 0$. Assume that the result is true for all

$\tau \in \Sigma^*$, such that $|\tau| = n - 1$ and $n > 0$. Suppose that $\sigma = \tau a$, $|\tau| = n - 1$, $\tau \in \Sigma^*$, and $a \in \Sigma$. Then,

$$\begin{aligned}
\delta(r) \wedge A^*(r, \sigma, q) &= \delta(r) \wedge A^*(r, \tau a, q) \\
&= \delta(r) \wedge \left(\bigvee \{A^*(r, \tau, s) \wedge A(s, a, q) | s \in Q\}\right) \\
&= \bigvee \{\delta(r) \wedge A^*(r, \tau, s) \wedge A(s, a, q)| \\
&\leq \bigvee \{\delta(r) \wedge A(s, a, q) | s \in Q\} \\
&\leq \delta(q).
\end{aligned}$$

This means that $\delta(q) \geq \delta(r) \wedge A^*(r, \sigma, q)$. The proof of the converse is trivial. □

Definition 8.3.8 Assume that $M = (Q, \Sigma, A)$ is a fuzzy finite state automaton and δ is a fuzzy subset of Q. Then, (Q, δ, Σ, A) is a *strong subsystem* of M if and only if for all $r, q \in Q$ and if there exists an $a \in \Sigma$ such that $A(r, a, q) > 0$, then $\delta(q) \geq \delta(r)$.

We can also compose automata.

Definition 8.3.9 Assume that $M_1 = (Q_1, \Sigma_1, A_1)$ and $M_2 = (Q_2, \Sigma_2, A_2)$ are two fuzzy finite state automata. Also, assume that $\Sigma_1 \cap \Sigma_2 = \emptyset$. Let

$$M_1 \cdot M_2 = (Q_1 \times Q_2, \Sigma_1 \cup \Sigma_2, A_1 \cdot A_2),$$

where

$$(A_1 \cdot A_2)((r_1, r_2), a, (q_1, q_2)) = \begin{cases} A_1(r_1, a, q_1), & \text{if } a \in \Sigma_1 \text{ and } r_2 = q_2, \\ A_2(r_2, a, q_2), & \text{if } a \in \Sigma_2 \text{ and } r_1 = q_1, \\ 0, & \text{otherwise.} \end{cases}$$

for all $(r_1, r_2), (q_1, q_2) \in Q_1 \times Q_2$, $a \in \Sigma_1 \cup \Sigma_2$. Then, $M_1 \cdot M_2$ is a fuzzy finite state automaton and it is called the *Cartesian composition* of M_1 and M_2.

Definition 8.3.10 Suppose that $M_1 = (Q_1, \Sigma_1, A_1)$ and $M_2 = (Q_2, \Sigma_2, A_2)$ are two fuzzy finite state automata such that $Q_1 \cap Q_2 = \emptyset$ and $\Sigma_1 \cap \Sigma_2 = \emptyset$. Then, the fuzzy finite state automaton

$$M_1 + M_2 = (Q_1 \cup Q_2, \Sigma_1 \cup \Sigma_2, A_1 + A_2),$$

where

$$(A_1 + A_2)(r, x, q) = \begin{cases} A_1(r, a, q), & \text{if } a \in \Sigma_1 \text{ and } r, q \in Q_1, \\ A_2(r, a, q), & \text{if } a \in \Sigma_2 \text{ and } r, q \in Q_2, \\ 0, & \text{otherwise.} \end{cases}$$

is called the *sum* of M_1 and M_2.

Mansoor Doostfatemeh and Stefan C. Kremer [104] had presented a new definition for fuzzy automata. A number of reasons (i.e. membership assignment, output mapping, multi-membership resolution, and the concept of acceptance for fuzzy automata) necessitated the introduction of this new definition:

Definition 8.3.11 An octuple $(Q, \Sigma, \tilde{R}, Z, \omega, \tilde{\delta}, F_1, F_2)$ is a *general fuzzy automaton* (GFA) $\tilde{F}$, where

- Q is a finite set of states, $Q = \{q_1, q_2, \dots, q_n\}$;
- Σ is a finite crisp set of input symbols, $\Sigma = \{a_1, a_2, \dots, a_m\}$;
- $\tilde{R} \subseteq \mathscr{F}(Q)$ is the set of fuzzy start states;
- Z is a finite crisp set of output symbols, $Z = \{b_1, b_2, \dots, b_k\}$;
- $\omega : Q \to Z$ is the output function that maps a (fuzzy) state to the output set;
- $F_1 : [0,1] \times [0,1] \to [0,1]$ is a fuzzy binary relation in μ, the set of membership values of a predecessor, and δ, the active states (it is called *membership assignment function*);
- $\tilde{\delta} : (Q \times [0,1]) \times \Sigma \times Q \xrightarrow{F_1(\mu,\delta)} [0,1]$ is the augment transition function; and
- $F_2 : [0,1]^* \to [0,1]$ is the multi-membership resolution function. If there are several simultaneous transitions to the active state q_m at time $t+1$, then F_2 is a function that specifies the *strategy* that resolves the multi-membership active states and assigns a single membership value to them.

The authors have presented a multi-membership resolution algorithm, but we are not going to present it here. But let us see how some simple fuzzy automata can be described as general fuzzy automata.

Example 8.3.4 A fuzzy finite automaton with final states is a GFA with $Z = \{\text{accept}, \text{reject}\}$ such that $\omega : Q \to Z$ is defined as follows:

$$\omega(q_i) = \begin{cases} \text{accept}, & \text{if } q_i \in Q_f, \\ \text{reject}, & \text{if } q_i \notin Q_f. \end{cases}$$

Here Q_f is the set of terminal states.

8.4 Fuzzy Turing Machines

The first general description of a fuzzy Turing machine was given by Zadeh [306]. He presumed that a *fuzzy algorithm* should contain vague commands (Zadeh used the term "fuzzy commands," but in order to be consistent with the terminology used in this book, we will call them "vague commands"). According to Zadeh, the following are examples of simple vague commands:

(i) Set y *approximately equal to* 10, if x is *approximately equal to* 5.
(ii) If x is *large*, increase y by *several* units.
(iii) If x is *large*, increase y by *several* units; if x is *small*, decrease y by *several* units; otherwise keep y unchanged.

This command is vague, because the text that appears slanted corresponds to fuzzy sets. For example, the numbers that are approximately equal to 10 or 5 are two different fuzzy sets. Based on this, Zadeh vaguely described a fuzzy Turing machine as one where instructions are performed to a degree. A number of concrete proposals followed (see Ref. [275] for a detailed presentation of all these proposals), however, we will only present the one by Jiří Wiedermann [298] since it is the most complete presentation and the most interesting of all previous attempts. Wiedermann's machine is a nondeterministic fuzzy Turing machine with hypercomputational capabilities. In a nutshell, Wiedermann has shown that his machine can solve problems no ordinary Turing machine can solve.

Definition 8.4.1 A nondeterministic fuzzy Turing machine with a unidirectional tape is a nonuple

$$\mathscr{F} = (Q, T, I, \Delta, \llcorner\!\lrcorner, q_0, q_f, \mu, *),$$

where:

- Q is a finite set of states;
- T is a finite set of tape symbols;
- I is a set of input symbols, where $I \subseteq T$;
- Δ is a transition relation and it is a subset of $Q \times T \times Q \times T \times \{L, N, R\}$. Each action that the machine takes is associated with an element $\delta \in \Delta$. In particular, for $\delta = (q_i, t_i, q_{i+1}, t_{i+1}, d)$ this means that, when the machine is in state q_i and the current symbol that has been read is t_i, then the machine will enter state q_{i+1}, the symbol t_{i+1} will be printed on the current cell and the scanning head will move according to the value of d, that is, if d is L, N, or R, then the head will move one cell to the left, will not move, or it will move one cell to the right, respectively.
- $\llcorner\!\lrcorner \in T \setminus I$ is the blank symbol;
- q_0 and q_f are the initial and the final state, respectively;
- $\mu : \Delta \to [0, 1]$ is a fuzzy relation on Δ; and
- $*$ is a t-norm.

Definition 8.4.2 When μ is a partial function from $Q \times T$ to $Q \times T \times \{L, N, R\}$ and T is a fuzzy subset of Q, then the resulting machine is called a deterministic fuzzy Turing machine.

A configuration gives

(i) the position of the scanning head,
(ii) what is printed on the tape, and
(iii) the current state of the machine.

If S_i and S_{i+1} are two configurations, then $S_i \vdash^{\alpha} S_{i+1}$ means that S_{i+1} is reachable in one step from S_i with a plausibility degree that is equal to α if and only if there is a $\delta \in \Delta$ such that $\mu(\delta) = \alpha$, and by which the machine goes from S_i to S_{i+1}. When a machine starts with input some string w, the characters of the string are printed on the tape starting from the leftmost cell; the scanning head is placed atop the leftmost cell, and the machine enters state q_0. If

$$S_0 \vdash^{\alpha_0} S_1 \vdash^{\alpha_1} S_2 \vdash^{\alpha_2} \cdots \vdash^{\alpha_{n-1}} S_n,$$

then S_n is reachable from S_0 in n steps. Assume that S_n is reachable from S_0 in n steps, then the plausibility degree of this *computational path* is

$$D((S_0, S_1, \dots, S_n)) = \begin{cases} 1, & n - 0, \\ D((S_0, S_1, \dots, S_{n-1})) * \alpha_{n-1}, & n > 0. \end{cases}$$

Obviously, the value that is computed with this formula depends on the specific path that is chosen. Since the machine is nondeterministic, it is quite possible that some configuration S_n can be reached via different computational paths. Therefore, when a machine starts from S_0 and finishes at S_n in n steps, the plausibility degree of this computational path, which is called a *computation*, should be equal to the maximum of all possible computation paths:

$$d(S_n) = \max[D((S_0, S_1, \dots, S_n))].$$

In other words, the plausibility degree of the computation is equal to the plausibility degree of the computational path that is most likely to happen.

Assume that a machine starts from configuration S_0 with w as input. Then, a computational path $S_0, S_1, \dots, S_m$ is an accepting path of configurations if the state of S_m is q_f. In addition, the string w is accepted with degree equal to $d(S_m)$.

Definition 8.4.3 Assume that $\mathscr{F}$ is a fuzzy nondeterministic Turing machine. Then, an input string w is accepted with plausibility degree $e(w)$ by $\mathscr{F}$ if and only if:

- there is an accepting configuration from the initial configuration S_0 on input w;
- $e(w) = \max_S\{d(S) | S \text{ is an accepting configuration reachable from } S_0\}$.

Also,

Definition 8.4.4 The fuzzy language accepted by some machine $\mathscr{F}$ is the fuzzy set that is defined as follows:

$$L(\mathscr{F}) = \{(w, e(w)) | w \text{ is accepted by } \mathscr{F} \text{ with plausibility degree } e(w)\}.$$

The class of all fuzzy languages accepted by a fuzzy Turing machine, in the sense just explained, with (classically) computable t-norms is denoted by Φ.

Theorem 8.4.1 $\Phi = \Sigma_1^0 \cup \Pi_1^0$, *that is,* Φ *is the union of the recursively enumerable and the co-recursively enumerable languages.*

Proof: (Sketch) In order to show that $\Phi = \Sigma_1^0 \cup \Pi_1^0$, one has to prove that $\Sigma_1^0 \cup \Pi_1^0 \subseteq \Phi$ and, at the same time, that $\Phi \subseteq \Sigma_1^0 \cup \Pi_1^0$. Assume that L is a language such that $L \in \Sigma_1^0 \cup \Pi_1^0$. Then either $L \in \Sigma_1^0$ or $L \in \Pi_1^0$. When L is recursively enumerable, there is a nonfuzzy machine $\mathscr{M}$ that is able to semidecide L. In what follows, it will be shown that for each constant $0 \leq c < 1$ there is a fuzzy Turing machine $\mathscr{F}$, whose acceptance criterion is given in Definition 8.4.4, such that $w \in L$ if and only if w is accepted by $\mathscr{F}$ with plausibility degree equal to 1, that is, $(w, 1) \in L(\mathscr{F})$, and $w \notin L$ if and only if $(w, c) \in L(\mathscr{F})$. Given any t-norm and $0 \leq c < 1$, one can specify $\mathscr{F}$ as follows: unless it is explicitly stated, by default all commands will have plausibility degree that is equal to 1. Suppose that w is the input to $\mathscr{F}$. Then, this machine will make a nondeterministic branch and one path will lead to the simulation of $\mathscr{M}$. In addition, when $\mathscr{M}$ enters an accepting state, $\mathscr{F}$ will enter an accepting state q with plausibility degree equal to one. Also, the other path leads to an accepting state q' with plausibility degree equal to c. Now, if $w \in L$, then both q and q' are reached. In addition, w will be accepted in q with plausibility degree equal to one (i.e. $(w, 1) \in L(\mathscr{F})$). If $w \notin L(\mathscr{F})$, then the machine will not enter any accepting state but r, and so $(w, c) \in L(\mathscr{F})$. Assume now that L is co-recursively enumerable and so $\overline{L}$ is recursively enumerable. This implies that there is a machine $\mathscr{M}'$ that recognizes $\overline{L}$. Following an argument similar to the one presented so far, one can show that for each constant $0 \leq c < 1$ there is a fuzzy Turing machine $\mathscr{G}$ such that $w \in \overline{L}$ if and only if w is accepted by $\mathscr{G}$ with plausibility degree equal to one (i.e. $(w, 1) \in L(\mathscr{G})$) and $w \notin \overline{L}$ if and only if $(w, c) \in L(\mathscr{G})$. In other words, $w \notin L$ if and only $(w, 1) \in L(\mathscr{G})$, and $w \in L$ if and only if $(w, c) \in L(\mathscr{G})$. And this concludes the first part of the proof. Let us now proceed with the second part.

Now it will be shown that for any fuzzy Turing machine $\mathscr{F} = (Q, T, I, \Delta, \sqcup, q_0, q_f, \mu, *)$ with a computable t-norm, there are crisp languages $L_1 \in \Sigma_1^0$ and $L_2 \in \Pi_1^0$ such that $(w, d) \in L(\mathscr{F})$ if and only if $w\#d \in L_1 \cap \in L_2 \Sigma_1^0 \cup \Pi_1^0$ for any $w \in (I \setminus \{\#\})^*$ and $\# \in T$. Assume that F is a fuzzy Turing machine, where all commands have plausibility degrees that are equal to one (essentially, this machine is a nondeterministic Turing machine). Also, assume that $ACF_{\mathscr{F}}(w)$ is a set that contains all accepting configurations of $\mathscr{F}$ on input w, and that e is the evaluation function that assigns to each accepting configuration its plausibility degree. In addition, consider the commutative ordered semigroup $G = ([0, 1], *, \leq)$, where $*$ is a t-norm and $\leq$ is the usual ordering relation. Furthermore, let

$\mathbb{Q} \supset J = \{\alpha_1, \alpha_2, \ldots, \alpha_k\} \subset [0,1]$, where $\alpha_1 < \alpha_2 < \cdots < \alpha_k$, be the range of μ and $G(J)$ be a subsemigroup of G generated by replacing $[0,1]$ with a set that has as elements all elements of J and all elements produced by the t-norm, that is, when α_m and α_n are elements of this set, then $\alpha_m * \alpha_n$ also belongs to this set. It is relatively easy to see that

$$\Sigma_1^0 \ni L_1 = \{w\#d | \exists a \in ACF_{\mathcal{F}}(w) : e(a) = d \in G(J)\},$$
$$\Pi_1^0 \ni L_2 = \{w\#d | \forall a \in ACF_{\mathcal{F}}(w) : e(a) \leq d \in G(J)\}.$$

In particular, in order to show that $L_1 \in \Sigma_1^0$, one should consider an ordinary nondeterministic machine $\mathcal{M}_1$, which on input $w\#d$, where $d = \alpha_n * \alpha_m$, first guesses the numbers α_n and α_m, then computes $\alpha_n * \alpha_m$, and checks whether $d \in G(J)$. Next, $\mathcal{M}_1$ guesses a and simulates F on w to see whether $a \in ACF_{\mathcal{F}}(w)$ and $e(a) = d$. In order to show that $L_2 \in \Pi_1^0$, one should consider an ordinary nondeterministic machine M_2 that accepts the language

$$\overline{L}_2 = \{w\#d \mid \exists a \in ACF_{\mathcal{F}}(w) : e(a) > d\}.$$

By a similar argument, one concludes that $\overline{L}_2 \in \Sigma_1^0$ and so $L_2 \in \Pi_1^0$. Furthermore, it is clear that $(w, d) \in L(\mathcal{F})$ if and only if $w\#d \in L_1 \cap L_2$, because the second condition is just a reformulation of conditions for $\mathcal{F}$ to accept w with plausibility degree equal to d according to 8.4.4. □

There have been some arguments against the validity of this proof, but we will not discuss it further (see Ref. [275] for details). In addition, there are some results (e.g. the existence of a universal fuzzy Turing machine) that are discussed in [275, 278].

8.5 Other Fuzzy Models of Computation

Fuzzy Turing machines and fuzzy automata are not the only fuzzy models of computation, however, they are the ones that keep busy most researchers. Indeed, there are models of computation that are based on *fuzzy multisets*.

Definition 8.5.1 Let X be a universe. A *multiset* A of the set X, is characterized by a function $A : X \to \mathbb{N}$. For every $x \in X$, the value $A(x)$ denotes the number of times x belongs to A.

Definition 8.5.2 Assume X is a set of elements. Then, a fuzzy multiset A drawn from X is characterized by a function $A : X \to \mathbb{N}^{[0,1]}$. The expression $\mathbb{N}^{[0,1]}$ is the set of all multisets of $[0,1]$.

It is not difficult to see that any fuzzy multiset A is actually characterized by a function

$$A : X \times [0,1] \to \mathbb{N},$$

which is obtained from the former function by *uncurrying* it. However, one can demand that for each element x, there is only one membership degree and one multiplicity. In other words, a "fuzzy multiset" A should be characterized by a function $X \to [0,1] \times \mathbb{N}$. To distinguish these structures from fuzzy multisets, we will call them *multi-fuzzy* sets [271]. Given a multi-fuzzy set A, the expression $A(x) = (i, n)$ denotes that there are n copies of x that belong to A with a degree that is equal to i. A generalization of this definition was presented in [273]:

Definition 8.5.3 An L-fuzzy hybrid set $\mathscr{A}$ is a mathematical structure that is characterized by a function $\mathscr{A} : X \to L \times \mathbb{Z}$, where L is a frame, and it is associated with an L-fuzzy set $A : X \to L$. More specifically, the equality $\mathscr{A}(x) = (\ell, n)$ means that $\mathscr{A}$ contains exactly n copies of x, where $A(x) = \ell$.

By substituting $\mathbb{Z}$ with $\mathbb{N}$ in the previous definition, the resulting structures will be called L-multi-fuzzy sets.

Assuming that $\mathscr{A}$ is an L-fuzzy hybrid set, then we can define the following two functions: the *multiplicity function* $\mathscr{A}_m : X \to \mathbb{Z}$ and the *degree function* $\mathscr{A}_\mu : X \to L$. Clearly, if $\mathscr{A}(x) = (\ell, n)$, then $\mathscr{A}_m(x) = n$ and $\mathscr{A}_\mu(x) = \ell$. Note that it is equally easy to define the corresponding functions for an L-multi-fuzzy set.

Definition 8.5.4 Assume that $\mathscr{A}$ is an L-fuzzy hybrid set that draws elements from a universe X. Then, its cardinality is defined as follows:

$$\text{card}\, \mathscr{A} = \sum_{x \in X} \mathscr{A}_\mu(x) \otimes \mathscr{A}_m(x),$$

where $\otimes : L \times \mathbb{Z} \to \mathbb{R}$ is a binary multiplication operator that is used to compute the product of $\ell \in L$ "times" $n \in \mathbb{Z}$.

Example 8.5.1 If $L = [0,1] \times [0,1]$ (i.e, when extending "intuitionistic" fuzzy sets), then $(i,j) \times n = in - jn$.

Remark 8.5.1 When L is the interval $[0,1]$, then $\otimes$ is the usual multiplication operator.

In order to give the basic properties of fuzzy hybrid sets, it is necessary to define the notion of subsethood. Before, going on with this definition, we will introduce the partial order $\ll$ over $\mathbb{Z}$, which is defined as follows, for all $n, m \in \mathbb{Z}$:

$$\begin{aligned} n \ll m \equiv\, & (n = 0) \vee \\ & ((n > 0) \wedge (m > 0) \wedge (n \leq m)) \vee \\ & ((n < 0) \wedge (m > 0)) \vee \\ & (|n| \leq |m|). \end{aligned}$$

Here $\wedge$ and $\vee$ denote the classical logical conjunction and disjunction operators, respectively. In addition, the symbols $\leq$ and $<$ are the well-known ordering operators, and $|n|$ is the absolute value of n.

Example 8.5.2 From Definition 8.5.4 it should be obvious that $0 \ll n$, for all $n \in \mathbb{Z}$. Also, $3 \ll 4$, $-3 \ll 4$, and $-4 \ll -3$.

Proposition 8.5.1 *The relation $\ll$ is a partial order.*

Proof: We have to prove that the relation $\ll$ is reflexive, antisymmetric, and transitive:

Reflexivity. Assume that $a \in \mathbb{Z}$. If $a = 0$, then $a \ll a$ from the first part of the disjunction. If $a < 0$, then $a \ll a$ from the fourth part of the disjunction, and if $a > 0$, then $a \ll a$ from the second part of the disjunction.

Antisymmetry. Assume that $a, b \in \mathbb{Z}$, $a \ll b$, and $b \ll a$. Then if $a = 0$, this implies that $b = 0$ and so $a = b$. If $a < 0$, then it follows that $b < 0$, $|a| \leq |b|$, and $|b| \leq |a|$, which implies that $a = b$. Similarly, if $a > 0$, then it follows that $b > 0$, $a \leq b$, and $b \leq a$, which implies that $a = b$.

Transitivity. Assume that $a, b, c \in \mathbb{Z}$, $a \ll b$, and $b \ll c$. If $a = 0$, then clearly $a \ll c$. If $a < 0$ and $b < 0$, then either $c < 0$ or $c > 0$, but since $|b| \leq |c|$, this implies that $a \ll c$. If $a > 0$ and $b > 0$, then $c > 0$ and since $b \leq c$ this implies that $a \ll c$. If $a < 0$ and $b > 0$, then since $b \ll c$, this implies that $c > 0$, which means that $a \ll c$. □

Note that $a \gg b$ is an alternative form of $b \ll a$, which will be used in the rest of this section. Let us now proceed with the definition of the notion of subsethood for L-fuzzy hybrid sets:

Definition 8.5.5 Assume that $\mathscr{A}, \mathscr{B} : X \to L \times \mathbb{Z}$ are two L-fuzzy hybrid sets. Then, $\mathscr{A} \subseteq \mathscr{B}$ if and only if $\mathscr{A}_\mu(x) \sqsubseteq \mathscr{B}_\mu(x)$ and $\mathscr{A}_m(x) \ll \mathscr{B}_m(x)$ for all $x \in X$.

Note that for all $\ell_1, \ell_2 \in L$, $\ell_1 \sqsubseteq \ell_2$ if ℓ_1 is "less than or equal" to ℓ_2 in the sense of the partial order defined over L. The definition of subsethood for L-multi-fuzzy sets is more straightforward:

Definition 8.5.6 Assume that $\mathscr{A}, \mathscr{B} : X \to L \times \mathbb{N}$ are two L-multi-fuzzy sets. Then $\mathscr{A} \subseteq \mathscr{B}$ if and only if $\mathscr{A}_\mu(x) \sqsubseteq \mathscr{B}_\mu(x)$ and $\mathscr{A}_m(x) \leq \mathscr{B}_m(x)$ for all $x \in X$.

Let us now present the definitions of union and sum of L-multi-fuzzy sets:

Definition 8.5.7 Assuming that $\mathscr{A}, \mathscr{B} : X \to L \times \mathbb{N}$ are two L-multi-fuzzy sets, then their union, denoted $\mathscr{A} \cup \mathscr{B}$, is defined as follows:

$$(\mathscr{A} \cup \mathscr{B})(x) = (\mathscr{A}_\mu(x) \sqcup \mathscr{B}_\mu(x), \max\{\mathscr{A}_m(x), \mathscr{B}_m(x)\}),$$

where $a \sqcup b$ is the join of $a, b \in L$.

Definition 8.5.8 Suppose that $\mathscr{A}, \mathscr{B} : X \to L \times \mathbb{N}$ are two L-multi-fuzzy sets. Then their sum, denoted $\mathscr{A} \uplus \mathscr{B}$, is defined as follows:

$$(\mathscr{A} \uplus \mathscr{B})(x) = (\mathscr{A}_\mu(x) \sqcup \mathscr{B}_\mu(x), \mathscr{A}_m(x) + \mathscr{B}_m(x)).$$

P system is a model of computation, inspired by the way cells live and function. The model is built around the notion of nested compartments surrounded by porous *membranes* (hence the term *membrane computing*). It is quite instructive to think of the membrane structure as a bubbles-inside-bubbles structure, where we have a bubble that contains bubbles, which, in turn, contain other bubbles, etc. Initially, each compartment contains a number of possible repeated objects (i.e. a multiset of objects). Once "computation" commences, the compartments exchange objects according to a number of multiset processing rules that are associated with each compartment; in the simplest case, these processing rules are just multiset rewriting rules. The activity stops when no rule can be applied anymore. The result of the computation is equal to the number of objects that reside within a designated compartment called the *output membrane*.

In [271], a fuzzified version of P systems was presented. The basic idea behind this particular attempt to fuzzify P systems is the substitution of one or all ingredients of a P system with their fuzzy counterparts. From a purely computational point of view, it turns out that only P systems that process multi-fuzzy sets are interesting, the reason being the fact that these systems are capable of computing (positive) real numbers. By replacing the multi-fuzzy sets employed in the first author's previous work with L-multi-fuzzy sets, the computational power of the resulting P systems will not be any "greater," nevertheless, these systems may be quite useful in modeling living organisms. But, things may get really interesting if we consider P systems with L-fuzzy hybrid sets, in general. Here we are going to give only the definition of these systems. For a full exposition see Ref. [275].

Definition 8.5.9 A general fuzzy P system is a construction

$$\Pi_{\text{FD}} = (O, \mu, w^{(1)}, \ldots, w^{(m)}, R_1, \ldots, R_m, i_0),$$

where:

(i) O is an alphabet (i.e. a set of distinct entities) whose elements are called *objects*;
(ii) μ is the membrane structure of degree $m \geq 1$; membranes are injectively labeled with succeeding natural numbers starting with one;
(iii) $w^{(i)} : O \to L \times \mathbb{Z}, 1 \leq i \leq m$, are L-fuzzy hybrid sets over O that are associated with each region i;
(iv) $R_i, 1 \leq i \leq m$, are finite sets of multiset rewriting rules (called *evolution rules*) over O. An evolution rule is of the form $u \to v$, $u \in O^*$ and $v \in O^*_{\text{TAR}}$, where $O_{\text{TAR}} = O \times \text{TAR}$, $\text{TAR} = \{\text{here}, \text{out}\} \cup \{\text{in}_j | 1 \leq j \leq m\}$. The effect of each rule is the removal of the elements of the left-hand side of each rule from the "current" compartment and the introduction of the elements of right-hand side to the designated compartments;
(v) $i_0 \in \{1, 2, \ldots, m\}$ is the label of an elementary membrane (i.e. a membrane that does not contain any other membrane), called the *output* membrane.

Fuzzy multisets have been also used to define *fuzzy multiset grammars* (see Ref. [259] and references therein). These grammars are recognized by special kinds of automata. In what follows, $\Sigma^{\uplus}$ will denote the set of all multisets whose universe is the set Σ. Also, given two multisets $A, B : X \to \mathbb{N}$, then $(A \uplus B)(x) = A(x) + B(x)$ and $(A \ominus B)(x) = \max(0, A(x) - B(x))$.

Definition 8.5.10 A *fuzzy multiset finite automaton* (FMFA) is a quintuple $M = (Q, \Sigma, \delta, \sigma, \tau)$, where

- Q and Σ are nonempty finite sets called the *state-set* and *input-set*, respectively;
- $\delta : Q \times \Sigma^{\uplus} \times Q \to [0, 1]$ is a map called *transition map*;
- $\sigma : Q \to [0, 1]$ is a map called the *fuzzy set of initial states*; and
- $\tau : Q \to [0, 1]$ is a map called the *fuzzy set of final states*.

A *configuration* of a fuzzy multiset finite automaton M is a pair (p, α), where p and α denote the current state and the current multiset, respectively. Transitions in a fuzzy multiset finite automaton are described with the help of configurations. The transition from configuration (p, α) leads to configuration (q, β)

with membership degree $k \in [0,1]$ if there exists a multiset $\gamma \in \Sigma^{\uplus}$ with $\gamma \subseteq \alpha$, $\delta(p,\gamma,q) = k$ and $\beta = \alpha \ominus \gamma$. This transition is written as $(p,\alpha) \xrightarrow{k} (q,\beta)$.

Definition 8.5.11 For a given set Σ, a *fuzzy multiset language* is a map $L : \Sigma^{\uplus} \to [0,1]$. Let $M = (Q,\Sigma,\delta,\sigma,\tau)$ be a fuzzy multiset (finite) automaton. Then, the set

$$L(M) = \{\alpha \in \Sigma^{\uplus} | \alpha \text{ is accepted by } M\}$$

is called the fuzzy multiset language of M. A fuzzy multiset language $L : \Sigma^{\uplus} \to [0,1]$ is *accepted* by an FMFA $M = (Q,\Sigma,\delta,\sigma,\tau)$ if

$$L(\alpha) = \bigvee\{\sigma(q) \wedge \mu_M((q,\alpha) \to\ast (p, 0_\Sigma)) \wedge \tau(p) : q,p \in Q\},$$

for all $\alpha \in \Sigma^{\uplus}$. Furthermore, a fuzzy multiset language $L : \Sigma^{\uplus} \to [0,1]$ is called *regular* if there exists a FMFA M such that $L = L(M)$.

9

Fuzzy Abstract Algebra

Abstract algebra (see Ref. [16] for an accessible overview) is a very important field of mathematics. For this reason, many researchers working on fuzzy mathematics tried to introduce various fuzzy versions of known algebraic structures such as groups, rings, fields, etc.

9.1 Groups, Rings, and Fields

As usually, we begin with a reminder of classical concepts and notions in abstract algebra. Let us start with groupoids and semigroups.

Definition 9.1.1 A nonempty set $\mathfrak{G}$ on which a binary operation $\circledcirc$ is defined, is said to form a groupoid with respect to this operation provided, for arbitrary $a, b \in \mathfrak{G}$, that $a \circledcirc b \in \mathfrak{G}$.

Definition 9.1.2 A nonempty set $\mathfrak{G}$ on which a binary operation $\circledcirc$ is defined, is said to form a semigroup with respect to this operation provided, for arbitrary $a, b, c \in \mathfrak{G}$, that $(a \circledcirc b) \circledcirc c = a \circledcirc (b \circledcirc c)$.

Definition 9.1.3 A nonempty set $\mathfrak{G}$ on which a binary operation $\circledcirc$ is defined, is said to form a monoid with respect to this operation provided, for arbitrary $a, b, c \in \mathfrak{G}$, that $(a \circledcirc b) \circledcirc c = a \circledcirc (b \circledcirc c)$ and there exists an element $e \in \mathfrak{G}$ such that for all $a \in \mathfrak{G}$, $e \circledcirc a = a \circledcirc e = a$.

Groups are more interesting structures.

Definition 9.1.4 A nonempty set $\mathfrak{G}$ on which a binary operation $\circledcirc$ is defined, is said to form a group with respect to this operation provided, for arbitrary $a, b, c \in \mathfrak{G}$, that the following properties hold:

A Modern Introduction to Fuzzy Mathematics, First Edition.
Apostolos Syropoulos and Theophanes Grammenos.

Associative law. $(a \circledcirc b) \circledcirc c = a \circledcirc (b \circledcirc c)$;

Existence of identity element. There is $u \in \mathfrak{G}$ such that $a \circledcirc u = u \circledcirc a = a$ for all $a \in \mathfrak{G}$; and

Existence of inverses. For each $a \in \mathfrak{G}$ there exists $a^{-1} \in \mathfrak{G}$ such that $a \circledcirc a^{-1} = a^{-1} \circledcirc a = u$.

Typically, we write $(\mathfrak{F}, \circledcirc)$ to denote a group. In case the operation is known, we can just write the set and mean the group. In a way, subgroups are to groups what subsets are to sets:

Definition 9.1.5 Suppose that $\mathfrak{G} = \{a, b, c, \dots\}$ is a group with respect to $\circledcirc$. Then, any nonempty subset $\mathfrak{G}'$ of $\mathfrak{G}$ is a *subgroup* of $\mathfrak{G}$ if $\mathfrak{G}'$ is a group with respect to $\circledcirc$.

Example 9.1.1 The set $\mathfrak{G} = \{1, -1, i, -i\}$ is a group with respect to multiplication (why?). Also, the set $\mathfrak{G}' = \{1, -1\}$ is a subgroup of $\mathfrak{G}$ with respect to multiplication.

Remark 9.1.1 An *additive* group is a group whose group operation has properties similar to addition. Similarly, a *multiplicative* group is a group whose group operation has properties similar to multiplication.

Definition 9.1.6 A subgroup $\mathfrak{N}$ of a group $\mathfrak{G}$ is called a *normal subgroup* if for all $g, h \in \mathfrak{G}, g \circledcirc h \in \mathfrak{N} \iff h \circledcirc g \in \mathfrak{N}$.

Definition 9.1.7 Assume that $\mathfrak{G}$ is a finite group with group operation $\circledcirc$, $\mathfrak{H}$ is a subgroup of $\mathfrak{G}$, and a an arbitrary element of $\mathfrak{G}$. Then, the *right coset* $\mathfrak{H}a$ of $\mathfrak{H}$ in $\mathfrak{G}$, generated by a, is the following subset of $\mathfrak{G}$:

$$\mathfrak{H}a = \{h \circledcirc a \,|\, h \in \mathfrak{H}\}.$$

Similarly, the *left coset* $a\mathfrak{H}$ of $\mathfrak{H}$ in $\mathfrak{G}$, generated by a, is the following subset of $\mathfrak{G}$:

$$a\mathfrak{H} = \{a \circledcirc h \,|\, h \in \mathfrak{H}\}.$$

Given an element a of a group, its *order*, which is also known as period length or period of a, is the smallest positive integer m such that

$$\underbrace{a \circledcirc a \circledcirc \cdots \circledcirc a}_{m \text{ times}} = e,$$

where e denotes the identity element of the group. If no such m exists, a is said to have infinite order. A group $(\mathfrak{G}, \circledcirc)$ is called *Abelian* if $a \circledcirc b = b \circledcirc a$ (i.e. the operator $\circledcirc$ is commutative). Otherwise, it is called *non-Abelian* or *non-commutative*.

Definition 9.1.8 A *torsion* subgroup $\mathfrak{A}_T$ of an Abelian group $\mathfrak{A}$ is the subgroup of $\mathfrak{A}$ that consists of all elements that have finite order (the *torsion elements* of $\mathfrak{A}$). An Abelian group $\mathfrak{A}$ is called a *torsion* (or periodic) group if every element of $\mathfrak{A}$ has finite order, and it is called *torsion-free* if every element of $\mathfrak{A}$ except the identity has infinite order.

An Abelian group $\mathfrak{G}$ that is neither torsion nor torsion-free is called a *mixed* Abelian group.

Definition 9.1.9 Assume that $(\mathfrak{G}, \circledcirc)$ and $(\mathfrak{F}, \boxtimes)$ are two groups. Then, a *homomorphism* of $\mathfrak{G}$ to $\mathfrak{F}$ is a mapping $\theta : \mathfrak{G} \rightarrow \mathfrak{F}$ such that

(i) every $g \in \mathfrak{G}$ has a unique image $f \in \mathfrak{F}$;
(ii) if $\theta(g_1) = f_1$ and $\theta(g_2) = f_2$, then $\theta(g_1 \circledcirc g_2) = f_1 \boxtimes f_2$.

Rings are sets equipped with two operations.

Definition 9.1.10 A nonempty set $\mathfrak{T}$ together with two binary operations $\boxplus$ and $\boxtimes$ is a *ring* if the following properties hold for all $a, b, c \in \mathfrak{T}$:

Associative law of addition. $(a \boxplus b) \boxplus c = a \boxplus (b \boxplus d)$;
Commutative law of addition. $a \boxplus b = b \boxplus a$;
Existence of an additive identity (zero). There is $0 \in \mathfrak{T}$ such that $a \boxplus 0 = a$;
Existence of additive inverses. For each $a \in \mathfrak{T}$ there is $-a \in \mathfrak{T}$ such that $a \boxplus (-a) = 0$;
Associative law of multiplication. $(a \boxtimes b) \boxtimes c = a \boxtimes (b \boxtimes c)$; and
Distributive laws. $a \boxtimes (b \boxplus c) = a \boxtimes b \boxplus a \boxtimes c$ and $(b \boxplus c) \boxtimes a = b \boxtimes a \boxplus c \boxtimes a$.

A ring is written as a triple $(\mathfrak{T}, \boxplus, \boxtimes)$.

Definition 9.1.11 Suppose that $(\mathfrak{T}, \boxplus, \boxtimes)$ is a ring. Then, if $\emptyset \neq \mathfrak{S} \subset \mathfrak{T}$ and $(\mathfrak{S}, \boxplus, \boxtimes)$ is a ring, then $(\mathfrak{S}, \boxplus, \boxtimes)$ is a *subring* of $(\mathfrak{T}, \boxplus, \boxtimes)$.

Definition 9.1.12 A homomorphism (isomorphism) of the additive group of a ring $\mathscr{R}$ into (onto) the additive group of a ring $\mathscr{R}'$ which also preserves the second operation, multiplication, is called a homomorphism (isomorphism) of $\mathscr{R}$ into (onto) $\mathscr{R}'$.

Definition 9.1.13 A ring $(\mathfrak{F}, \boxplus, \boxtimes)$ whose non-zero elements form an Abelian multiplicative group (i.e. $a \boxtimes b = b \boxtimes a$) is called a *field*.

Many interesting algebraic structures are special cases of vector spaces.

Definition 9.1.14 Assume that $\mathscr{F}$ is a field and V is an Abelian additive group such that there is a scalar multiplication of V by $\mathscr{F}$ which associates with each $s \in \mathscr{F}$ and $\xi \in V$ the element $s\xi \in V$. Then, V is called a *vector space* (or *linear space*) over $\mathscr{F}$, with 1 the unity of $\mathscr{F}$, and

(i) $s(\xi + \eta) = s\xi + s\eta$ (iii) $s(t\xi) = (st)\xi$
(ii) $(s + t)\xi = s\xi + t\xi$ (iv) $1\xi = \xi$

hold for all $s, t \in \mathscr{F}$ and $\xi, \eta \in V$.

Example 9.1.2 The set $\mathbb{R}^n$ together with the following operations

$$\begin{aligned}\mathbf{x} + \mathbf{y} &= (x_1, x_2, \dots, x_n) + (y_1, y_2, \dots, y_n)\\ &= (x_1 + y_1, x_2 + y_2, \dots, x_n + y_n),\\ \kappa\mathbf{x} &= (\kappa x_1, \kappa x_2, \dots, \kappa x_n),\end{aligned}$$

where $\mathbf{x}, \mathbf{y} \in \mathbb{R}^n$ and $\kappa \in \mathbb{R}$, is a real vector space.

Definition 9.1.15 A nonempty subset U of a vector space V over $\mathscr{F}$ is a subspace of V, provided U is itself a vector space over $\mathscr{F}$ with respect to the operations defined on V.

A *linear transformation* of a vector space V into a vector space W over the same field $\mathscr{F}$ is a map $T : V \to W$ for which

(i) $T(\xi_i + \xi_j) = T(\xi_i) + T(\xi_j)$ for all $\xi_i, \xi_j \in V$; and
(ii) $T(s\xi_i) = sT(\xi_i)$ for all $\xi_i \in V$ and $s \in \mathscr{F}$.

Assume that V, W, and Y are vector spaces over the same field $\mathscr{F}$. Then, a function $F : V \times W \to Y$ is a *bilinear transformation* if the following conditions hold:

(i) for each $\mathbf{w} \in W$, the function from V to Y given by $\mathbf{v} \mapsto F(\mathbf{v}, \mathbf{w})$ is a linear transformation;
(ii) for each $\mathbf{v} \in V$, the function from W to Y given by $\mathbf{w} \mapsto F(\mathbf{v}, \mathbf{w})$ is a linear transformation.

A bilinear transformation F from $V \times W$ to $\mathscr{F}$ is called a *bilinear form*. A multiplication or product on a vector space V is a bilinear map from $V \times V$ to V.

Definition 9.1.16 Assume that V is a vector space over $\mathbb{R}$. Then, a *norm* $\|\cdot\|$ on V is a function that assigns a real number to each element $v \in V$, satisfying

(i) $\|v\| \geq 0$ and $v = 0$ if and only if $v = 0$;
(ii) $\|av\| = |a|\,\|v\|$ for any scalar $a \in \mathbb{R}$; and
(iii) for all $v, w \in V$,

$$\|v + w\| \leq \|v\| + \|w\|, \quad \text{(triangle inequality)}.$$

The $(V, \|\cdot\|)$ is called a *normed vector space*. Obviously, different norms define different normed vector spaces.

Definition 9.1.17 Given a normed vector space $(V, \|\cdot\|)$, then the function defined by

$$d(v, w) = \|v - w\|, \quad \text{for all } v, w, \in V,$$

is a *metric* on V called the *metric generated by* $\|\cdot\|$ (see Section 10.1 for more details about metrics).

Suppose that V is a vector space over the field $\mathbb{K}$ ($\mathbb{K}$ is either $\mathbb{R}$ or $\mathbb{C}$, the set of complex numbers). Then, a function $\phi : V \times V \to \mathbb{K}$ is called *inner product* if it satisfies the following four properties:

(i) for all $\mathbf{x}, \mathbf{y} \in V$, $\overline{\phi(\mathbf{x}, \mathbf{y})} = \phi(\mathbf{y}, \mathbf{x})$, where $\overline{x}$ is the complex conjugate of x and obviously, for any real number x, it holds that $\overline{x} = x$;
(ii) for all $\mathbf{x}, \mathbf{y}, \mathbf{z} \in V$, $\phi(\mathbf{x} + \mathbf{y}, \mathbf{z}) = \phi(\mathbf{x}, \mathbf{z}) + \phi(\mathbf{y}, \mathbf{z})$;
(iii) for all $\kappa \in \mathbb{K}$ and all $\mathbf{x}, \mathbf{y} \in V$, $\phi(\kappa\mathbf{x}, \mathbf{y}) = \kappa\phi(\mathbf{x}, \mathbf{y})$; and
(iv) for all $\mathbf{x} \in V$, $\phi(\mathbf{x}, \mathbf{x}) \geq 0$ and $\phi(\mathbf{x}, \mathbf{x}) =$ if and only if $\mathbf{x} = \mathbf{0}$.

The pair (V, ϕ) is called *inner product space*. Usually, the inner product of $\mathbf{x}$ and $\mathbf{y}$ is written as $\langle \mathbf{x}, \mathbf{y} \rangle$.

Theorem 9.1.1 *Let $(V, \langle\ ,\ \rangle)$ be an inner product space. Then, the function $\|\ \| : V \to \mathbb{R}$, where for every $\mathbf{x} \in V$, $\|\mathbf{x}\| = \sqrt{\langle \mathbf{x}, \mathbf{x} \rangle}$, is a norm on V.*

The norm described in the previous theorem is known as the *naturally defined norm*.

Banach spaces, Hilbert spaces, and Lie algebras [254] are special cases of vector spaces. However, we will not discuss Banach and Hilbert spaces in this chapter since one needs to have a basic understanding of metric spaces. So they will be presented in Chapter 10.

Definition 9.1.18 A *Lie algebra* consists of a (finite dimensional) vector space, over a field $\mathscr{F}$, and a multiplication on the vector space (denoted by [ ], pronounced "bracket," the image of a pair $(\mathbf{x}, \mathbf{y})$ denoted by $[\mathbf{xy}]$ or $[\mathbf{x}, \mathbf{y}]$) with the properties

(i) $[\mathbf{x}, \mathbf{x}] = \mathbf{0}$ (ii) $[\mathbf{x}, [\mathbf{y}, \mathbf{z}]] + [\mathbf{y}, [\mathbf{z}, \mathbf{x}]] + [\mathbf{z}, [\mathbf{x}, \mathbf{z}]] = \mathbf{0}$

9.2 Fuzzy Groups

Rosenfeld [249] introduced fuzzy groupoids and fuzzy groups (see Ref. [223] for a thorough presentation of fuzzy group theory). Let us start with his definition of fuzzy groupoids.

Definition 9.2.1 Suppose that $(X, \circledcirc)$ is a groupoid and $A : X \to [0, 1]$ is a fuzzy subset of X. Then, A is a *fuzzy subgroupoid* of X if, for all $x, y \in X$,

$$A(x \circledcirc y) \geq \min(A(x), A(y)).$$

A will be called a *fuzzy left ideal*, if $A(x \circledcirc y) \geq A(y)$; a *fuzzy right ideal*, if $A(x \circledcirc y) \geq A(x)$; and a *fuzzy ideal*, if it is a fuzzy left and a fuzzy right ideal, that is, if

$$A(x \circledcirc y) \geq \max(A(x), A(y)).$$

Fuzzy subgroups are defined similarly:

Definition 9.2.2 Suppose that $(X, \circledcirc)$ is a group and A is a fuzzy subgroupoid of X. Then, A will be called a *fuzzy subgroup* of X if

$$A(x^{-1}) \geq A(x), \quad \text{for all } x \in X.$$

The condition of the previous definition can be restated as follows:

$$A(x \circledcirc y^{-1}) \geq \min(A(x), A(y)), \quad \text{for all } x, y \in X.$$

By replacing min with some other fuzzy t-norm, one gets a more general structure called a t-fuzzy group [11]. In fact, this generalization is necessary since the fuzzy subgroups generated with min have a number of problems (see Ref. [11] for details).

Definition 9.2.3 Suppose that $(X, \circledcirc)$ is a group and A is a fuzzy subgroup of X. If

$$A(x) = A(y \circledcirc x \circledcirc y^{-1}), \quad \text{for all } x, y \in X,$$

then A is called a *normal fuzzy subgroup* of X.

Proposition 9.2.1 *Assume that A is a fuzzy subgroup of a group $\mathfrak{G}$. Then, A is a normal fuzzy subgroup of $\mathfrak{G}$ if and only if ${}^{\alpha}A$ is a normal subgroup of $\mathfrak{G}$ for all $\alpha \in [0, 1]$.*

Definition 9.2.4 Assume that A is a fuzzy subgroup of $\mathfrak{G}$ and $x \in \mathfrak{G}$. Then, xA is the *left fuzzy coset* of A in $\mathfrak{G}$, where

$$xA(u) = \begin{cases} \bigvee_{z\in\mathfrak{G}} A(z), & x \circledcirc z = u, \; u \in \mathfrak{G}, \\ 0, & \text{if there is no such } z. \end{cases}$$

Similarly, Ax is the *right fuzzy coset* of A in $\mathfrak{G}$.

So far we have presented fuzzy subgroups that are generated from the set of the group. However, it is quite possible to start again from an ordinary group and fuzzify the binary operation. Indeed, Mustafa Demirci [94] followed this path to define his *vague groups*. In order to introduce vague groups, we have to first introduce a few new notions.

Definition 9.2.5 A mapping $E_X : X \times X \to [0,1]$ is called a *fuzzy equality* on X if and only if the following properties hold:

(i) $E_X(x,y) = 1 \iff x = y$, for all $x, y \in X$;
(ii) $E_X(x,y) = E_X(y,x)$, for all $X, y \in X$; and
(iii) $E_X(x,y) \wedge E_X(y,z) \leq E_X(x,z)$, for all $x, y, z \in X$.

For $x, y \in X$, the real number $E_X(x,y)$ is the degree to which x is equal to y. We can use this definition to give a functional definition of crisp equality for all $x, y \in X$:

$$E_X^*(x,y) = \begin{cases} 1, & \text{if } x = y, \\ 0, & \text{otherwise.} \end{cases}$$

Definition 9.2.6 Suppose that X and Y are two nonempty sets, and E_X and E_Y are two fuzzy equalities on X and Y, respectively. Then, a fuzzy relation f in X and Y is called a *fuzzy function* from X to Y with respect to the fuzzy equalities E_X and E_Y if and only if the following properties hold:

(i) for all $x \in X$ there is a $y \in Y$ such that $f(x,y) > 0$; and
(ii) for all $x, y \in X$ and for all $z, w \in Y, f(x,z) \wedge f(y,w) \wedge E_x(x,y) \leq E_Y(z,w)$.

Typically a fuzzy function f from X to Y is denoted by $f : X \to Y$. In addition, a fuzzy function f is called a *strong* fuzzy function if and only if, in addition, the following property holds:

(iii) for all $x \in X$, there is a $y \in Y$ such that $f(x,y) = 1$.

Remark 9.2.1 If E_X, E_Y, and f are such that $E_X = E_X^*$, $E_Y = E_Y^*$, and for all $x, y \in X, f(x,y) \in \{0,1\}$, then obviously f is a crisp function.

Definition 9.2.7 A strong fuzzy function $f : X \times X \to X$ with respect to a fuzzy equality $E_{X \times X}$ in $X \times X$ and a fuzzy equality E_X in X is a *vague binary operation* on X with respect to $E_{X \times X}$ and E_X. Such an operation is *transitive of the first order* if and only if for all $a, b, c, d \in X$

$$f(a,b,c) \wedge E_X(c,d) \leq f(a,b,d).$$

A vague binary operation f is *transitive of the second order* if and only if for all $a, b, c, d \in X$

$$f(a,b,c) \wedge E_X(b,d) \leq f(a,d,c).$$

A binary operation $\circledcirc : X \times X \to X$ can be thought of as a special vague binary operation $\circledcirc$ on X with respect to $E^*_{X\times X}$ and E^*_X satisfying the condition $\circledcirc(X \times X \times X) \subseteq \{0, 1\}$. This simply means that $a \circledcirc b = c$ can be expressed as $\circledcirc(a, b, c) = 1$.

Definition 9.2.8 Assume that $\circledcirc$ is a vague binary operation on X with respect to a fuzzy equality $E_{X\times X}$ in $X \times X$ and a fuzzy equality E_X in X. Then,

(i) X together with $\circledcirc$ is the *vague semigroup* $(X, \circledcirc)$ if and only if for all $a, b, c, d, m, q, w \in X$, the condition that follows is satisfied:

VG.1 $\quad \circledcirc(b, c, d) \wedge \circledcirc(a, d, m) \wedge \circledcirc(a, b, q) \wedge \circledcirc(q, c, w) \leq E_X(m, w)$;

(ii) a vague semigroup $(X, \circledcirc)$ is a *vague monoid* if and only if there exists an identity element $e \in X$ such that for all $a \in X$

VG.2 $\quad \circledcirc(e, a, a) \wedge \circledcirc(a, e, a) = 1$;

(iii) a vague monoid $(X, \circledcirc)$ is a *vague group* if and only if for all $a \in X$, there is an inverse element a^{-1} such that

VG.3 $\quad \circledcirc(a^{-1}, a, e) \wedge \circledcirc(a, a^{-1}, e) = 1$; and

(iv) a vague semigroup $(X, \circledcirc)$ is Abelian if and only if for all $a, b, n, w \in X$

VG.4 $\quad \circledcirc(a, b, m) \wedge \circledcirc(b, a, w) \leq E_X(m, w)$.

Example 9.2.1 Suppose that $(X, \boxtimes)$ is a crisp group, and α, β, θ are real numbers such that $0 < \theta \leq \alpha \leq \beta < 1$. In addition, for all $x, y, z, w \in X$, we define the fuzzy equalities in X and $X \times X$ as follows:

$$E_X(x, y) = \begin{cases} 1, & \text{if } x = y, \\ \beta, & \text{otherwise,} \end{cases} \quad \text{and } E_{X\times X}((x, y), (z, w)) = \begin{cases} 1, & \text{if } (x, y) = (z, w), \\ \alpha, & \text{otherwise,} \end{cases}$$

and the fuzzy relation $\circledcirc$ in $X \times X \times X$ is defined as follows:

$$\circledcirc(x, y, z) = \begin{cases} 1, & \text{if } z = x \boxtimes y, \\ \theta, & \text{otherwise.} \end{cases}$$

Then, we can easily prove that $(X, \circledcirc)$ is a vague semigroup.

The example that follows is borrowed from [255].

Example 9.2.2 Assume that $\alpha \in [0, 1)$ and

$$x^\bullet = \min\left\{\frac{1}{\max(x, 1)}, \min(x, 1)\right\}$$

for all $x \in \mathbb{R}^+$ (i.e. the set of positive reals including zero). Also, for all $x, y, u, v, z \in \mathbb{R}^+$,

$$E_{\mathbb{R}^+}(x,y) = \begin{cases} 1, & \text{when } x = y, \\ \max\left(\alpha, \min\left(\dfrac{1}{x^\bullet}, \dfrac{1}{y^\bullet}\right)\right), & \text{otherwise,} \end{cases}$$

and

$$E_{\mathbb{R}^+\times\mathbb{R}^+}((x,y),(u,v)) = \begin{cases} 1, & \text{if } x = u \text{ and } y = v, \\ \max\{\alpha, \min[(x^2)^\bullet, (y^2)^\bullet, (u^2)^\bullet, (v^2)^\bullet]\}, & \text{otherwise,} \end{cases}$$

are fuzzy equalities in $\mathbb{R}^+$ and $\mathbb{R}^+ \times \mathbb{R}^+$, respectively. In addition, for $x, y, z, u, v, \in \mathbb{Q}^+$,

$$E_{\mathbb{Q}^+}(x,y) = E_{\mathbb{R}^+}(x,y)|\mathbb{Q}^+$$
$$E_{\mathbb{Q}^+\times\mathbb{Q}^+}((x,y),(u,v)) = E_{\mathbb{R}^+\times\mathbb{R}^+}((x,y),(u,v))|\mathbb{Q}^+$$

as fuzzy equalities in $\mathbb{Q}^+$ and $\mathbb{Q}^+ \times \mathbb{Q}^+$, respectively. For $n \in \mathbb{N}^+$, we get the fuzzy operators $\circledast$ and $\odot$ in $\mathbb{R}^+ \times \mathbb{R}^+ \times \mathbb{R}^+$ and $\mathbb{Q}^+ \times \mathbb{Q}^+ \times \mathbb{Q}^+$, respectively, defined by

$$\circledast(x,y,z) = \begin{cases} 1, & \text{if } z = x \cdot y, \\ \alpha \cdot \min\{x^\bullet, y^\bullet, z^\bullet\}, & \text{otherwise} \end{cases}$$

and

$$\odot_n(x,y,z) = \begin{cases} 1, & \text{if } z = x \cdot y, \\ \dfrac{\alpha}{n} \cdot \min\{x^\bullet, y^\bullet, z^\bullet\}, & \text{otherwise} \end{cases}$$

are vague binary operators in $\mathbb{R}^+$ and $\mathbb{Q}^+$, respectively. In addition, $(\mathbb{R}^+, \circledast)$ and $(\mathbb{Q}^+, \odot)$ are vague groups.

Proposition 9.2.2 *Assume that $(X, \circledcirc)$ is a vague group. Then, there is a crisp binary operator $\sigma_\circledcirc$ on X such that $(X, \sigma_\circledcirc)$ is a crisp group.*

Assume that $(X, \circledcirc)$ is a vague semigroup. Then, if there is either $e_L \in X$ or $e_R \in X$ such that for all $a \in X$

$$\circledcirc(e_L, a, a) = 1 \quad \text{or} \quad \circledcirc(a, e_R, a) = 1,$$

we say that e_L and e_R are the left or the right identity element of $(X, \circledcirc)$, respectively. In addition, given a vague semigroup $(X, \circledcirc)$ that has either a left identity element e_L or a right identity element e_R, then if for each $a \in X$, there is either an element $a_L^{-1} \in X$ or an element $a_R^{-1} \in X$ such that

$$\circledcirc(a_L^{-1}, a, e_L) = 1 \quad \text{or} \quad \circledcirc(a, a_R^{-1}, e_R) = 1,$$

respectively, where a_L^{-1} or a_R^{-1} are the left or right inverse of a, respectively.

Proposition 9.2.3 *Suppose that $(X, \circledcirc)$ is a vague semigroup with respect to fuzzy equalities $E_{X \times X}$ in $X \times X$ and E_X in X. If $(X, \circledcirc)$ has either a left identity element e_L or a right identity element e_R, and for each $a \in X$, there is either a left inverse a_L^{-1} or a right inverse a_R^{-1} of a, then for each $c \in X$*

$$\circledcirc(c, c, c) \leq E_X(c, e_L) \quad or \quad \circledcirc(c, c, c) \leq E_X(c, e_R),$$

respectively.

Proof: Suppose that $(X, \circledcirc)$ has a left identity element e_L and for each $a \in X$, its left inverse a_L^{-1} exists. Then, since $\circledcirc(e_L^{-1}, c, e_L) = \circledcirc(e_L, c, c) = 1$ for all $c \in X$, and using property VG.1 from Definition 9.2.8, we observe that

$$\circledcirc(c, c, c) = \circledcirc(c, c, c) \wedge \circledcirc(c_L^{-1}, c, e_L) \wedge \circledcirc(c_L^{-1}, c, e_L) \wedge \circledcirc(c_L, c, e) \leq E_X(c, e_L).$$

In the case of e_R, we can obtain the corresponding inequality in a similar way. □

Let us proceed with vague subgroups and vague homomorphisms.

Assume that $(X, \circledcirc)$ is a vague group and $Y \subset X$, and $\otimes$ a vague binary operation on X. Then, Y is *closed* under $\otimes$ if and only if

$$\otimes(a, b, c) = 1 \Rightarrow c \in B, \quad \forall a, \forall b, \in B, \; \forall c \in X.$$

Definition 9.2.9 Suppose that $(X, \circledcirc)$ is a vague group with respect to a fuzzy equality $E_{X \times X}$ in $X \times X$ and a fuzzy equality E_X in X. Also, suppose that H is a nonempty, crisp subset of X that is vague closed under $\circledcirc$. Then, H is a *vague subgroup* of X if and only if $(H, \otimes|_{H \times H \times H})$ is itself a vague group.

We state the following theorem without proof.

Theorem 9.2.1 *Suppose that $(X, \circledcirc)$ is a vague group with respect to a fuzzy equality $E_{X \times X}$ in $X \times X$ and a fuzzy equality E_X in X. Then, $X \subset H \neq \emptyset$ is a vague subgroup of X if for all $a, b \in H$ and for all $c \in X$, $\circledcirc(a, b^{-1}, c) = 1$ implies that $c \in H$.*

Definition 9.2.10 Assume that $(X, \circledcirc)$ and $(Y, \otimes)$ are two vague semigroups. A crisp function $f : X \to Y$ is a *vague homomorphism* if and only if

$$\circledcirc(a, b, c) \leq \otimes(f(a), f(b), f(c)), \quad \text{for all } a, b, c \in X.$$

9.3 Abelian Fuzzy Subgroups

Abelian fuzzy subgroups were introduced by Lu Tu and Gu Wenxiang [289].

Definition 9.3.1 Suppose that A is a fuzzy subset of a group $\mathfrak{G}$ and ${}^{\alpha}A$ is an Abelian subgroup of $\mathfrak{G}$ for every $\alpha \in [0, 1]$. Then, A is called an *Abelian fuzzy subgroup of* $\mathfrak{G}$.

Corollary 9.3.1 *A fuzzy subgroup A of a group G is an Abelian fuzzy subgroup if and only if G is an Abelian group.*

Proof: If A is an Abelian fuzzy group, then ${}^0A = \mathfrak{G}$ is an Abelian group of $\mathfrak{G}$. Conversely, if $\mathfrak{G}$ is an Abelian group, then for any ${}^\alpha A$, $\alpha \in [0,1]$, ${}^\alpha A \subset {}^0A$ is the subgroup of ${}^0A$, thus ${}^\alpha A$ is an Abelian subgroup of ${}^0A$. □

Corollary 9.3.2 *If A is an Abelian fuzzy subgroup of group G, then ${}^\alpha A$ is a normal subgroup of $\mathfrak{G}$ for every $\alpha \in [0,1]$, thus A is a normal fuzzy subgroup of $\mathfrak{G}$.*

Proof: From group theory we know that the subgroup of an Abelian group $\mathfrak{G}$ is the normal subgroup of G. □

Corollary 9.3.3 *A product of finite Abelian fuzzy subgroups of a group G is an Abelian fuzzy subgroup of G.*

Proof: Suppose that $A_1, A_2, \ldots, A_n$ are Abelian fuzzy subgroups of $\mathfrak{G}$. Then, the algebraic product $A_1A_2\cdots A_n$ is a fuzzy subgroup of G. Since $\mathfrak{G}$ is an Abelian group, this means that the algebraic product $A_1A_2\cdots A_n$ is an Abelian fuzzy subgroup of $\mathfrak{G}$. □

Definition 9.3.2 Suppose that A is an Abelian fuzzy subgroup of a group $\mathfrak{G}$, for all $\alpha \in [0,1]$. Then,

(i) if the elements of ${}^\alpha A$ have a finite order, then A is called a *torsion* fuzzy subgroup of $\mathfrak{G}$;
(ii) if the elements of ${}^\alpha A$ have infinite order, then A is called a *torsion-free* fuzzy subgroup of $\mathfrak{G}$; and
(iii) if A is neither a torsion fuzzy subgroup of $\mathfrak{G}$ nor a torsion-free fuzzy subgroup of $\mathfrak{G}$, then it is called a *mixed* fuzzy subgroup of $\mathfrak{G}$.

Proposition 9.3.1 *Let A be an Abelian fuzzy subgroup of a group $\mathfrak{G}$. Then,*

(i) A is a torsion (torsion-free, mixed, respectively) fuzzy subgroup if and only if $\mathfrak{G}$ is a torsion (torsion-free, mixed, respectively) Abelian group; and
(ii) if A is a mixed fuzzy subgroup of $\mathfrak{G}$, then there exists $i \in [0,1]$, and ${}^iA \subset \mathfrak{G}$ is a mixed Abelian group, while ${}^jA$ is a mixed Abelian group for every $j \in [0,1]$.

In particular, if ${}^1A$ is a mixed Abelian group, then ${}^\alpha A$ is a mixed Abelian group for every $\alpha \in [0,1]$.

Proof: Since A is a torsion (torsion-free, mixed, respectively) fuzzy subgroup of $\mathfrak{G}$, ${}^0A = \mathfrak{G}$ is a torsion (torsion-free, mixed, respectively) Abelian group. Conversely, if

$\mathfrak{G}$ is a torsion (torsion-free, mixed, respectively) Abelian group, for any $\alpha \in [0, 1]$, ${}^{0}A \subset \mathfrak{G}$ is a torsion (torsion-free, mixed, respectively) Abelian group. Thus, A is a torsion (torsion-free, respectively) Abelian fuzzy group. A is a mixed Abelian fuzzy group when $\mathfrak{G}$ is a mixed Abelian group. □

Definition 9.3.3 Assume that A is a fuzzy subset of a nonempty set X, $Y \subset X$, and for every $x \in X$

$$A|_Y(x) = \begin{cases} A(x), & \text{when } x \in Y, \\ \text{undefined}, & \text{otherwise}, \end{cases} \quad \text{and}$$

$$(A|_Y)^0(x) = \begin{cases} A(x), & \text{when } x \in Y, \\ 0, & \text{otherwise.} \end{cases}$$

Then, $A|_Y$ is the restriction of A in Y and $(A|_Y)^0$ is the zero-extension of $A|_Y$.

Definition 9.3.4 Suppose that A is a fuzzy subgroup of a group $\mathfrak{G}$ and B is a fuzzy subgroup of a subgroup $\mathfrak{H}$ of the group $\mathfrak{G}$. Then, for all $\alpha \in [0, 1]$

(i) if ${}^{\alpha}B$ is a subgroup of ${}^{\alpha}A$, then B is a subgroup of A; and
(ii) if ${}^{\alpha}B$ is a normal subgroup of ${}^{\alpha}A$, then B is a normal subgroup of A.

Proposition 9.3.2 *If A is a fuzzy subgroup of $\mathfrak{G}$ and $\mathfrak{H}$ is a subgroup of $\mathfrak{G}$, then*

(a) $A|_{\mathfrak{H}}$ is a fuzzy subgroup of $\mathfrak{H}$ and $(A|_{\mathfrak{H}})^0$ is a fuzzy subgroup of $\mathfrak{G}$;
(b) both $A|_{\mathfrak{H}}$ and $(A|_{\mathfrak{H}})^0$ are subgroups of A; and
(c) when $\mathfrak{G}$ is an Abelian group, both $A|_{\mathfrak{H}}$ and $(A|_{\mathfrak{H}})^0$ are normal subgroups of A.

Proof:

(a) For any $\alpha \in [0, 1]$,

$${}^{\alpha}A|_{\mathfrak{H}} = \{x \mid x \in \mathfrak{H} \text{ and } A(x) \geq \alpha\} \subset {}^{\alpha}A,$$

and $e = {}^{\alpha}A|_{\mathfrak{H}} \neq \emptyset$, where e is the identity of $\mathfrak{G}$. For any $x, y \in {}^{\alpha}A|_{\mathfrak{H}}$, $x \circledcirc y^{-1} \in \mathfrak{H}$ since $\mathfrak{H}$ is a subgroup of $\mathfrak{G}$, and

$$A|_{\mathfrak{H}}(x \circledcirc y^{-1}) = A(x \circledcirc y^{-1}) \geq \min\{A(x), A|_{\mathfrak{H}}(y)\} = \alpha.$$

Therefore, $x \circledcirc y^{-1} \in {}^{\alpha}A|_{\mathfrak{H}}$, that is, ${}^{\alpha}A|_{\mathfrak{H}}$ is a subgroup of $\mathfrak{H}$. Thus, $A|_{\mathfrak{H}}$ is a fuzzy subgroup of $\mathfrak{H}$ because of Proposition 9.2.1.

(b) For all $\alpha \in [0, 1]$, both ${}^{\alpha}A|_{\mathfrak{H}}$ and ${}^{\alpha}(A|_{\mathfrak{H}})^0$ are subgroups of ${}^{\alpha}A$.

(c) If $\mathfrak{G}$ is an Abelian group, then for any $\alpha \in [0, 1]$ both ${}^{\alpha}A|_{\mathfrak{H}}$ and ${}^{\alpha}(A|_{\mathfrak{H}})^0$ are normal subgroups of ${}^{\alpha}A$.

□

9.4 Fuzzy Rings and Fuzzy Fields

Wang-Jin Liu [193] introduced fuzzy subrings back in 1982.

Definition 9.4.1 Assume that "$\times$" is a binary operation on a set X, and $A, B : X \to [0,1]$ are fuzzy subsets of X. Then, the binary operator "$\boxtimes$" induced by "$\times$" is defined as follows:

$$(A \boxtimes B)(x) = \begin{cases} \bigvee\limits_{y\times z=x} \min\{A(y), B(z)\}, & \text{if } y \times z = x \quad \text{for } y, z \in X, \\ 0, & \text{if } y \times z \neq x \quad \text{for any } y, z \in X. \end{cases}$$

Clearly, $A \boxtimes B$ is a fuzzy subset of X.

Definition 9.4.2 Suppose that $(X, +, \times)$ is a ring and $A : X \to [0,1]$ is not the empty fuzzy subset of X. Then, A is a *fuzzy subring* of X if $(A, \boxplus)$ is a fuzzy subgroup and $(A, \boxtimes)$ is a fuzzy subgroupoid, where $\boxplus$ and $\boxtimes$ are binary operations induced by $+$ and $\times$, respectively.

Proposition 9.4.1 *Suppose that $(X, +, \times)$ is a ring and A a fuzzy subset of X different from the empty fuzzy subset. Then, A is a fuzzy subring of X if and only if for all $x, y \in X$,*

$$A(x + y) \geq \min\{A(x), A(y)\} \text{ and}$$
$$A(x \times y) \geq \min\{A(x), A(y)\}.$$

Definition 9.4.3 Suppose that $(X, +, \times)$ is a ring. Then, a fuzzy subring A is a *fuzzy left ideal* of X, if $A(x \times y) \geq A(y)$ for all $x, y \in X$. A is a *fuzzy right ideal* of X, if $A(x \times y) \geq A(x)$ for all $x, y \in X$. A is a *fuzzy ideal* of X, if it is both a fuzzy left and a fuzzy right ideal.

Let X and Y be two rings and let $f : X \to Y$ be a ring homomorphism. If A is a fuzzy subring (ideal) of X, then so is $f^{\rightarrow}(A)$, and if B is a fuzzy subring (ideal) of Y, then so is $f^{\leftarrow}(B)$.

Proposition 9.4.2 *Suppose that X is a ring and A is a nonempty fuzzy subset of X. Then, A is a fuzzy left or right ideal of X if and only if for all $x, y \in X$*

(i) $A(x + y) \geq \min\{A(x), A(y)\}$;
(ii) $A(x \times y) \geq \max\{A(x), A(y)\}$, *that is,* $A(x \times y) \geq A(y)$ *or* $A(x \times y) \geq A(x)$.

Vague rings [255] are to rings what vague groups are to groups.

Definition 9.4.4 Assume that $E_{H\times H}$ and E_H are fuzzy equalities in $H \times H$ and H, respectively. Also, assume that $\lozenge$, $\blacklozenge$ are two vague binary operations in H. Then, the triple $(H, \lozenge, \blacklozenge)$ is a *vague ring* with respect to $E_{H\times H}$ and E_H if the following three conditions are satisfied:

(VR.1) $(H, \lozenge)$ is an Abelian vague group;
(VR.2) $(H, \blacklozenge)$ is a vague semigroup; and
(VR.3) $(H, \lozenge, \blacklozenge)$ satisfies distributive laws, that is, for all $a, b, c, d, t, x, y, z \in H$,

$$\blacklozenge(x,y,a) \wedge \blacklozenge(x,z,b) \wedge \lozenge(a,b,c) \wedge \lozenge(y,z,d) \wedge \blacklozenge(x,d,t) \leq E_H(t,c),$$
$$\blacklozenge(x,z,a) \wedge \blacklozenge(y,z,b) \wedge \lozenge(a,b,c) \wedge \lozenge(x,y,d) \wedge \blacklozenge(d,z,t) \leq E_H(t,c).$$

In addition,

(VR.4) A vague ring $(H, \lozenge, \blacklozenge)$ is a vague ring which identifies if there is an element $e_{\blacklozenge} \in H$ such that

$$\blacklozenge(x, e_{\blacklozenge}, x) \wedge \blacklozenge(e_{\blacklozenge}, x, x) = 1$$

for all $x \in H$.

(VR.5) A vague ring $(H, \lozenge, \blacklozenge)$ is Abelian if

$$\blacklozenge(x,y,s) \wedge \blacklozenge(y,x,t) \leq E_H(s,t),$$

for all $x, y, s, t \in H$.

Example 9.4.1 Assume that $(H, \boxplus, \boxtimes)$ is a ring. Also, for all $x, y, a, b \in H$ and $\alpha, \beta, \gamma, \nu \in \mathbb{R}$ such that $0 \leq \nu \leq \gamma \leq \beta \leq \alpha < 1$, assume that

$$E_H(a,b) = \begin{cases} 1, & \text{when } a = b, \\ \alpha, & \text{otherwise,} \end{cases} \quad \text{and}$$

$$E_{H\times H}((a,b),(x,y)) = \begin{cases} 1, & \text{when } (a,b) = (x,y), \\ \beta, & \text{otherwise,} \end{cases}$$

are fuzzy equalities in H and $X \times H$, respectively. In addition, assume that $\lozenge : H \times H \to H$ and $\blacklozenge : H \times H \to H$ are vague binary operations defined as follows:

$$\lozenge(a,b,c) = \begin{cases} 1, & \text{when } a \boxplus b = c, \\ \gamma, & \text{otherwise,} \end{cases} \quad \text{and} \quad \blacklozenge(a,b,c) = \begin{cases} 1, & \text{when } a \boxtimes b = c, \\ \nu, & \text{otherwise,} \end{cases}$$

Now, it is easy to show that $(H, \lozenge, \blacklozenge)$ is a vague ring.

Proposition 9.4.3 *If $(H, \lozenge, \blacklozenge)$ is a vague ring, then $(H, \boxplus, \boxtimes)$ is a ring.*

Definition 9.4.5 Suppose that $(H, \lozenge, \blacklozenge)$ is a vague ring and $H \supset A \neq \emptyset$. Also, assume that $\oplus$ and $\odot$ are two vague binary operations in A, such that

$$\oplus(a,b,c) \leq \lozenge(a,b,c) \quad \text{and} \quad \odot(a,b,c) \leq \blacklozenge(a,b,c) \qquad \text{for all } a,b,c \in A.$$

If $(A, \oplus, \odot)$ is a vague ring with respect to $E_{A\times A}$ and E_A, then $(A, \oplus, \odot)$ is a vague subring of $(H, \lozenge, \blacklozenge)$.

We conclude the presentation of vague rings with the definition of vague ideals.

Definition 9.4.6 Let $(H, \lozenge, \blacklozenge)$ be a vague ring and $(A, \oplus, \odot)$ be a vague subring of $(H, \lozenge, \blacklozenge)$. If for all $a \in A$ and for all $h, t, s \in H$

$$\blacklozenge(a, h, t) = 1 \Rightarrow r \in A \quad \text{and} \quad \blacklozenge(h, a, s) = 1 \Rightarrow s \in A,$$

then $(A, \oplus, \odot)$ is a *vague ideal* of $(H, \lozenge, \blacklozenge)$.

Fuzzy fields have been introduced by Sudarsan Nanda,[1] however, Nanda's definitions were improved by Ranjit Biswas [31]:

Definition 9.4.7 Suppose that X is a field and F is a fuzzy subset of X. Then, F is a *fuzzy field* in X if and only if the following conditions are satisfied:

(i) $F(x \boxplus y) \geq \min\{F(x), F(y)\}$ for all $x, y \in X$;
(ii) $F(-x) \geq F(x)$ for all $x \in X$;
(iii) $F(x \boxtimes y) \geq \min\{F(x), F(y)\}$ for all $x, y \in X$;
(iv) $F(x^{-1}) \geq F(x)$ for all $\mathbf{0} \neq x \in X$

9.5 Fuzzy Vector Spaces

Athanasios Katsaras and D.B. Liu [173] introduced fuzzy vector spaces. For the rest of this short presentation, E will stand for a vector space over K, which is either $\mathbb{R}$ or $\mathbb{C}$.

Definition 9.5.1 Suppose that $A_1, \dots, A_n$ are fuzzy subsets of E. Then, the product $A = A_1 \times \cdots \times A_n$ is a fuzzy subset of E^n such that

$$A(x_1, \dots, x_n) = \min\{A_1(x_1), \dots, A_n(x_n)\}.$$

Assume that $f : E^n \to E$ is a function such that $f(x_1, \dots, x_n) = x_1 + \cdots + x_n$. Then, $A_1 + \cdots + A_n = f(A)$. Let λ be a scalar and B a fuzzy subset of E. Then, $\lambda B = g(B)$, where $g : E \to E, g(x) = \lambda x$.

Definition 9.5.2 A fuzzy subset F of E is a fuzzy subspace if

(i) $F + F \subset F$ and
(ii) $\lambda F \subset F$ for all scalars λ.

---

1 A number of papers cite a paper by this author entitled "Fuzzy fields and fuzzy linear space," nevertheless, we could not locate this paper in any bibliographic database!

9.6 Fuzzy Normed Spaces

Fuzzy normed spaces have been introduced by Reza Saadati and S. Mansour Vaezpour [252] (but see also [75]).

Definition 9.6.1 A triple $(X, N, *)$ is a *fuzzy normed space* if X is a vector space, $*$ is a continuous t-norm, and N is a fuzzy subset of $X \times (0, \infty)$ that satisfies the following conditions for all $x, y \in X$ and $t, s > 0$:

(FN1) $N(x, t) > 0$;
(FN2) $N(x, t) = 1$ if and only if $x = 0$;
(FN3) $N(\alpha x, t) = N(x, \frac{t}{|\alpha|})$ for any $\alpha \neq 0$;
(FN4) $N(x, t) * N(y, s) \leq N(x + y, t + s)$;
(FN5) $N(x, \cdot) : (0, \infty) \to [0, 1]$ is continuous; and
(FN6) $\lim_{t\to\infty} N(x, t) = 1$ and $\lim_{t\to 0} N(x, t) = 0$.

The fuzzy subset N is a *fuzzy norm.*

Example 9.6.1 Suppose that $(X, \|\cdot\|)$ is a vector space and for all $x \in X$ and $t > 0$

$$N(x, t) = \frac{t}{t + \|x\|}$$

is a mapping. Then, $(X, N, \min)$ is a fuzzy normed space.

Definition 9.6.2 A mapping f from a fuzzy normed space $(X, N, *)$ to a fuzzy normed space $(Y, M, \circledast)$ is called *uniformly continuous* if, for all $r \in (0, 1)$ and $t > 0$, there exists $r_0 \in (0, 1)$ and $t_0 > 0$ such that

$$N(x - y, t_0) > 1 - r_0 \text{ implies that } M(f(x) - f(y), t) > 1 - r.$$

There are more general versions of fuzzy normed spaces.

Definition 9.6.3 A *fuzzy φ-normed space* is a triple $(X, N, *)$, where X is a vector space, $*$ is a continuous t-norm, and N is a fuzzy set in $X \times (0, \infty)$ that satisfies the following conditions:

(ϕ-FN1) $N(x, t) < 0$ for all $x \in X$ and $t > 0$;
(ϕ-FN2) $N(x, t) = 1$ for all $t > 0$ and $x = 0$;
(ϕ-FN3) $N(\alpha x, t) = N(x, \frac{t}{\phi(\alpha)})$ for any $\alpha \neq 0$, for all $x \in X$, and $t > 0$;
(ϕ-FN4) $N(x, t) * N(y, s) \leq N(x + y, t + s)$ for all $x, y \in X$, $t > 0$, and $s > 0$;
(ϕ-FN5) $N(x, \cdot) : (0, \infty) \to [0, 1]$ is continuous for all $x \in X$; and
(ϕ-FN6) $\lim_{t\to\infty} N(x, t) = 1$ and $\lim_{t\to 0} N(x, t) = 0$ for all $x \in X$.

The most general form of fuzzy normed spaces is given in the following definition.

Definition 9.6.4 The triple $(V, P, \circledast)$ is an *L-fuzzy normed space* if V is a vector space, $\circledast$ is a continuous t-norm on a lattice L, and $P : V \times [9, +\infty) \to L$ is an L-fuzzy subset if, for all $x, y \in V$ and $t, s \in [0, +\infty)$,

(LFN1) $\mathbf{0} < P(x, t)$;
(LFN2) $P(x, t) = \mathbf{1}$ if and only if $x = 0$;
(LFN3) $P(\alpha x, t) = P\left(x, \frac{t}{|\alpha|}\right)$ for all $\alpha \neq 0$;
(LFN4) $P(x, t) \circledast P(y, s) \leq P(x + y, t + s)$;
(LFN5) $P(x, \cdot) : (0, +\infty) \to L$ is continuous; and
(LFN6) $\lim_{t \to 0} P(x, t) = \mathbf{0}$ and $\lim_{t \to +\infty} P(x, t) = \mathbf{1}$.

In this case, $\circledast$ is called an L-fuzzy norm.

The following example is borrowed from [75].

Example 9.6.2 Suppose that $L = [0, 1] \times [0, 1]$ and the operation $\leq$ is defined as follows:

$$L = \{(a_1, a_2) | (a_1, a_2) \in [0, 1] \times [0, 1] \text{ and } a_1 + a_2 \leq 1\},$$
$$(a_1, a_2) \leq (b_1, b_2) \iff a_1 \leq b_1 \text{ and } a_2 \geq b_2.$$

Now, it can be proved that $(L, \leq)$ is a complete lattice (see Ref. [96]).

Let $(X, \|\cdot\|)$ be a normed vector space and let $a \circledast b = (\min\{a_1, b_1\}, \max\{a_2, b_2\})$ for all $a = (a_1, a_2)$ and $b = (b_1, b_2)$ such that $a, b \in [0, 1] \times [0, 1]$. Also, suppose that N is a mapping defined by

$$N(x, t) = \left(\frac{t}{t + \|x\|}, \frac{\|x\|}{t + \|x\|} \right)$$

for all $t \in \mathbb{R}^+$. Then, $(X, N, \circledast)$ is an L-fuzzy normed space.

9.7 Fuzzy Lie Algebras

Fuzzy Lie algebras are covered in Muhammad Akram's [6] monograph.

Definition 9.7.1 A fuzzy set $A : L \to [0, 1]$ is a *fuzzy Lie subalgebra* of a vector space L over a field F if it is a fuzzy subspace of L such that for all $x, y \in L$ the following condition is satisfied:

$$A([x, y]) \geq \min\{A(x), A(y)\}.$$

Example 9.7.1 The real vector space $\mathbb{R}^3$ with $[x, y] = x \times y$, where $x, y \in \mathbb{R}^3$, is a real Lie algebra. Let us define a fuzzy subset A of $\mathbb{R}^3$ as follows:

$$A(x) = \begin{cases} 0.9, & \text{if } x = (0, 0, 0), \\ 0.6, & \text{if } x = (c, 0, 0) \text{ and } c \neq 0, \\ 0.2, & \text{otherwise.} \end{cases}$$

We can easily verify that A is a fuzzy Lie algebra.

Definition 9.7.2 Given a vector space L over a field F, a fuzzy subset $A : L \to [0, 1]$ of L is a *fuzzy Lie ideal* of L if the following conditions are satisfied for all $x, y \in L$ and $\alpha \in F$:

(i) $A(x \boxplus y) \geq \min\{A(x), A(y)\}$;
(ii) $A(\alpha x) \geq A(x)$; and
(iii) $A([x, y]) \geq A(x)$.

Proposition 9.7.1 *Every fuzzy Lie ideal is a fuzzy Lie subalgebra.*

The following lemma is obvious, so we omit its proof.

Lemma 9.7.1 *Suppose that A is a fuzzy ideal of L. Then, for all $x, y \in L$*

(i) $A(\mathbf{0}) \geq A(x)$;
(ii) $A([x, y]) \geq \max\{A(x), A(y)\}$;
(iii) $A([x, y]) = A(-[y, x]) = A([y, x])$;
(iv) $A(x \boxplus y) = A(\mathbf{0}) \Rightarrow A(x) = A(y)$; *and*
(v) $A(x \boxplus y) = A(x) = A(y \boxplus x)$ *if* $A(x) < A(y)$.

Definition 9.7.3 Assume that L is a Lie algebra. Then, a fuzzy subset Λ of L is an *anti-fuzzy Lie ideal* of L if for all $x, y \in L$ and $\alpha \in F$, the following conditions are satisfied:

(i) $\Lambda(x \boxplus y) \leq \max\{\Lambda(x), \Lambda(y)\}$;
(ii) $\Lambda(\alpha x) \leq \Lambda(x)$; and
(iii) $\Lambda([x, y]) \leq \Lambda(x)$.

Example 9.7.2 Suppose that $\mathbb{R}^2 = \{(x, y) \mid x, y \in \mathbb{R}\}$ is the set of two-dimensional real vectors. Then, $\mathbb{R}^2$ with $[x, y] = x \times y$ is a real Lie algebra. Next, we define a fuzzy subset of $\mathbb{R}^2$ as follows:

$$\Lambda(x, y) = \begin{cases} 0, & \text{when } x = y = 0, \\ 1, & \text{otherwise.} \end{cases}$$

Now, we can easily check that Λ is an anti-fuzzy Lie ideal of $\mathbb{R}^2$.

The following lemma is obvious:

Lemma 9.7.2 *Suppose that Λ is an anti-fuzzy Lie ideal of L. Then, for all $x, y \in L$, the following conditions are true:*

(i) $\Lambda(\mathbf{0}) \leq \Lambda(x)$;
(ii) $\Lambda([x, y]) \leq \min\{\Lambda(x), \Lambda(y)\}$; *and*
(iii) $\Lambda([x, y]) = \Lambda(-[y, x]) = \Lambda([y, x])$.

Exercises

9.1 Prove that every ring is an Abelian additive group.

9.2 Give the complete definition of the right fuzzy coset (see Definition 9.2.4).

9.3 Prove that $(X, \circledcirc)$ from Example 9.2.1 is a vague semigroup.

9.4 Prove Proposition 9.2.2.

9.5 Find a vague subgroup of the vague group $(\mathbb{Q}^+, \odot)$ defined in Example 9.2.2.

9.6 Define a vague homomorphism from the vague group $(\mathbb{Q}^+, \odot)$ to the vague group $(\mathbb{R}^+, \circledast)$, which are defined in Example 9.2.2.

9.7 Let $(\mathbb{R}, +, \cdot)$ denote the ring of real numbers under the usual operations of addition and multiplication. Then, prove that the following fuzzy subset A of $\mathbb{R}$ (see Ref. [102])

$$A(x) = \begin{cases} t, & \text{when } x \text{ is rational,} \\ t', & \text{when } x \text{ is irrational,} \end{cases}$$

where $t, t' \in [0, 1]$ and $t > t'$, is a fuzzy subring but not a fuzzy ideal of the ring $\mathbb{R}$.

9.8 Show that $(X, N, \cdot)$ is a fuzzy normed space, where X and N are as in Example 9.7.1 and $\cdot$ is the product t-norm.

10

Fuzzy Topology

Geometry and topology are two different branches of mathematics that deal with the same objects. However, as K.D. Joshi [167, p. 67] notes "…if two objects are equivalent for a geometer, they are certainly so for a topologist" and "…two objects which look distinct to a geometer may look the same to a topologist." In this chapter, we discuss fuzzy topology, and in the next chapter, we explore fuzzy geometry. Since metrics induce topologies, we will start with fuzzy metric spaces.

10.1 Metric and Topological Spaces[1]

On the line $\mathbb{R}$, the *distance* from a to b is equal to $|a - b|$. More generally, the distance from $\mathbf{x} \in \mathbb{R}^n$ to $\mathbf{y} \in \mathbb{R}^n$ is denoted by $\|\mathbf{x} - \mathbf{y}\|$ and it is equal to

$$d(\mathbf{x}, \mathbf{y}) = \|\mathbf{x} - \mathbf{y}\| = \sqrt{\sum_{i=1}^{n} (x_i - y_i)^2}.$$

Using the notion of distance, we can define continuous functions from $\mathbb{R}^n$ to $\mathbb{R}^m$ as follows:

Definition 10.1.1 Map $f : \mathbb{R}^n \to \mathbb{R}^m$ is continuous at $\mathbf{x} \in \mathbb{R}^n$ if, given $\epsilon > 0$, there is a $\delta > 0$ such that $d(\mathbf{x}, \mathbf{y}) < \delta$ implies that $d(f(\mathbf{x}), f(\mathbf{y})) < \epsilon$.

This distance is called the *Euclidean metric.*

If we want to be more general, we can talk about any "space" X that has a suitable notion of distance. The new spaces are known as metric spaces.

Definition 10.1.2 A *metric space* is a set X together with a function

$$d : X \times X \to \mathbb{R},$$

1 The overview of metric spaces and topological spaces that follows is based on [38, 167].

A Modern Introduction to Fuzzy Mathematics, First Edition.
Apostolos Syropoulos and Theophanes Grammenos.

called a *metric*, that assigns a real number $d(x, y)$ to every pair $x, y \in X$ having the properties:

(i) $d(x, y) \geq 0$ for all $x, y \in X$ (positivity);
(ii) $d(x, y) \geq 0$ and $d(x, y) = 0 \Leftrightarrow x = y$ (identity);
(iii) $d(x, y) = d(y, x)$ (symmetry); and
(iv) $d(x, y) + d(y, z) \geq d(x, z)$ (triangle inequality).

Given a metric space X, the ϵ-ball, $\epsilon > 0$, about a point $x \in X$ is the set

$$B_\epsilon(x, t) = \{t | t \in X \text{ and } d(x, t) < \epsilon\}.$$

An ϵ-ball about a point x is also called an *open ball* with center the point x and radius ϵ. The *closed ball* with center the point x and radius ϵ is the set

$$\overline{B}_\epsilon(x, t) = \{t | t \in X \text{ and } d(x, t) \leq \epsilon\}.$$

A subset $U \subset X$ is *open* if, for each point $x \in U$, there is an ϵ-ball about x completely contained in U. A subset $U' \subset X$ is *closed* if its complement (i.e. $(U')^{\complement} = X \setminus U'$) is open. Also, if $y \in B_\epsilon(x, t)$ and $\delta = \epsilon - d(x, y)$, then $B_\epsilon(y, t) \subset B_\epsilon(x, t)$ by the third property of the previous definition. This shows that all ϵ-balls are open sets. The following proposition shows that for continuity we need only open sets:

Proposition 10.1.1 *A map $f : X \to Y$ between metric spaces is continuous if and only if $f^{-1}(U)$ is open in X for each open subset U of Y.*

Example 10.1.1 The set $\mathbb{R}^2$ together with the *taxi cab* metric

$$d((x_1, y_1), (x_2, y_2)) = |x_1 - x_2| + |y_1 - y_2|$$

is a metric space. Also, the set of all continuous real valued functions on $[0, 1]$ together with the function

$$d(f, g) = \int_0^1 |f(x) - g(x)| \, dx$$

that defines a metric, form a metric space.

A *sequence* is a list of objects (usually numbers) that are in some order. Typically, we can write its elements like a set, that is, $\{a_0, a_1, a_2, \ldots\}$, or using a rule, for example $x_n = 2n + 1$, which means that the nth element of the sequence is the number $2n + 1$.

Definition 10.1.3 A sequence $\{x_n\} = \{x_1, x_2, \ldots\}$ in a metric space (X, d) is said to *converge* to a limit $x \in X$ if, given $\epsilon > 0$, there exists a positive integer N such that

$$d(x_n, x) < \epsilon \quad \text{for all } n > N.$$

We write $\lim_n x_n = x$ or $x_n \to x$.

Definition 10.1.4 A sequence $\{x_n\}$ in a metric space (X, d) is a *Cauchy sequence* if, given $\epsilon > 0$, there exists a natural number N such that

$$d(x_m, x_n) < \epsilon \quad \text{for all } m, n > N.$$

If every Cauchy sequence in (X, d) converges to a point in X, then we say that (X, d) is a *complete metric space.*

Definition 10.1.5 A complete normed vector space is called a *Banach space.*

Definition 10.1.6 A complete inner product space, that is a Banach space with the naturally defined norm, is called a *Hilbert space.*

The inner product of two elements $\mathbf{x}$ and $\mathbf{y}$ of some Hilbert space H is written as $\langle \mathbf{x} \mid \mathbf{y} \rangle$.

Since continuity can be expressed in terms of open sets alone, and since some interesting constructions of spaces do not rely on the metric, it is very useful to forget about metrics and to focus on the basic properties of open sets that are required in order to talk about continuity. Quite naturally, this leads to the notion of a general topological space.

Definition 10.1.7 A *topological space* is a set X together with a collection of subsets of X, τ, called *open* sets that have the following properties:

(i) the intersection of two open sets is open;
(ii) the union of any collection of open sets is open; and
(iii) the empty set $\emptyset$ and the whole space X are open.

In addition, a subset $C \subset X$ is called *closed* if its complement $C^{\complement}$ is open. The collection τ is called a *topology on* X.

Definition 10.1.8 Assume that (X, τ) is a topological space, $x_0 \in X$, and $N \subset X$. Then, N is a neighborhood of x_0 if there is an open set V such that $x_0 \in V$ and $V \subset N$.

The following can be found in [219]:

Remark 10.1.1 Suppose that (X, τ) is a topological space. Then, the family of open neighborhoods $T(x)$ of point x, where

$$T(x) = \{V | V \in \tau \ \text{ and } \ x \in V\},$$

may be considered as the soft set $(T(x), \tau)$.

Definition 10.1.9 Assume that (X, τ) is a topological space and that $A \subset X$. Then, the *interior* of A is the set of all interior points of A, that is, the set

$$\mathrm{Int}(A) = \{x | x \in A \text{ and } A \text{ is a neighborhood of } x\}.$$

Proposition 10.1.2 *Suppose that (X, τ) is a topological space and that $A \subset X$. Then,* Int(A) *is the union of all open sets contained in A. It is also the largest open subset of X contained in A.*

Definition 10.1.10 The *closure* of a subset A of a topological space (X, τ) is the intersection of all closed subsets containing it, that, is the set

$$\mathrm{Cl}(A) = \bigcap \{C | C \text{ is closed and } C \supset A\}.$$

Definition 10.1.11 Assume that A is a subset of a space (X, τ). Then, A is said to be *dense* in X if $\mathrm{Cl}(A) = X$.

Proposition 10.1.3 *A subset A of space (X, τ) is* dense *in X if and only if for every nonempty open subset B of (X, τ), $A \cap B \neq \emptyset$.*

Continuity is a basic notion in topology and neighborhoods can also be used to define continuity:

Definition 10.1.12 A function $f : X \to Y$ between the topological spaces (X, τ_X) and (Y, τ_Y) is *continuous* at $x \in X$, if, given any neighborhood $\mathcal{N}$ of $f(x)$ in Y, there is a neighborhood $\mathcal{M}$ of x in X such that $f(\mathcal{M}) \subset \mathcal{N}$.

Definition 10.1.13 A function $f : X \to Y$ between topological spaces is continuous if and only if it is continuous at each point $x \in X$.

Proposition 10.1.4 *Assume that (X, τ_X) and (Y, τ_Y) are topological spaces and that $f : X \to Y$ is a function. Then, f* is continuous *if $f^{-1}(U)$ is open for each open set $U \subset Y$.*

The following definition is borrowed from [38, p. 7]:

Definition 10.1.14 If X is a set and some condition is given on subsets of X which may or may not hold for any particular subset, then if there is a topology τ whose open sets satisfy the condition, and such that, for any topology τ' whose open sets satisfy the condition, then the τ-open sets are also τ'-open (i.e. $\tau \subset \tau'$), then τ is called the *smallest* (or *weakest* or *coarsest*) topology satisfying the condition. If, instead, for any topology τ' whose open sets satisfy the condition, any τ'-open sets are also τ-open, then τ is called the *largest* (or *strongest* or *finest*) topology satisfying the condition.

Definition 10.1.15 Assume that $(X_j)_{j\in J}$ is any family of topological spaces. Then, we say that a topology τ on X is *initial* with respect to the family of mappings $f_j : X \to X_j, j \in J$, if τ is the coarsest topology on X which makes all f_j's continuous.

Definition 10.1.16 Suppose that (X, τ) is a topological space. Then, a family $\mathfrak{B}$ of τ is a *base* for τ if every member of τ can be expressed as the union of some members of $\mathscr{B}$.

Definition 10.1.17 A family $\mathfrak{S}$ of subsets of X is said to be a *sub-base* for a topology τ on X, if the family of all finite intersections of members of $\mathfrak{S}$ is a base of τ.

Definition 10.1.18 Suppose that $T = (X, \tau)$ is a topological space and $A, B \subseteq X$. Then, A and B are *separated* (*in* T) if and ony if

$$\mathrm{Cl}(A) \cap B = A \cap \mathrm{Cl}(B) = \emptyset.$$

The so-called separation axioms are used to describe topological spaces that have further restrictions in addition to the "standard" ones.

Definition 10.1.19 The separation axioms:

(T_0) A topological space (X, τ) is called a T_0-*space* if for any two points $x \neq y$ there is an open set containing one of them but not the other.

(T_1) A topological space (X, τ) is called a T_1-*space* or *Fréchet* if for any two points $x \neq y$ there is an open set containing x but not y and another open set containing y but not x.

(T_2) A topological space (X, τ) is called a T_2-*space* or *Hausdorff* if for any two points $x \neq y$ there are disjoint open sets U and V with $x \in U$ and $y \in V$.

(T_3) A T_1-space (X, τ) is called a T_3-*space* or *regular* if for any point x and closed set F not containing x there are disjoint open sets U and V with $x \in U$ and $F \subset V$.

(T_4) A T_1-space (X, τ) is called a T_4-*space* or *normal* if for any two disjoint closed sets F and G there are disjoint open sets U and V with $F \subset U$ and $G \subset V$.

(T_5) A T_1-space (X, τ) is called a T_5-*space* or *completely normal* if for two separated sets $A, B \subseteq S$ there exist disjoint open sets $U, V \in \tau$ containing A and B, respectively.

A normal space has a number of important properties. One of them is that normal spaces admit a lot of continuous functions:

Theorem 10.1.1 ***(Urysohn's Lemma)*** *If A and B are disjoint closed sets in a normal space (X, τ), then there is a continuous function $f : X \to [0, 1]$ such that for all $a \in A, f(a) = 0$ and for all $b \in B, f(b) = 1$.*

Definition 10.1.20 A family $\mathscr{U}$ of sets is a *cover* of a set A if A is contained in the union of members of $\mathscr{U}$. A *subcover* of $\mathscr{U}$ is a subfamily $\mathscr{V}$ of $\mathscr{U}$ which itself is a cover of A. If we are in a topological space, then a cover is open if all its members are open.

Definition 10.1.21 A subset A of a space X is *compact* (*Lindelöf*) subset of X if every cover of A by open subsets of X has a finite (respectively countable) subcover. A space X is *compact* (*Lindelöf*) if X is a compact (respectively Lindelöf) subset of itself.

Definition 10.1.22 A topological space is *countably compact* if every countable open cover of it has a finite sub-cover.

In a connected space, it is possible to go from one point to any other point without jumps. Intuitively, one could say that if Greece and Switzerland were topological spaces, then Greece is not connected while Switzerland is connected.

Definition 10.1.23 A topological space X is *connected* if it is not the disjoint union of two nonempty open subsets.

Definition 10.1.24 A subset of a topological space X is called *clopen* if it is both open and closed in X.

Proposition 10.1.5 *A topological space X is connected if and only if its only clopen subsets are X and $\varnothing$.*

10.2 Fuzzy Metric Spaces

Fuzzy metric spaces were introduced by Ivan Kramosil and Jiří Michálek [181]. In their approach, the distance between two points is not just a real number but something vague. For example, a real number with some plausibility degree. In what follows, the set E contains such degrees.

Lemma 10.2.1 *A metric d on the set X is uniquely determined by the relation $R_d \subset X \times X \times E$, where for all $x, y \in X$ and $\lambda \in E$, $R_d(x, y, \lambda)$ is valid if and only if $d(X, y) < \lambda$.*

Proof: Suppose that d_1 and d_2 are two different metrics on X. Then, there is at least one pair $(x, y) \in X \times X$ such that $d_1(x, y) \neq d_2(x, y)$. Suppose that $d_1(x, y) < d_2(x, y)$. Then, $(x, y, d_1(x, y)) \in R_{d_1}$ but $(x, y, d_2(x, y)) \notin R_{d_2}$, that is, $R_{d_1} \neq R_{d_2}$. □

This lemma leads to the definition that follows.

Definition 10.2.1 A fuzzy metric R on a set X is the fuzzy relation F_R in $X \times X \times E$ that has the following properties:

(i) $F_R(x, y, \lambda) = 0$ for all $x, y \in X$ and all $\lambda \leq 0$;
(ii) $F_R(x, y, \lambda) = 1$ for $\lambda < 0$ if and only if $x = y$;
(iii) $F_R(x, y, \lambda) = F_R(y, x, \lambda)$ for all $x, y \in X$ and all $\lambda \in E$;
(iv) $F_R(x, z, \lambda + \mu) \geq S(F_R(x, y, \lambda), F_R(y, z, \mu))$, where $S : [0, 1] \times [0, 1] \rightarrow [0, 1]$ is a *measurable*[2] binary real function such that $S(1, 1) = 1$; and
(v) $F_R(x, y, \lambda)$ is for every pair $(x, y) \in X \times X$ a left-continuous and non-decreasing function of λ such that $\lim_{\lambda \rightarrow \infty} F_R(x, y, \lambda) = 1$.

A simplified formulation of this definition has been provided in [134]:

Definition 10.2.2 The triple $(X, M, *)$ is a fuzzy metric space if X is an arbitrary set, $*$ is a continuous t-norm, and M is a fuzzy subset of $X \times X \times [0, \infty)$ having the following properties:

(i) $M(x, y, 0) = 0$;
(ii) $M(x, y, t) = 1$ for all $t > 0$ if and only if $x = y$;
(iii) $M(x, y, t) = M(y, x, t)$;
(iv) $M(x, y, t) * M(y, z, s) \leq M(x, z, t + s)$, $t, s > 0$ and
(v) for fixed $x, y \in X$, the map $M'(t) = M(x, y, t) : [0, \infty) \rightarrow [0, 1]$ is left continuous

2 A nonempty class of sets **F** is called a *σ-ring* if and only if
(a) for all $E, F \in \mathbf{F}$, $E \setminus F \in \mathbf{F}$ and
(b) for all $E_i \in \mathbf{F}$, $i = 1, 2. \dots, \cup_{i=1}^{\infty} E_i \in \mathbf{F}$.
A σ-algebra is a σ-ring that contains X, where all elements of class **F** are subsets of X. Consider the class of all bounded, left closed, and right open intervals of the real line. These sets have certain properties and are known as Borel sets. Assume that $(X, \mathbf{F})$ is a *measurable space* (i.e. **F** is a σ-algebra). Then, a real-valued function $f : X \rightarrow (-\infty, \infty)$ on X is *measurable* if and only if $f^{\leftarrow}(B) = \{x | f(x) \in B\} \in \mathbf{F}$, for any Borel set B.

Example 10.2.1 Suppose that $X = \mathbb{R}$, $a * b = ab$, and

$$M(x,y,t) = \left[\exp\left(\frac{|x-y|}{t}\right)\right]^{-1}$$

for all $x, y \in X$ and $t \in (0, \infty)$, Then, $(X, M, *)$ is a fuzzy metric space.

Before presenting another approach to the definition of fuzzy metric spaces, which was formulated by Phil Diamond and Peter Kloeden [99, 100], we need to give some definitions.

Definition 10.2.3 The *Hausdorff distance*[3] between nonempty subsets A and B of $\mathbb{R}^n$ is defined by

$$d_H(A,B) = \max\left\{\bigvee_{b\in B}\bigwedge_{a\in A}\|b-a\|, \bigvee_{a\in A}\bigwedge_{b\in B}\|a-b\|\right\}.$$

The definition that follows is borrowed from [270, p. 120]

Definition 10.2.4 Assume that S is a nonempty subset of $\mathbb{R}^n$. Then, a real-valued function $f : S \to \mathbb{R}$ is said to be *upper semi-continuous* at a point $x \in S$ if for any $\epsilon > 0$, there exists a $\delta > 0$ such that

$$A(y) < A(x) + \epsilon,$$

for all $y \in S$, and $\|y - x\| < \delta . f : S \to \mathbb{R}^1$ is said to be upper semi-continuous if it is upper semi-continuous at each point of S. Here A is a fuzzy number of $\mathbb{R}^n$.

Alternatively, one can use the definition that follows.[4]

Definition 10.2.5 Consider a function $f : \mathbb{R} \to \mathbb{R}$ and a point $x_0 \in \mathbb{R}$. The function f is said to be upper (resp. lower) semi-continuous at the point x_0 if

$$f(x_0) \geq \lim_{x\to x_0} \bigwedge f(x) \quad \left(\text{resp. } f(x_0) \leq \lim_{x\to x_0} \bigvee f(x)\right).$$

This definition can be easily extended to functions defined on subdomains of $\mathbb{R}$ and taking values in the extended real line $[-\infty, \infty]$. If a function is upper (resp.

3 The authors defined the Hausdorff distance in [99] using min, while in [100] they defined it using max. Apparently, the definition using max is the correct one.
4 See Ref. [117].

lower) semi-continuous at every point of its domain of definition, then it is simply called an upper (resp. lower) semi-continuous function.

Definition 10.2.6 Assume that S is a Euclidean space, or more generally, a vector space over the real numbers. A set C in S is said to be convex if, for all $x, y \in C$ and all $t \in (0, 1)$, the point $(1 - t)x + ty$ also belongs to C. In other words, every point on the line segment connecting x and y is in C.

Diamond and Kloeden examined the class $\mathscr{E}^n$ of fuzzy sets $A : \mathbb{R}^n \to [0, 1]$ that are

(i) normal;
(ii) convex;
(iii) upper semi-continuous; and
(iv) the closure of ${}^{0+}A$ is compact.

Note that, according to the Heine-Borel theorem, a subspace of $\mathbb{R}^n$ (with the usual topology) is compact if and only if it is closed and bounded.[5] These properties imply that for each $0 < x \leq 1$, the set ${}^{x}A$ is a nonempty compact convex subset of $\mathbb{R}^n$, as is the support set ${}^{0+}A$.

Definition 10.2.7 For each $1 \leq p \leq \infty$ and $A, B : \mathbb{R}^n \to [0, 1]$ we define

$$d_p(A, B) = \left(\int_0^1 d_H({}^{\alpha}A, {}^{\alpha}B)^p d\,\alpha \right)^{1/p} \quad \text{and} \quad d_{\infty}(A, B) = \bigvee_{0 \leq \alpha \leq 1} d_H({}^{\alpha}A, {}^{\alpha}B).$$

We state the following result without proof:

Theorem 10.2.1 *$(\mathscr{E}^n, d_p)$, $1 \leq p < \infty$, is a metric space.*

Osmo Kaleva and Seppo Seikkala [171] introduced their own version of fuzzy metric space. The distance between two points in this alternative fuzzy metric space is a nonnegative, upper semi-continuous fuzzy number. An upper semi-continuous fuzzy number has α-cuts of the form $[a^{\alpha}, b^{\alpha}]$, where it is possible to have $a^{\alpha} = -\infty$ and $b^{\alpha} = \infty$. Thus, when $b^{\alpha} = \infty$, then $[a^{\alpha}, b^{\alpha}]$ is the interval $[a^{\alpha}, \infty)$. The set of all upper semi-continuous fuzzy numbers will be denoted by E. Also, the set of all nonnegative fuzzy numbers of E is denoted by G.

5 See Ref. [297].

The arithmetic operations between upper semi-continuous fuzzy numbers are defined as follows:

$$(A+B)(x) = \bigvee_{y\in\mathbb{R}} \min\{A(y), B(x-y)\},$$

$$(A-B)(x) = \bigvee_{y\in\mathbb{R}} \min\{A(y), B(y-x)\},$$

$$(A\cdot B)(x) = \bigvee_{\substack{y\in\mathbb{R}\\ y\neq 0}} \min\{A(y), B(x/y)\},$$

$$(A/B)(x) = \bigvee_{y\in\mathbb{R}} \min\{A(xy), B(y)\},$$

for all $x \in \mathbb{R}$. In addition, the neutral elements of addition and multiplication in E are denoted by $\overline{0}$ and $\overline{1}$, respectively, and are defined as follows:

$$\overline{0}(x) = \begin{cases} 1, & \text{if } x = 0 \\ 0, & \text{if } x \neq 0 \end{cases} \quad \text{and} \quad \overline{1}(x) = \begin{cases} 1, & \text{if } x = 1 \\ 0, & \text{if } x \neq 1. \end{cases}$$

Definition 10.2.8 Assume that X is a set such that $X \neq \emptyset$, d is a mapping from $X \times X$ into G, and the mappings $L, R : [0,1] \times [0,1] \to [0,1]$ are symmetric, non-decreasing in both arguments and satisfy $L(0,0) = 0$ and $R(1,1) = 1$. Let

$${}^{\alpha}d(x,y) = [\lambda_\alpha(x,y), \rho_\alpha(x,y)] \quad \text{for } x, y \in X \text{ and } 0 < \alpha \leq 1.$$

The quadruple (X, d, L, R) is called a fuzzy metric space and d a fuzzy metric, if

(i) $d(x,y) = \overline{0}$ if and only if $x = y$;
(ii) $d(x,y) = d(y,x)$ for all $x, y, \in X$;
(iii) for all $x, y, z \in X$,
  (a) $d(x,y)(s+t) \geq L(d(x,z)(s), d(z,y)(t))$ whenever $s \leq \lambda_1(x,z)$, $t \leq \lambda_1(z,y)$, and $s+t \leq \lambda_1(x,y)$; and
  (b) $d(x,y)(s+t) \leq R(d(x,z)(s), d(z,y)(t)$ whenever $s \geq \lambda_1(x,z)$, $t \geq \lambda_1(z,y)$, and $s+t \geq \lambda_1(x,y)$,

There is a connection between fuzzy normed spaces and fuzzy metric spaces, which is shown by the following result.

Lemma 10.2.2 *Suppose that $(X, N, *)$ is a fuzzy normed space. If we define*

$$M(x,y,t) = N(x-y,t)$$

for all $x, y \in X$ and $t > 0$, then M is a fuzzy metric on X.

The metric M is a *fuzzy metric generated by the fuzzy norm N.*

10.3 Fuzzy Topological Spaces[6]

Fuzzy topological spaces have been introduced by C.L. Chang [66].

Definition 10.3.1 A fuzzy topology is a family τ of fuzzy sets in X that has the following properties:

(i) $\emptyset, \chi_X \in \tau$, where χ_X is the characteristic function of X (i.e. $\chi_X(x) = 1$, for all $x \in X$);
(ii) if $A, B \in \tau$, then $A \cap B \in \tau$; and
(iii) if $A_i \in \tau$ for each $i \in I$, then $\bigcup_{i \in I} A_i \in \tau$.

All members of τ are called *open fuzzy sets*, and a fuzzy set is *closed* if and only if its complement is open.

The pair (X, τ) is called a *fuzzy topological space.*

Example 10.3.1 Suppose that $X = \{\alpha, \beta\}$ and that A is a fuzzy subset of X defined as $A(\alpha) = 0.8$ and $A(\beta) = 0.5$. Then, $\tau = \{\chi_\emptyset, A, \chi_X\}$ is a fuzzy topology and (X, τ) is a fuzzy topological space.

Robert Lowen [197] has given an alternative definition of fuzzy topology that is based on his work on fuzzy-compactness:

Definition 10.3.2 Assume that E is some set (e.g., $E = [0, 1]$). Then, $\delta \subset [0, 1]^E$ is a fuzzy topology on E if and only if

(i) for all *constant* fuzzy sets C, $C \in \delta$[7];
(ii) for all $A, B \in \delta, A \wedge B \in \delta$; and
(iii) for all $\{M_j\}_{j \in J} \subset \delta, (\bigvee_{j \in J} M_j) \in \delta$.

The members of δ are called *open sets*. A fuzzy set $A \in [0, 1]^E$ is called closed if and only if $A^{\complement}$ is open.

In what follows, we will use Chang's definition since it is the one that is used the most in the literature.

Bases and sub-bases are defined as follows:

Definition 10.3.3 Suppose that (X, τ_X) is a fuzzy topological space. Then, a family $\mathfrak{B}$ of τ_X is called a *base* for τ_X if and only if for each $A \in \tau_X$, there exists

6 The text that follows is based on [230] unless explicitly stated otherwise.
7 Obviously, for a constant fuzzy set $\mathbf{r} : X \to [0, 1]$ we have that $\mathbf{r}(x) = r$ for all $x \in X$ and $r \in [0, 1]$.

$(A_j)_{j\in J} \subset \mathfrak{B}$ such that

$$A = \bigvee_{j\in J} A_j.$$

A subfamily $\mathfrak{S}$ of τ_X is called a *sub-base* for τ_X if and only if the family of finite intersections of members of $\mathfrak{S}$ forms a base for τ_X.

In Lowen's [197] version, bases and sub-bases are defined as follows:

Definition 10.3.4 A subset $\sigma \subset \delta$, where δ is a fuzzy topology on a set E, is a base for δ if and only if for all $A \in \delta$, there exists a family $(A_i)_{j\in J} \subset \sigma$ such that

$$A = \vee_{j\in J} A_j.$$

A subset $\sigma' \subset \delta$ is a sub-base of δ if and only if the family of finite infima of members of σ' is a base of δ.

Chang defined the neighborhood of a fuzzy set as follows:

Definition 10.3.5 A fuzzy set U of X, where (X, τ) is a fuzzy topological space, is a neighborhood of a fuzzy subset A if and only if there exists an open fuzzy set O such that $A \subset O \subset U$.

Note that here we are not talking about the neighborhood of a point. Assume that $A: X \to [0,1]$ and $B : X \to [0,1]$ are two fuzzy subsets of X. If $\min[A(x), B(x)] \neq 0$, for some $x \in X$, and if $A(x) > B^{\complement}(x)$ or $A(x) + B(x) > 1$, then A is *quasi-coincident with* B. Usually, we write AqB to denote that A is quasi-coincident with B.

Definition 10.3.6 Suppose that (X, τ) is a fuzzy topological space and that A is a fuzzy subset of X. Then, the neighborhood system $\mathcal{N}$ of A is the set of all neighborhoods of A.

In order to speak about the neighborhood of a point we need to define fuzzy points.

Definition 10.3.7 A fuzzy subset P of X is called a *fuzzy point* if and only if it has a membership function defined by

$$P(x) = \begin{cases} \lambda, & x = y, \\ 0, & x \neq y, \end{cases}$$

where $0 < \lambda < 1$. P has support y, value λ, and it is denoted by P_y^{λ}.

Given a fuzzy subset A of X, then a fuzzy point P_y^λ is in A, $P_y^\lambda \in A$, if and only if $\lambda < A(y)$. If $\lambda \leq A(x)$, then P_y^λ is *contained* in A.

Definition 10.3.8 Let (X, τ) be a fuzzy topological space. A fuzzy subset A of X is a neighborhood of a fuzzy point P_y^λ if and only if there exists a $B \in \tau$ such that $P_y^\lambda \in B \subseteq A$. A neighborhood A is said to be *open* if and only if A is open. The family made up of all the neighborhoods of P_y^λ is called the *system of neighborhoods* of P_y^λ.

Definition 10.3.9 A fuzzy set A in (X, δ) is called a *quasi-coincident neighborhood* of P_x^λ or just a *Q-neighborhood* of P_x^λ if $P_x^\lambda qA$.

Definition 10.3.10 A fuzzy point P_x^λ is called a δ-cluster point of a fuzzy set S in a fuzzy topological space (X, τ) if and only if every regular Q-neighborhood of P_x^λ is quasi-coincident with S. The set of δ-cluster points of S is written as $[S]_\delta$. If $S = [S]_\delta$, then S is *δ-closed*. The complement of a δ-closed set is *δ-open*.

The definition that follows is from [300].

Definition 10.3.11 Assume that A is a fuzzy set in a fuzzy topological space (X, τ). Then, the *closure* of A, which is written as $\mathrm{Cl}(A)$, and its *interior*, which is denoted by $\mathrm{Int}(A)$, are defined as follows:

$$\mathrm{Cl}(A) = \bigcap \{B | B \supset A \text{ and } B^{\mathrm{C}} \in \tau\},$$

$$\mathrm{Int}(A) = \bigcup \{B | B \subset A \text{ and } B \in \tau\}.$$

The interior and the closure of a fuzzy set in Lowen's [197] formulation are defined as follows:

Definition 10.3.12 The closure and interior of a fuzzy set $A \in [0,1]^E$ are defined, respectively

$$\mathrm{Cl}(A) = \bigwedge \{B | B \geq A \text{ and } B^{\mathrm{C}} \in \delta\},$$

$$\mathrm{Int}(A) = \bigvee \{B | B \leq A \text{ and } B \in \delta\}.$$

Essentially, the two definitions are similar.

Definition 10.3.13 Suppose that (X, τ) is an ordinary topological space and that

$$F(\tau) = \{f | f \in [1,0]^X \text{ and } f \text{ is lower semi-continuous}\},$$

then $(X, F(\tau))$ is called the *induced fuzzy topological space* of (X, τ).

A complete theory of fuzzy topological spaces should include a definition of continuity.

Definition 10.3.14 Assume that (X, τ_X) and (Y, τ_Y) are fuzzy topological spaces. Then, a function f from X into Y is F-continuous if $f^{\leftarrow}(U) \in \tau_X$ for every $U \subset Y$ that is an element of τ_Y.

Theorem 10.3.1 *A function f from a fuzzy topological space (X, τ_X) to a fuzzy topological space (Y, τ_Y) is F-continuous if and only if the complement of $f^{\leftarrow}(B^{\complement})$, where $B \in \tau_Y$, is an open set of τ_X.*

We conclude this section with a definition of subspaces.

Definition 10.3.15 Suppose that (X, τ) is a fuzzy topological space and $Y \subset X$. Then, the family

$$U = \{A | A \in [1, 0]^Y \text{ and } A \in \tau\},$$

which is a fuzzy topology for Y, is the *relative fuzzy topology*, or the relativization of τ to Y. The topological space (Y, U) is called a *subspace* of (X, τ). A U-open (resp. U-closed) set is also called a *relative open* (resp. *relative closed*) set on Y.

10.4 Fuzzy Product Spaces

Let us first define the product of fuzzy topological spaces.

Definition 10.4.1 Assume that (X, τ_X) and (T, τ_Y) are fuzzy topological spaces. Then, the *fuzzy product space* of X and Y is the Cartesian product $X \times Y$ with the topology generated by the family

$$\{(\pi_X^{\leftarrow}(A_j), \pi_Y^{\leftarrow}(B_k)) | A_j \in \tau_X, B_k \in \tau_Y, \text{ and } \pi_X, \pi_Y \text{ are projections maps}\}.$$

Since $\pi_X^{\leftarrow}(A_j) = A_j \times 1$, $\pi_Y^{\leftarrow}(B_k) = 1 \times B_k$, and $(A_j \times 1) \wedge (1 \times B_k) = A_j \times B_k$, the family

$$\{A_j \times B_k | A_j \in \tau_X, B_k \in \tau_Y\}$$

forms a base for the product topology on $X \times Y$.

Proposition 10.4.1 *For a family $(A_j)_{i \in I}$ of fuzzy sets of a fuzzy topological space (X, τ_X),*

$$\bigvee \text{Cl}(A_j) \leq \text{Cl}\left(\bigvee A_j\right).$$

If j takes only finite values, then

$$\bigvee \mathrm{Cl}(A_j) = \mathrm{Cl}\left(\bigvee A_j\right).$$

Also,

$$\bigvee \mathrm{Int}(A_j) \leq \mathrm{Int}\left(\bigvee A_j\right).$$

Proposition 10.4.2 *For a fuzzy set A of a fuzzy topological space X,*

(i) $1 - \mathrm{Int}\, A = \mathrm{Cl}(1 - A)$, *and*
(ii) $1 - \mathrm{Cl}\, A = \mathrm{Int}(1 - A)$.

Proposition 10.4.3 *Suppose that A is a fuzzy closed set of a fuzzy topological space X and B is a fuzzy closed set of a fuzzy topological space Y, then $A \times B$ is a fuzzy closed set of the fuzzy product topological space $X \times Y$.*

θ-topologies are very foundational and important.

Definition 10.4.2 Assume that A is a fuzzy set in X. Then, the subset

$$A^S = \{(x, a) | P_x^a \in A \text{ and } a \in (0, 1)\}$$

of the product set $X \times (0, 1)$ is called the *shape* of the fuzzy set A. The family

$$\{A^S | A \in [0, 1]^X\}$$

of all shapes of fuzzy sets in X is denoted by $\mathscr{G}$.

Theorem 10.4.1 *The operator* $^S : [0, 1]^X \to \mathscr{G}$ *is an isomorphism for "∪" and finite "∩."*

Proof: It should be clear that the operator ^{S} is injective. Assume that $(A_j)_{j \in J}$ is a family of fuzzy subsets of X. Then,

$$\begin{aligned}
\bigcup_{j \in J} A_j^S &= \bigcup_{i \in J} \{(x, \alpha) | x \in X \text{ and } 0 < \alpha < A_j(x)\} \\
&= \left\{ (x, \alpha) | x \in X \text{ and } 0 < \bigvee_{j \in J} A_j(x) \right\} \\
&= \left(\bigcup_{j \in J} A_j \right)^S .
\end{aligned}$$

Similarly, we can prove

$$\bigcap_{j=1,\dots,n} A_j^S = \left(\bigcap_{j=1,\dots,n} A_j\right)^S .$$

□

Definition 10.4.3 A topology for $(0,1)$ is called *θ-topology* if and only if the family of open sets consists of some open intervals $(0,\alpha)$, where $\alpha \in [0,1]$.

One can easily show that

$$\theta_0 = \{\emptyset, (0,1)\} \quad \text{and} \quad \theta_I = \{(0,\alpha) | \alpha \in [0,1]\}$$

are θ-topologies.

Definition 10.4.4 Assume that $\mathfrak{J}$ is a topology for X and θ is a θ-topology for $(0,1)$. Then, the family of fuzzy sets of X

$$\mathsf{F}_{\mathfrak{J}\times\theta} = \{A | A \in [0,1]^X \text{ and } A^S \in \mathfrak{J} \times \theta\}$$

is a fuzzy topology for X that is called *product-induced topology* and $(X, \mathsf{F}_{\mathfrak{J}\times\theta})$ is the product-induced space.

Theorem 10.4.2 *The product-induced space $(X, \mathsf{F}_{\mathfrak{J}\times\theta})$ and the topological space $(X \times (0,1), \mathfrak{J} \times \theta)$ are topologically isomorphic, that is, there exists an injection correspondence $^S : \mathsf{F}_{\mathfrak{J}\times\theta} \times \mathfrak{J} \times \theta$ such that*

(i) $\chi_\emptyset^S = \emptyset$, $\chi_X^S = X \times (0,1)$,
(ii) $(A \cap B)^S = A^S \cap B^S$, *and*
(iii) $(\bigcup_{j\in J} A_j)^S = \bigcup_{j\in J} A_j^S$,

for all $A, B, A_j \in \mathsf{F}_{\mathfrak{J}\times\theta}$, $j \in J$.

The proof of the theorem is left as an exercise.

10.5 Fuzzy Separation

Let $\mathfrak{L}(X)$ be the set of all topologies on X and $W(X)$ the set of all fuzzy topologies on X. On $\mathbb{R}$ we consider the topology

$$\mathfrak{L}_r = \{(a,\infty) | a \in \mathbb{R}\} \cup \{\emptyset\}.$$

Given a topological space I, the induced topology will be denoted by I_r. We define the mapping $\iota : W(X) \to \mathfrak{L}(X)$, which maps the family of "functions" δ to the initial topology $\iota(\delta)$ on X.

Assume that (X, σ) is a fuzzy topological space and $U \in \tau = \iota(\sigma)$. Then, the fuzzy set with shape $U \times (0, \alpha)$ is called a *fundamental* fuzzy set for (X, σ) and is denoted by N_U^α.

Definition 10.5.1 A fuzzy topological space (X, τ) is a *fuzzy quasi-T_0-space* if and only if for every $x \in X$, and $\lambda \neq \mu$, $\lambda, \mu \in [0, 1]$, either $P_x^\lambda \notin \mathrm{Cl}(P_x^\mu)$ or $P_x^\mu \notin \mathrm{Cl}(P_x^\lambda)$.

When $\mu < \lambda$, $P_x^\mu \in \mathrm{Cl}(P_x^\lambda)$, then (S, ρ) is a fuzzy quasi-T_0-space if and only if for every $x \in S$ and $0 < \mu < \lambda \leq 1$, $P_x^\lambda \in \mathrm{Cl}(P_x^\mu)$.

Definition 10.5.2 A fuzzy topological space (X, τ) is a *fuzzy T_0-space* if and only if for any two distinct points in (X, τ), at least one of them has a neighborhood that is not a neighborhood of the other.

Theorem 10.5.1 *If the fuzzy topological space (X, τ) is a fuzzy T_0-space, then the initial space $(X, \iota(\tau))$ is a T_0-space.*

Proof: Suppose that x and y are two distinct points in X. Now, consider the fuzzy points P_x^α and P_y^α, $0 < \alpha < 1$. Since (X, τ) is a fuzzy T_0-space, we can suppose that P_x^α has an open neighborhood B such that $P_y^\alpha \notin B$. This means that $x \in {}^{\alpha+}B \in \iota(\tau)$ but $y \notin {}^{\alpha+}B$, and so $(X, \iota(\tau))$ is a T_0-space. □

Definition 10.5.3 A fuzzy topological space (X, τ) is a *fuzzy T_1-space* if and only if $P_y^\beta \not\subset P_x^\alpha$ and the dual point $P_y^{1-\beta}$ has a neighborhood that is not quasi-coincident with P_x^α.

Theorem 10.5.2 *If the fuzzy topological space (X, τ) is a fuzzy T_1-space, then the initial space $(X, \iota(\tau))$ is a T_1-space.*

Proof: Assume that $x \neq y$ and consider the points $P_x^{0.5}$ and $P_y^{0.5}$. Then, $P_y^{0.5}$ has a neighborhood A such that A and $P_x^{0.5}$ are not quasi-coincident. Thus, $P_x^{0.5} \notin A$ and so $x \notin {}^{0.5+}A$. Hence, $(X, \iota(\tau))$ is T_1. □

We state the following result without proof:

Theorem 10.5.3 *A fuzzy topological space (X, τ) is a T_1-space if and only if every point P_x^α is fuzzy closed.*

Theorem 10.5.4 *A fuzzy topological space (X, τ) is a fuzzy T_1-space if and only if for any fuzzy set A, we have*

$$A = \bigcap\{B | A \subset B \in \tau\}.$$

Proof: ⇒: For any P_x^α and $P_y^\beta \not\subset P_x^\alpha$, we have

$$P_y^{1-\beta} = \bigcap\{B|P_y^{1-\beta} \subset B \in \tau\}.$$

This means that $P_y^{1-\beta}$ has an open neighborhood B such that B and P_x^α are not quasi-coincident and so $P_y^\beta \not\subset P_x^{\alpha^C}$. Hence, $\mathrm{Cl}(P_x^\alpha) = P_x^\alpha$. Because of Theorem 10.5.3, the fuzzy topological space (X, τ) is a fuzzy T_1-space.

⇐: For any fuzzy set A and point $P_x^\alpha \subset A$, there is a $\beta < \alpha$ such that $P_x^\beta \not\subset A$. The fuzzy topological space (X, τ) is a fuzzy T_1-space and the point $P_x^{1-\beta}$ is a closed fuzzy set, so $P_x^{1-\beta C}$ is an open fuzzy set and $A \subset P_x^{1-\beta C}$, $P_x^\alpha \not\subset P_x^{1-\beta C}$. Suppose that $B = P_x^{1-\beta C}$. Then,

$$A = \bigcap\{B|A \subset B \in \tau\}.$$

□

Definition 10.5.4 A fuzzy topological space (X, τ) is a fuzzy T_2-space if and only if for any points P_x^α and $P_y^\beta \not\subset P_x^\alpha$ there are neighborhoods of P_x^α and $P_y^{1-\beta}$ that are not quasi-coincident to each other.

Theorem 10.5.5 *If (X, τ) is a fuzzy T_2-space, then $(X, \iota(\tau))$ is a T_2-space.*

Theorem 10.5.6 *A fuzzy topological space is a fuzzy T_2-space if and only if for any P_x^α we have*

$$P_x^\alpha = \bigcap\{C|P_x^\alpha \in \mathrm{Int}(C) \text{ and } C^C \in \tau\}.$$

Proof: ⇒: For any points P_x^α and $P_y^\beta \not\subset P_x^\alpha$, there is a closed neighborhood C of P_x^α such that $P_y^\beta \not\subset C^C$ and so $P_y^{1-\beta} \in C^C$. Thus, $\mathrm{Int}(C)$ and C^C are the neighborhoods of P_x^α and $P_y^{1-\beta}$, respectively, that are not quasi-coincident to each other. So, the fuzzy topological space (X, τ) is a fuzzy T_2-space.

⇐: Suppose that $P_y^\beta \not\subset P_x^\alpha$, then P_x^α and $P_y^{1-\beta}$ have neighborhoods A and B, respectively, that are not quasi-coincident to each other. That is, $P_x^\alpha \in A \subset B^C$ and $P_y^\beta \not\subset B^C$. Assume that $C = B^C$. Then, CE is a closed neighborhood of P_x^α and we have

$$P_x^\alpha = \bigcap\{C|P_x^\alpha \in \mathrm{Int}(C) \text{ and } C^C \in \tau\}.$$

□

It should be clear that a fuzzy T_2-space implies a fuzzy T_1-space, and a fuzzy T_1-space implies a fuzzy T_0-space.

Definition 10.5.5 A fuzzy topological space (X, τ) is a fuzzy T_3-space or fuzzy *regular* if and only if for any point P_x^α and its open neighborhood A, there is a fuzzy set B such that

$$P_x^\alpha \in \mathrm{Int}(B) \subset \mathrm{Cl}(B) \subset A,$$

and A is a neighborhood of B.

A fuzzy set A is a neighborhood of B if and only if A is a neighborhood of every point $P_x^{B(x)}$, $B(x) > 0$, and $x \in X$.

Theorem 10.5.7 *If (X, τ) is a fuzzy T_3-space, then the initial space $(X, \iota(\tau))$ is a T_3-space.*

Proof: Since the following family

$$\{{}^{\alpha}A | A \in \tau \text{ and } \alpha \in (0, 1)\}$$

forms a base of $\iota(\tau)$, we have to show that, for any $x \in X$ and its open neighborhood

$$U = \bigcap_{i=1,\dots,n} \{{}^{\alpha_i+}A_i | A_{\in}\tau \text{ and } \alpha_i \in (0, 1)\},$$

there is an open set V and a closed set W in $(X, \iota(\tau))$ such that $x \in V \subset W \subset U$. In fact, since (X, τ) is a fuzzy T_3-space, then for any fuzzy point $P_V^{\alpha_i}$ and any of its open neighborhoods A_i there exists an open fuzzy set B_i such that

$$P_V^{\alpha_i} \in B_i \subset \text{Cl}(B^{\mathsf{C}}) \subset A_i,$$

and A_i is a neighborhood of $\text{Cl}(B_i)$. So we have

$$x \in {}^{\alpha_i+}B_i \subset {}^{\alpha_i}\text{Cl}(B_i) \subset {}^{\alpha_i+}A_i.$$

Assume that

$$V = \bigcap_{i=1,\dots,n} {}^{\alpha_i+}B_i \quad \text{and} \quad W = \bigcap_{i=1,\dots,n} {}^{\alpha_i}\text{Cl}(B_i).$$

Clearly, they are open and closed sets in $(X, \iota(\tau))$, respectively, and we have that $x \in V \subset W \subset U$. Therefore, $(X, \iota(\tau))$ is a T_3-space. □

Definition 10.5.6 The fuzzy sets A and B in (X, τ) are *strong non-quasi-coincident* (or *strong non-quasi-coincident*) if and only if for any $x \in X$, $A(x) + B(x) \leq 1$, and if $A(X) + B(x) = 1$, then either $A(x) = 1$ or $B(x) = 1$.

Theorem 10.5.8 *A fuzzy topological space (X, τ) is a fuzzy T_3-space if and only if for any point P_x^{α} and any closed set C that is strong non-quasi-coincident with P_x^{α}, there are neighborhoods A and B of P_x^{α} and C, respectively, such that A and B are not quasi-coincident.*

Proof: $\Rightarrow$: Suppose that C is a closed fuzzy set and point $P_x^{\alpha} \in C^{\mathsf{C}} \in \tau$. Then, P_x^{α} and C are strong non-quasi-coincident and there are neighborhoods A and B of P_x^{α} and C, respectively, such that A and B are not quasi-coincident, so

$$P_x^{\alpha} \in A \subset B^{\mathsf{C}} \subset C^{\mathsf{C}},$$

and C^{C} is a neighborhood of B^{C}. Hence (X, τ) is a fuzzy T_3-space.

$\Leftarrow$: Assume that the fuzzy point P_x^α and the closed fuzzy set C are strong non-quasi-coincident, that is, if $P_x^\alpha \in C^{\mathrm{C}}$, then by the fuzzy T_3-space property there is an open fuzzy set A such that

$$P_x^\alpha \in A \subset \mathrm{Cl}(A) \subset C^{\mathrm{C}},$$

and C^{C} is a neighborhood of B^{C}. Suppose that $B = \mathrm{Cl}(A^{\mathrm{C}})$, then A and B are neighborhoods of P_x^α and C, respectively, and they are non-quasi-coincident. □

It is easy to show that a fuzzy T_3-space is a fuzzy T_2-space. Suppose that (X, τ) is a fuzzy T_3-space. Then, any two $P_y^\beta \not\subset P_x^\alpha$ are different because of the basic property of fuzzy T_3-spaces. Now, we can assume that $A \in \tau$, $P_x^\alpha \in A$, and $P_y^\beta \notin A$. Then, there is an open fuzzy set B such that

$$P_x^\alpha \in B \subset \mathrm{Cl}(B) \subset A.$$

Assume that $C = B^{\mathrm{C}}$. Then, C is a neighborhood of $P_y^{1-\beta}$ and is not quasi-coincident with B. That is, the fuzzy topological space (X, τ) is a fuzzy T_2-space.

Definition 10.5.7 A fuzzy topological space (X, τ) is a fuzzy T_4-space if and only if for any closed fuzzy set C and any of its open neighborhoods B in (X, τ), there is a fuzzy set A such that

$$C \subset \mathrm{Int}(A) \subset \mathrm{Cl}(A) \subset B,$$

where $\mathrm{Int}(A)$ is a neighborhood of C and B is a neighborhood of $\mathrm{Cl}(A)$.

Theorem 10.5.9 *A fuzzy topological space (X, τ) is a fuzzy T_4-space if and only if for any two strong non-quasi-coincident closed sets C and D there exist open neighborhoods A and B of C and D, respectively, such that A and B are non-quasi-coincident.*

Definition 10.5.8 Suppose that (X, τ) is a fuzzy topological space and D a dense subset of the interval $[0, 1]$. Then, the family $\{O_{d_{-d \in D}}\}$ of open fuzzy sets in (X, τ) is called a *scale* of fuzzy sets if and only if $\mathrm{Cl}(O_{d_1}) \subset O_{d_2}$ and O_{d_2} is a neighborhood of $\mathrm{Cl}(O_{d_1})$ for any pair $d_1 < d_2$ in D.

Now we can generalize Urysohn's lemma (see Theorem 10.1.1) on usual topologies to the case of fuzzy topologies.

Theorem 10.5.10 *A fuzzy topological space (X, τ) is a fuzzy T_4-space if and only if for any closed fuzzy set B and any of its open neighborhoods A, there is a scale of fuzzy open sets $\{O_d\}_{d \in D}$ such that $B \subset O_d \subset A$, for all $d \in D$.*

Proof: ⇒: Obvious.

⇐: Suppose that B is a closed set and A its open neighborhood in (X, τ). Then, there is an open fuzzy set O_0 such that

$$B \subset O_0 \subset \mathrm{Cl}(O_0) \subset A.$$

In addition, O_0 and A are the neighborhoods of B and $\mathrm{Cl}(O_0)$, respectively.

Assume that D is the set of *dyadic rationals*[8], and

$$D_n = \left\{ \frac{m}{2^n} | m < 2^b \text{ and } m = 2k-1,\ k = 1, 2, \ldots \right\}.$$

Then, $D = \bigcup_{n=1,2,\ldots} D_n$. Let us construct a scale of fuzzy open sets. First we set $O_1 = A$. Next, we choose an open set $O_{1\ 2}$ such that

$$\mathrm{Cl}(O_0) \subset O_{1\ 2} \subset \mathrm{Cl}(O_{1\ 2}) \subset O_1,$$

where $O_{1\ 2}$ and O_1 are the neighborhoods of $\mathrm{Cl}(O_0)$ and $\mathrm{Cl}(O_{1\ 2})$, respectively. Now, suppose that for all $d \in D_1 \cup D_2 \cup \cdots \cup D_n$, O_d is defined such that $\mathrm{Cl}(O_{d'}) \subset O_{d''}$ and $O_{d''}$ is a neighborhood of $\mathrm{Cl}(O_{d'})$ for any pair $d' < d''$. For any $d \in D_{n+1}$, there are adjacent elements $d', d'' \in D_1 \cup D_c \cup \cdots \cup D_n$ such that $d = (d' + d'')/2$. Using the properties of the fuzzy T_4-space, we can choose an open fuzzy set O_d satisfying

$$\mathrm{Cl}(O'_d) \subset O_d \subset \mathrm{Cl}(O_d) \subset D_{d''},$$

where O_d and $O_{d''}$ are the neighborhoods of $\mathrm{Cl}(O_{d'})$ and $\mathrm{Cl}(O_d)$, respectively. By induction, we get a scale of fuzzy open sets $\{O_d\}_{d\in D}$ such that $B \subset O_d \subset A$, where O_d and A are the neighborhoods of B and $\mathrm{Cl}(O_d)$, respectively. □

Definition 10.5.9 A fuzzy topological space (X, τ) is a fuzzy T_5-space if and only if for any set A and one of its neighborhoods B having the property $P_x^{A(x)} \in B$, there is an open fuzzy set O such that

$$A \subset O \subset \mathrm{Cl}(O) \subset B.$$

In addition, O is a neighborhood of A and $P_x^{(\mathrm{Cl}(O))(x)} \in B, x \in X$.

Theorem 10.5.11 *The product-induced space $(X, \mathbf{F}_{\mathfrak{F}\times\theta})$ has any of the fuzzy separation properties mentioned above if and only if the topological space $(X, \mathfrak{F})$ has the corresponding properties.*

---

8 A dyadic rational is a fraction whose denominator, when expressed in canonical form (i.e. an expression of the form $\frac{p}{q}$, where $p, q \in \mathbb{Z}$ and $p \perp q$, that is, they have no common divisor except 1) is an integral power of 2. For example, 1/2 or 3/8 are dyadic rationals, but 1/3 is not.

10.5.1 Separation

Definition 10.5.10 Two fuzzy sets A_1 and A_2 in (X, τ) are called *separated* if and only if there is $\mathcal{U}_i \in \tau, i = 1, 2$, such that $\mathcal{U}_i \supset A_i, i = 1, 2$, and $\mathcal{U}_1 \wedge A_2 = \emptyset = \mathcal{U}_2 \wedge A_1$.

Definition 10.5.11 Two fuzzy sets A_1 and A_2 in a fuzzy topological space (X, τ) are *Q-separated* if and only if there is a δ-closed set H_i, $i = 1, 2$, such that $H_i \supset A_i$, $i = 1, 2$, and $H_1 \wedge A_2 = \emptyset = H_2 \wedge A_1$. Obviously, A_1 and A_2 are Q-separated if and only if $\text{Cl}(A_1) \wedge A_2 = \emptyset = \text{Cl}(A_2) \wedge A_1$.

10.6 Fuzzy Nets

Assume that X is a set and $\geq$ is a binary relation on X. Elements x and y are said to be incomparable, written here as $x \sim y$, if neither $x \geq y$ nor $y \geq x$ is true. Then, the pair $(X, \geq)$ is a *semiorder* if it has the following properties:

(i) For all x and y, the expressions $x \geq y$ and $y \geq x$ cannot be simultaneously true.
(ii) For all x, y, z, and w, if the expressions $x \geq y$, $y \sim z$, and $z \geq w$ are true, then it must also be true that $x \geq w$.
(iii) For all x, y, z, and w, if the expressions $x \geq y$, $y \geq z$, and $y \sim w$ are true, then it cannot also be true that $x \sim w$ and $z \sim w$ simultaneously.

Definition 10.6.1 Assume that D is a nonempty set and that "$\geq$" is a semiorder on D. Then, the pair $(D, \geq)$ is called a *directed set*, directed by $\geq$, if and only if for every pair $m, n \in D$, there is a $p \in D$ such that $p \geq m$ and $p \geq n$.

Definition 10.6.2 Suppose that $(D, \geq)$ is a directed set, X a crisp set, and $\mathcal{X}$ is the collection of all fuzzy points in X. Then, the function $S : D \to \mathcal{X}$ is called a *fuzzy net* in X. In other words, a fuzzy net is a pair $(S, \geq)$ such that S is a function from D to $\mathcal{X}$ and $\geq$ directs D (the domain of S). For $n \in D$, $S(n)$ is often denoted by S_n and so the net S is often denoted by $\{S_n, n \in D\}$.

Definition 10.6.3 Let $\{S_n, n \in D\}$ be a fuzzy net in X. Then, S is *quasi-coincident* with A if and only if for each $n \in D$, S_n is quasi-coincident with A. S is *eventually quasi-coincident* with A if and only if there is an element m of D such that, if $n \in D$ and $n \geq m$, then S_n is quasi-coincident with A. S is *frequently quasi-coincident* with A if and only if for each $m \in D$ there is an $n \in D$ such that $n \geq m$ and S_n is quasi-coincident with A. S is in A if and only if for each $n \in D$, $S_n \in A$.

10.7 Fuzzy Compactness

As in the case of compact topological spaces, we need to define the notion of a fuzzy cover:

Definition 10.7.1 A family $\mathcal{U}$ of fuzzy sets is a *cover* of a fuzzy set B if and only if

$$B \leq \bigvee \{A | A \in \mathcal{U}\},$$

It is an *open cover* if and only if each member of $\mathcal{U}$ is an open fuzzy set.

Definition 10.7.2 A fuzzy topological space (X, τ) is *quasi fuzzy compact* if and only if each open cover has a finite subcover.

Definition 10.7.3 A fuzzy topological space (X, τ) is *quasi fuzzy countable com pact* if and only if each open countable cover has a finite subcover.

In the previous definition, ϵ is both a number and a constant fuzzy set.

Definition 10.7.4 A fuzzy topological space (X, τ) is called *weakly fuzzy compact* if and only if for every open cover

$$\mathcal{U} = \{U_j | j \in J\}$$

of fuzzy open sets of X, that is,

$$\bigvee \{U_j | j \in J\} = \chi_X$$

and for every $\epsilon > 0$, there exists a finite subfamily $\{U_{j_1}, \dots, U_{j_m}\}$ of $\mathcal{U}$ such that

$$\bigvee \{U_{j_i} | i = 1, \dots, m\} \geq \mathbf{1} - \epsilon.$$

Suppose that (X, τ) is a fuzzy topological space (à la Chang or à la Lowen).

Definition 10.7.5 A fuzzy set $B \in [0, 1]^X$ is fuzzy compact if and only if for all families

$$\mathcal{D} \subset \tau \text{ such that } \bigvee_{D \in \mathcal{D}} D \geq B$$

and for all $\epsilon > 0$, there exists a finite subfamily $\mathcal{D}_0 \subset \mathcal{D}$ such that

$$\bigvee_{A \in \mathcal{D}_0} A \geq B - \epsilon.$$

Definition 10.7.6 A fuzzy topological space (X, τ) is *fuzzy compact* if and only if each constant fuzzy set in (X, τ) is fuzzy compact.

An alternative definition of à la Chang fuzzy compact spaces follows:

Definition 10.7.7 A fuzzy topological space (X, τ) is called *fuzzy compact* if and only if for every family $\mathscr{D}$ of fuzzy open sets of X and for every $a \in [0, 1]$ such that

$$\bigvee\{D|D \in \mathscr{D}\} \geq \mathbf{a}$$

and for every $\varepsilon \in (0, a]$, there exists a finite subfamily $\mathscr{D}_1$ of $\mathscr{D}$ such that

$$\bigvee\{D|U \in \mathscr{D}_1\} \geq \mathbf{a} - \epsilon.$$

10.8 Fuzzy Connectedness

In what follows, we will briefly describe the notion of fuzzy connectedness.

Definition 10.8.1 A fuzzy set D in (X, δ) is *disconnected* if and only if there are two nonempty sets A and B in the subspace D_0 (i.e. supp(D)) such that A and B are Q-separated and $D = A \vee B$. A fuzzy set is *connected* if and only if it is not disconnected.

The proof of the following result is omitted since it is based on a number of results that we have intentionally omitted – obviously, we did not plan to cover the whole field of fuzzy topology in a chapter.

Lemma 10.8.1 *A fuzzy set D is disconnected if and only if there are relative closed sets in the subspace D_0 such that $A \wedge D \neq \emptyset$, $B \wedge D \neq \emptyset$, $A \wedge B = \emptyset$, and $A \vee B \supset D$.*

10.9 Smooth Fuzzy Topological Spaces

In Section 10.3, we presented two different formulations of fuzzy topological spaces. Both of these formulations are crisp, mainly because fuzzy sets either belong or do not belong to the collection of fuzzy open sets. However, this seems a bit unnatural and so Šostak [264] introduced his own version of fuzzy topological spaces where an open set is open to some degree.

Definition 10.9.1 Assume that X is a nonempty crisp set and $\tau : [0, 1]^X \to [0, 1]$ is a function that has the following three properties:

(i) if $A, B \in [0,1]^X$, then $\tau(A \wedge B) \geq \tau(A) \wedge \tau(B)$;
(ii) if $A_i \in [0,1]^X$, $i \in J$, then

$$\tau\left(\bigvee_{i \in J} A_i\right) \geq \bigvee_{i \in J} \tau(A_i); \text{ and}$$

(iii) $\tau(\emptyset) = \tau(\chi_X) = 1$.

Then, (X, τ) is a *smooth*[9] fuzzy topological space and τ is a smooth fuzzy topology on X. Also, for any fuzzy set $C \in [0,1]^X$, $\tau(C)$ is the *degree of openness* of C.

If we assume that $\tau : L_1^X \to L_2$, where L_1 and L_2 are lattices, then we get more general structures that are called *smooth topological spaces*. These structures have been introduced by Ahmed A. Ramadan [246]. However, there are many generalizations of fuzzy topological spaces and we had no intention to cover all of them. After all, it is quite possible to define any kind of extension.

Definition 10.9.2 Suppose that τ_1 and τ_2 are smooth fuzzy topologies on X. Then, τ_1 is *finer* than τ_2 (or τ_2 is *coarser* than τ_1) if and only if $\tau_1(A) \geq \tau_2(A)$, for all $A \in [0,1]^X$.

Definition 10.9.3 Given a smooth fuzzy topological space (X, τ), the function $\tau^* : [0,1]^X \to [0,1]$ where $\tau^*(A) = \tau(A^{\complement})$, for all $A \in [0,1]^X$, defines the *degree of closedness* of a fuzzy set A.

The following result can be easily proved using the definitions above.

Proposition 10.9.1 *The mapping τ^* has the following properties:*

(i) if $A, B \in [0,1]^X$, then $\tau^*(A \vee B) \geq \tau^*(A) \wedge \tau^*(B)$;
(ii) if $A_1 \in [0,1]^X$ for all $i \in I$, then $\tau^*(\wedge_i A_i) \geq \vee_i \tau^*(A_i)$; and
(iii) $\tau^*(\emptyset) = \tau^*(\chi_X) = 1$.

Definition 10.9.4 Suppose that (X, τ) and (Y, σ) are smooth fuzzy topological spaces and $f : X \to Y$ is a map. Then, this mapping is called *fuzzy continuous* if $\tau(f^{\leftarrow}(B)) \geq \sigma(B)$, for all $B \in [0,1]^Y$.

In simple words, a mapping is a fuzzy continuous mapping if it does not decrease the degree of openness of fuzzy subsets in the opposite direction (i.e. the direction of the preimage operator).

---

9 Šostak did use the term "fuzzy topological spaces," however, later on the term "smooth fuzzy topological spaces" was put in wide use in the literature and so we have adopted the later term.

Proposition 10.9.2 *A map $f : X \to Y$ of smooth fuzzy topological spaces (X, τ) and (Y, σ) is fuzzy continuous if and only if $\tau^*(f^{\leftarrow}(B)) \geq \sigma^*(B)$ for all $B \in [0,1]^Y$.*

Proposition 10.9.3 *Suppose that (X, τ), (Y, σ), and (Z, δ) are smooth fuzzy topological spaces and $f : X \to Y$ and $g : Y \to Z$ are fuzzy continuous mappings. Then, the composition $g \circ f : X \to Z$ is also fuzzy continuous.*

The proof is obvious and so it is omitted.

Remark 10.9.1 It is easy to define a category of smooth fuzzy topological spaces. First, the objects would be the collection of smooth fuzzy topological spaces and the arrows would denote maps that are fuzzy continuous. Second, we note that composition is associative and the identity map $e : X \to X$ is fuzzy continuous for any space (X, τ). Thus, we have all ingredients to form a category.

10.10 Fuzzy Banach and Fuzzy Hilbert Spaces

In this section, we introduce some topological notions of fuzzy normed spaces (see Section 9.6).

Definition 10.10.1 Suppose that $(X, N, *)$ is a fuzzy normed space. Then, we define the ϵ-ball $B_\varepsilon(x, t)$ and the closed ϵ-ball $\overline{B}_\varepsilon(x, t)$ with center $x \in X$ and radius $0 < \varepsilon < 1$ for all $t > 0$ as follows:

$$B_\varepsilon(x, t) = \{y | y \in X \text{ and } N(x - y, t) > 1 - \varepsilon\},$$
$$\overline{B}_\varepsilon(x, t) = \{y | y \in X \text{ and } N(x - y, t) \geq 1 - \varepsilon\},$$

respectively.

Theorem 10.10.1 *Assume that $(X, N, *)$ is a fuzzy normed space. Then, every ϵ-ball $B_\varepsilon(x, t)$ is an open set.*

Proof: Suppose that $B_\varepsilon(x, t)$ is an ϵ-ball whose center is x and its radius is ε for all $t > 0$. Also, suppose that $y \in B_\varepsilon(x, t)$. Then, $N(x - y, t) > 1 - \varepsilon$, which means that there is a $t_0 \in (0, t)$ such that $N(x - y, t_0) > 1 - \varepsilon$. Let $\varepsilon_0 = N(x - y, t_0)$. Then, since $\varepsilon_0 > 1 - \varepsilon$, there exists $s \in (0, 1)$ such that $\varepsilon_0 > 1 - s > 1 - \varepsilon$. Furthermore, for any ε_0 and s such that $\varepsilon_0 > 1 - s$, there exists $\varepsilon_1 \in (0, 1)$ such that $\varepsilon_0 * \varepsilon_1 > 1 - s$. Now, we consider the ϵ-ball $B_{1-\varepsilon_1}(y, t-t_0)$ and we claim that $B_{1-\varepsilon_1}(y, t-t_0) \subset B_\varepsilon(x, t)$. Let $z \in B_{1-\varepsilon_1}(y, t-t_0)$. Then, $N(y-z, t-t_0) > \varepsilon_1$ and so

$$\begin{aligned} N(x-z, t) &\geq N(x-y, t_0) * N(y-z, t-t_0) \\ &\geq \varepsilon_0 * \varepsilon_1 \\ &\geq 1-s \\ &> 1-\varepsilon. \end{aligned}$$

Thus, $z \in B_\varepsilon(x, t)$ and hence $B_{1-\varepsilon_1}(y, t-t_0) \subset B_\varepsilon(x, t)$. □

Definition 10.10.2 Suppose that $(X, N, *)$ is a fuzzy normed space. Then, a subset A of X is *F-bounded* if there is $t > 0$ and $\varepsilon \in (0, 1)$ such that $N(x-y, t) > 1-\varepsilon$ for all $x, y \in A$.

Definition 10.10.3 Suppose that $(X, N, *)$ is a fuzzy normed space. Then,

(i) A sequence $\{x_n\}$ in X *converges* to a point $x \in X$ for all $\epsilon > 0$ and $\lambda > 0$ if there exists a positive integer M, such that

$$N(x_n - x, \epsilon) > 1 - \lambda$$

whenever $n \geq M$.

(ii) A sequence $\{x_n\}$ in X is called a *Cauchy sequence* for all $\epsilon > 0$ and $\lambda > 0$ if there exists a positive integer M, such that

$$N(x_n - x_m, \epsilon) > 1 - \lambda$$

whenever $n \geq m \geq M$.

(iii) A fuzzy normed space $(X, N, *)$ is *complete*.

Definition 10.10.4 A fuzzy normed space $(X, N, *)$ is a *fuzzy Banach space* if X is complete with respect to the fuzzy metric generated by the fuzzy norm (see Definition 9.1.17).

Fuzzy Hilbert spaces have been introduced by Manizheh Goudarzi and Vaezpour [147]. In what follows, we will make use of the following real step function:

$$H(t) = \begin{cases} 1, & \text{if } t > 0, \\ 0, & \text{otherwise.} \end{cases}$$

Definition 10.10.5 A fuzzy inner product space is a triplet $(X, F, *)$, where X is a real vector space, $*$ is a continuous t-norm, and F is a fuzzy subset on $X^2 \times \mathbb{R}$ such that for all $x, y, z \in X$ and $s, t, r \in \mathbb{R}$, the following conditions are satisfied:

(FI-1) $F(x, x, 0) = 0$ and $F(x, x, t) > 0$ for all $t > 0$;
(FI-2) $F(x, x, t) \neq H(t)$ for all $t \in \mathbb{R}$ if and only if $x \neq 0$;
(FI-3) $F(x, y, t) = F(y, x, t)$;
(FI-4) For all real numbers α,

$$F(\alpha x, y, t) = \begin{cases} F(x, y, \frac{t}{\alpha}), & \text{if } \alpha > 0, \\ H(t), & \text{if } \alpha = 0, \\ 1 - F\left(x, y, \dfrac{t}{-\alpha}\right), & \text{if } \alpha < 0; \end{cases}$$

(FI-5) $F(x, x, t) * F(y, y, s) \leq F(x + y, x + y, t + s)$;
(FI-6) $\bigvee_{s+r=t}(F(x, z, s) * F(y, z, r)) = F(x + y, z, t)$;
(FI-7) $F(x, y, \cdot) : \mathbb{R} \to [0, 1]$ is continuous in $\mathbb{R} \setminus \{0\}$; and
(FI-8) $\lim_{t\to\infty} F(x, y, t) = 1$.

Example 10.10.1 Suppose that $(X, \langle\ ,\ \rangle)$ is a crisp inner product space. Let us define a mapping $F : X^2 \times \mathbb{R} \to [0, 1]$ as follows

$$F(\alpha x, y, t) = \begin{cases} \dfrac{kt^{n/2}}{kt^{n/2} + m|\langle \alpha x|y\rangle|^{n/2}}, & \text{if } t > 0,\ \alpha > 0, \text{ and } k, m, n \in \mathbb{R}^+ \\ 1 - \dfrac{kt^{n/2}}{kt^{n/2} + m|\langle \alpha x|y\rangle|^{n/2}}, & \text{if } t < 0 \text{ and } \alpha < 0, \\ 0, & \text{if } t \leq 0. \end{cases}$$

Assume that $k = m = n = 1$ and that the t-norm is min. Then, it can be easily shown that the $(X, F', \min)$ is a fuzzy inner product space, where

$$F'(\alpha x, y, t) = \begin{cases} \dfrac{t^{1/2}}{kt^{1/2} + |\langle \alpha x|y\rangle|^{1/2}}, & \text{if } t > 0 \text{ and } \alpha > 0 \\ 1 - \dfrac{t^{1/2}}{t^{1/2} + |\langle \alpha x|y\rangle|^{1/2}}, & \text{if } t < 0 \text{ and } \alpha < 0, \\ 0, & \text{if } t \leq 0. \end{cases}$$

This mapping will be called the *standard fuzzy inner product* induced by the inner product $\langle\ ,\ \rangle$.

Lemma 10.10.1 *If $(X, F, *)$ is a fuzzy inner product space, then $F(p, q, t)$ is nondecreasing with respect to t for each $p, q \in X$.*

Theorem 10.10.2 *Suppose that $(X, F, *)$ is a fuzzy inner product space, where $*$ is a strong t-norm and for each x.y $\in$, $\bigvee\{t|t \in \mathbb{R}$ and $F(x, y, t) < 1\} < \infty$. We define $\langle \mid \rangle : X \times X \to \mathbb{R}$ by*

$$\langle x|y\rangle = \bigvee\{t|t \in \mathbb{R} \text{ and } F(x, y, t) < 1\}.$$

Then, $(X, \langle \mid \rangle)$ is an inner product space.

Definition 10.10.6 Assume that $(X, F, *)$ is a fuzzy inner product space, where $*$ is a strong t-norm and for each $x, y \in X$, $\bigvee\{t|t \in \mathbb{R}$ and $F(x, y, t) < 1\} < \infty$. Next, we define $\langle \mid \rangle : X \times X \to \mathbb{R}$ by

$$\langle x|y\rangle = \bigvee\{t|t \in \mathbb{R} \text{ and } F(x, y, t) < 1\},$$

and $\|x\| = \sqrt{\langle x|x\rangle}$. We say that $(X, F, *)$ is a *fuzzy Hilbert space* if $(X, \|\cdot\|)$ is a complete normed space.

10.11* Fuzzy Topological Systems

The Dialectica categories construction (see for example [93]) can be instantiated using any lineale and the basic category **Set**. A lineale is a structure defined as follows:

Definition 10.11.1 The quintuple $(L, \leq, \circ, 1, \multimap)$ is a lineale if:

- $(L, \leq)$ is poset,
- $\circ : L \times L \to L$ is an order-preserving multiplication, such that $(L, \circ, 1)$ is a symmetric monoidal structure (i.e. for all $a \in L$, $a \circ 1 = 1 \circ a = a$).
- if for any $a, b \in L$ there exists a largest $x \in L$ such that $a \circ x \leq b$, then this element is denoted as $a \multimap b$ and is called the pseudo-complement of a with respect to b.

As was discussed in [272], the unit interval, since it is a Heyting algebra, has all the properties of a lineale structure. In particular, one can prove that the quintuple $([0, 1], \leq, \wedge, 1, \Rightarrow)$, $a \wedge b = \min\{a, b\}$, and $a \Rightarrow b = \bigvee\{c : c \wedge a \leq b\}$ $(a \vee b = \max\{a, b\})$, is a lineale.

Definition 10.11.2 The category $\mathsf{Dial}_{[0,1]}(\mathbf{Set})$ has as objects triples $A = (U, X, \alpha)$, where U and X are sets and α is a map $U \times X \to [0, 1]$. Thus, each object is a fuzzy relation. A map from $A = (U, X, \alpha)$ to $B = (V, Y, \beta)$ is a pair of **Set** maps $(f, g), f : U \to V, g : Y \to X$ such that

$$\alpha(u, g(y)) \leq \beta(f(u), y),$$

or in pictorial form:

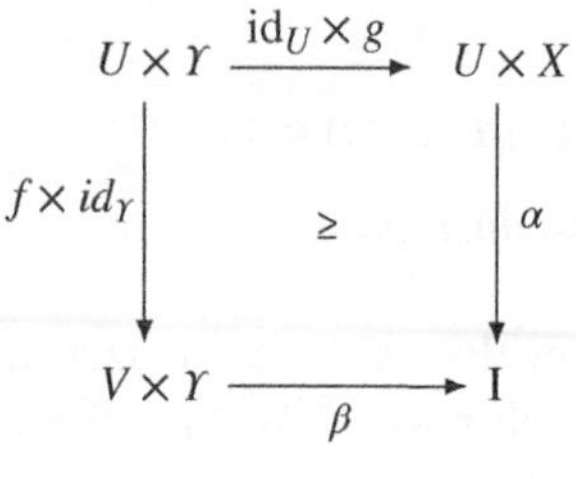

$$\begin{array}{ccc} U\times Y & \xrightarrow{id_U\times g} & U\times X \\ {\scriptstyle f\times id_Y}\downarrow & \geq & \downarrow{\scriptstyle \alpha} \\ V\times Y & \xrightarrow[\beta]{} & I \end{array}$$

Remark 10.11.1 When one uses "=" instead of "≤," then the resulting category is a Chu category.

Assume that (f, g) and (f', g') are the following arrows:

$$(U,X,\alpha) \overset{(f,g)}{\rightarrow} (V,Y,\beta) \overset{(f',g')}{\rightarrow} (W,Z,\gamma).$$

Then, $(f, g) \circ (f', g') = (f \circ f', g' \circ g)$ is such that

$$\alpha(u, (g' \circ g)(z)) \leq \gamma((f \circ f')(u), z).$$

Tensor products and the internal-hom in $\mathsf{Dial}_{[0,1]}(\mathbf{Set})$ are given as in the Girard-variant of the Dialectica construction [89]. Given the objects $A = (U, X, \alpha)$ and $B = (V, Y, \beta)$, the tensor product $A \otimes B$ is $(U \times V, X^V \times Y^U, \alpha \times \beta)$, where $\alpha \times \beta$ is the relation that, using the lineale structure of I, takes the minimum of the membership degrees. The linear function-space or internal-hom is given by $A \to B = (V^U \times Y^X, U \times X, \alpha \to \beta)$, where again the relation $\alpha \to \beta$ is given by the implication in the lineale. With this structure, we obtain:

Theorem 10.11.1 *The category* $\mathtt{Dial}_{[0,1]}$*(**Sets**) is a monoidal closed category with products and coproducts.*

Products and coproducts are given by $A \times B = (U \times V, X + Y, \gamma)$ and $A \oplus B = (U + V, X \times Y, \delta)$, where $\gamma : U \times V \times (X + Y) \to [0, 1]$ is the fuzzy relation that is defined as follows:

$$\gamma((u, v), z) = \begin{cases} \alpha(u, x), & \text{if } z = (x, 0), \\ \beta(v, y), & \text{if } z = (y, 1). \end{cases}$$

Similarly, for the coproduct $A \oplus B$.

Steven Vickers [290] introduced *topological systems*. These structures are triples $(U, X, \models)$, where X is a frame whose elements are called *opens* and U is a set whose elements are called *points*. Also, the relation $\models$ is a subset of $U \times X$, and when

$u \models x$, where $u \in U$ and $x \in X$, we say that u *satisfies* x. In addition, the following must hold:

- if S is a finite subset of X, then

$$u \models \bigwedge S \Leftrightarrow u \models x \text{ for all } x \in S.$$

- if S is any subset of X, then

$$u \models \bigvee S \Leftrightarrow u \models x \text{ for some } x \in S.$$

Given two topological systems $(U, X, \models)$ and $(V, Y, \models')$, a map from (U, X) to (V, Y) consists of a function $f : U \to V$ and a frame homomorphism $\phi : Y \to X$, if $u \models \phi(y) \Leftrightarrow f(u) \models' y$. Topological systems and continuous maps between them form a category, which we write as **TopSystems**.

Fuzzy topological systems have been introduced by Syropoulos and de Paiva [279].

Definition 10.11.3 A *fuzzy topological system* is an object (U, X, α) of $\mathsf{Dial}_{[0,1]}(\mathbf{Set})$ such that X is a frame and α satisfies the following conditions:

(i) If S is a finite subset of X, then

$$\alpha(u, \bigwedge S) \leq \alpha(u, x) \text{ for all } x \in S.$$

(ii) If S is any subset of X, then

$$\alpha(u, \bigvee S) \leq \alpha(u, x) \text{ for some } x \in S.$$

(iii) $\alpha(u, \top) = 1$ and $\alpha(u, \bot) = 0$ for all $u \in U$.

To see that fuzzy topological systems also form a category, we need to show that, given morphisms $(f, F) : (U, X) \to (V, Y)$ and $(g, G) : (V, Y) \to (W, Z)$, the obvious composition $(g \circ f, F \circ G) : (U, X) \to (W, Z)$ is also a morphism of fuzzy topological systems. But we know that $\mathsf{Dial}_{[0,1]}(\mathbf{Set})$ is a category, and conditions (i), (ii), and (iii) do not apply to morphisms. Identities are given by $(id_U, id_X) : (U, X) \to (U, X)$.

The collection of objects of $\mathsf{Dial}_{[0,1]}(\mathbf{Set})$ that are fuzzy topological systems and the arrows between them, form the category **FTopSystems**, which is a subcategory of $\mathsf{Dial}_{[0,1]}(\mathbf{Set})$.

Proposition 10.11.1 *Any topological system $(U, X, \models)$ is a fuzzy topological system (U, X, ι), where*

$$\iota(u, x) = \begin{cases} 1, & \text{when } u \models x, \\ 0, & \text{otherwise.} \end{cases}$$

Proof: Consider the first property of the relation "$\models$"

$$u \models \bigwedge S \Leftrightarrow u \models x \text{ for all } x \in S.$$

This will be translated to

$$\iota(u, \bigwedge S) \leq \iota(u, x) \text{ for all } x \in S.$$

The inequality is in fact an equality since, whenever $u \models x$, $\iota(u, x) = 1$. Therefore, we can transform this condition into the following one:

$$\iota(u, \bigwedge S) = \iota(u, x) \text{ for all } x \in S.$$

A similar argument holds true for the second property. □

The following result is based on the previous one:

Theorem 10.11.2 *The category of topological systems is a full subcategory of* $\mathrm{Dial}_{[0,1]}(\mathbf{Set})$.

Obviously, it is not enough to provide a generalization of structures – one needs to demonstrate that these new structures have some usefulness. The following example provides an interpretation of these structures in a "real-life" situation.

Example 10.11.1 Vickers [290, p. 53] gives an interesting physical interpretation of topological systems. In particular, he considers the set U to be a set of programs that generate bit streams and the opens to be assertions about bit streams. For example, if u is a program that generates the infinite bit stream 010101010101...and "**starts** 01010" is an assertion that is satified if a bit stream starts with the digits "01010," then this is expressed as follows:

$$x \models \textbf{starts } 01010.$$

Assume now that x' is a program that produces bit streams that look like the following:

0 1 0 1 0 1 0 1 0

The individual bits are not identical to either "1" or "0," but rather similar to these. One can speculate that these bits are the result of some interaction of x' with its environment and this is the reason why they are not identical. Then, we can say that x' satisfies the assertion "**starts** 01010" to some degree, since the elements that make up the stream produced by x' are not identical, but rather similar.

Bart Kosko [180] has argued that fuzziness "measures the degree to which an event occurs, not whether it occurs. Randomness describes the uncertainty of event occurrence." This idea can be used to provide a different physical interpretation of fuzzy topological systems. Again, let U be a set of programs that generate bit streams and the opens be assertions about bit streams. Then, the expression $\alpha(u, x)$ would denote the degree to which program u will output a bitstream that will satisfy assertion x. When $\alpha(u, x) = 0$, we know that program u will not generate a bitstream that satisfies assertion x. Also, when $\alpha(u, x) = 1$, then the program will definitely generate a bitstream that satisfies assertion x. In addition, this new physical interpretation is based on the assumption that the programs behave dynamically and nondeterministically. Thus, one could say that some of the programs in U are interactive.

Exercises

10.1 Explain why the fuzzy metric defined in Example 10.2.1 generalizes the conditions that each crisp metric must satisfy.

10.2 Show that the triple that is described in Example 10.2.1 is a metric space.

10.3 Prove that the usual metric space is a special case of the fuzzy metric space defined in Definition 10.2.8.

10.4 Prove that U in Definition 10.3.15 is a fuzzy topological space.

10.5 Prove Proposition 10.4.1.

10.6 Prove Proposition 10.4.2.

10.7 Prove Proposition 10.4.3.

10.8 Prove Theorem 10.4.2.

10.9 Prove Theorem 10.5.5.

10.10 Prove Theorem 10.5.9.

10.11 Prove Theorem 10.5.11 for any kind of a fuzzy T_i-space, where $i = 1, 2, 3, 4, 5$.

10.12 Give a formal definition of smooth topological spaces.

10.13 Prove Proposition 10.9.2.

10.14 Show that $(X, F', \min)$ from Example 10.10.1 is a fuzzy inner product space.

11

Fuzzy Geometry

The theorems and propositions of classical Euclidean geometry which are based on the corresponding axiomatic corpus, study idealized shapes and figures made by crisp points and having exact dimensions on the plane or in three-dimensional space. However, natural reality appears different. Already on 27 January 1921, in his lecture on "Geometry and Experience" at a public session of the Prussian Academy of Sciences [114, 115], Albert Einstein, made the famous statement: "As far as the propositions of mathematics refer to reality they are not certain, and as far as they are certain they do not refer to reality". Einstein's aim was to underline the realistic possibility of adopting non-Euclidean geometry as the true geometry of the universe. In the context of fuzzy mathematics, one follows a different route in geometry and its relation to reality. Indeed, when one studies or measures real objects, she soon realizes that the very process of measurement will often render expectations and theoretically exact predictions as improbable. Points, lines, lengths, areas, shapes, and other geometrical concepts and notions become rather uncertain. It is here where, since the late 1970s, a steadily increasing interest, coming mostly from areas such as pattern recognition and image processing and analysis where the use of fuzzy mathematics has already been established, for a new subject that has become known as geometry based on vagueness or *fuzzy geometry*, has come into play.

11.1 Fuzzy Points and Fuzzy Distance

The first studies on the geometric properties of fuzzy sets appeared in the late 1970s, thus constituting the beginnings of fuzzy geometry (for a review see, e.g. [251]). In fact, fuzzy geometry has been studied from different perspectives, here however we will follow and in fact review the fine line of thought initiated by J.J. Buckley and E. Eslami [42, 43] who focus on the notion of membership and where all the standard basic geometric notions, such as lengths (heights,

A Modern Introduction to Fuzzy Mathematics, First Edition.
Apostolos Syropoulos and Theophanes Grammenos.

widths, diameters, etc.) and areas of fuzzy subsets, are fuzzy real numbers. This approach is in the spirit of the present book. We, further, point out that the concepts introduced in [42, 43] are based on the sup-min composition of fuzzy sets resembling Zadeh's extension principle (see Section 2.8). The aforementioned work of Buckley and Eslami is complemented by certain formulas for membership functions given for some of the introduced geometric concepts in [304].

In order to deal with fuzzy geometry in the Euclidean plane, it is necessary to keep in mind that fuzzy points on the plane may be visualized as surfaces in space. Now, one could imagine that an ordered pair of fuzzy (real) numbers (see Definition 3.1.1) would give a natural definition of a fuzzy point in the plane. However, this kind of definition turns out to be not useful in the study of fuzzy lines, thus we adopt the following definition of a fuzzy point in $\mathbb{R}^2$:

Definition 11.1.1 Let $(x_0, y_0) \in \mathbb{R}^2$ with respect to a Cartesian coordinate system and let P be a fuzzy subset of $\mathbb{R}^2$. Then, P is called a *fuzzy point* at (x_0, y_0) if it satisfies the following properties:

(i) $P(x_0, y_0)$ is upper semicontinuous,
(ii) $P(x, y) = 1 \Longleftrightarrow (x, y) = (x_0, y_0), \forall (x, y) \in \mathbb{R}^2$,
(iii) The α-cut ${}^{\alpha}P$ is a compact and convex subset of $\mathbb{R}^2$ $\forall \alpha \in [0, 1]$.

If P is a fuzzy point at $(x_0, y_0) \in \mathbb{R}^2$, then it can be visualized as a surface in Euclidean space through the graph of the equation $z = P(x, y)$, for all $(x, y) \in \mathbb{R}^2$.

Example 11.1.1 Suppose P_1 and P_2 are two fuzzy (real) numbers such that $P_1(x) = 1 \Longleftrightarrow x = x_0$ and $P_2(y) = 1 \Longleftrightarrow y = y_0$. Then, the fuzzy subset of $\mathbb{R}^2$ defined by $P(x, y) = P_1(x) \wedge P_2(y)$ for all $(x, y) \in \mathbb{R}^2$ is a fuzzy point at (x_0, y_0).

The next example is adopted from [135].

Example 11.1.2 Assume that the point $(x_0, y_0) \in \mathbb{R}^2$ is the vertex of a right elliptical cylinder with the points of its base satisfying the equation $(\frac{x-x_0}{a})^2 + (\frac{y-y_0}{b})^2 \leq 1$. Then, the point $P(x_0, y_0)$ is a fuzzy point with the membership function

$$\mu(x, y) | P(x_0, y)$$
$$= \begin{cases} 1 - \sqrt{\left(\frac{x - x_0}{a}\right)^2 + \left(\frac{y - y_0}{b}\right)^2}, & \text{if } \left(\frac{x - x_0}{a}\right)^2 + \left(\frac{y - y_0}{b}\right)^2 \leq 1, \\ 0, & \text{otherwise.} \end{cases} \tag{11.1}$$

We know that the Euclidean distance between two points $P_1(x_1, \ldots, x_n)$ and $P_2(y_1, \ldots, y_n)$ in $\mathbb{R}^n$ is given by $d(P_1, P_2) = (\sum_{i=1}^{n} |x_i - y_i|^2)^{1/2}$, a relation expressing the *generalized Pythagorean theorem* in n dimensions. However, things are not that intuitive in fuzzy geometry. In order to define the *fuzzy distance* ${}^{\alpha}D$ between two fuzzy points, we consider two arbitrary fuzzy points P_1 and P_2 and the set

$$\Omega(\alpha) = \{d(x,y)|x \in {}^{\alpha}P_1 \text{ and } y \in {}^{\alpha}P_2 \text{ and } \alpha \in [0,1]\},$$

with d denoting the Euclidean distance. Then, we have the following definition:

Definition 11.1.2 Let $D(P_1, P_2)$ be the fuzzy subset of $\mathbb{R}$ given by

$$D(P_1,P_2)(t) = \bigvee\{t \in \Omega(\alpha)\} \quad \forall t \in \mathbb{R}, \qquad \forall \alpha \in [0,1].$$

Then, $\Omega(\alpha) = \{t \in \mathbb{R}|\exists x \in {}^{\alpha}P_1,\ \exists y \in {}^{\alpha}P_2 \ : \ t = d(x,y)\}$.

With the above definition, we have the next theorem for the fuzzy distance ${}^{\alpha}D$ between two fuzzy points P_1 and P_2:

Theorem 11.1.1

(i) ${}^{\alpha}D(P_1, P_2) = \Omega(\alpha)$, $\forall \alpha \in [0,1]$, *and*
(ii) $D(P_1, P_2$ *is a fuzzy (real) number.*

Proof:

(i) Let $d \in \Omega(\alpha)$. Then $D(d) \geq \alpha$, so $\omega(\alpha) \subseteq {}^{\alpha}D$. Now, we will prove that ${}^{\alpha}D \subseteq \Omega(\alpha)$. Suppose $d \in {}^{\alpha}D$. Then, $D(\alpha) \geq \alpha$. Let us set $D(d) = \beta$. We will examine two cases: $\alpha < \beta$ and $\alpha = \beta$. First, suppose $\alpha < \beta$. Then, $\exists \gamma \ : \ \alpha < \gamma \leq \beta$ with $d \in \Omega(\gamma)$. But $\Omega(\gamma) \subseteq \Omega(\alpha)$, so $d \in \Omega(\alpha)$ and this implies that ${}^{\alpha}D \subseteq \Omega(\alpha)$. Now, suppose $\alpha = \beta$ and let $F = \{\delta | d \in \Omega(\delta)\}$. Then, ${}_{\vee}F = \beta = \alpha = D(d)$. There exists a sequence $\gamma_n \in F \ : \ \lim \gamma_n \to \alpha$. Then, for an arbitrary $\varepsilon > 0$, $\exists N \in \mathbb{R} \ : \ \alpha - \varepsilon < \gamma_n \ \forall n \geq N$. From $d \in \Omega(\gamma_n) \ \forall n \Rightarrow d \in \Omega(\alpha - \varepsilon) \ \forall \varepsilon > 0$. So, $d = d(x,y)$ for some $x \in P_1(x_1,y_1)(\alpha - \varepsilon)$ and for some $y \in P_2(x_2,y_2)(\alpha - \varepsilon)$. Hence, $P_1(x_1,y_1) \geq (\alpha - \varepsilon)$ and $P_2(x_2,y_2) \geq (\alpha - \varepsilon)$. But, since ε is arbitrary, we have $P_1(x_1,y_1) \geq \alpha$ and $P_2(x_2,y_2) \geq \alpha$ and this implies that $d \in \Omega(\alpha)$, so that ${}^{\alpha}D \subseteq \Omega(\alpha)$. Since we have seen above that $\Omega(\alpha) \subseteq {}^{\alpha}D$, we conclude that ${}^{\alpha}D = \Omega(\alpha) \ \forall \alpha \in (0,1]$, and because of ${}^{0}D = \Omega(0)$, we have proven that ${}^{\alpha}D = \Omega(\alpha) \ \forall \alpha \in [0,1]$.

(ii) The α-cuts of $P_1(x_1,y_1)$ and $P_2(x_2,y_2)$ are compact and this implies that $\Omega(\alpha)$ is a bounded and closed interval $\forall \alpha \in [0,1]$. Further, since the α-cuts of a fuzzy number are closed sets, then the membership function of this fuzzy number is upper semicontinuous [49]. But ${}^{\alpha}D \subseteq \Omega(\alpha)$ is a closed interval $\forall \alpha \in [0,1]$. Thus D is upper semicontinuous. Now, suppose $\Omega(\alpha) = [k_1(\alpha), k_2(\alpha)] = {}^{\alpha}D \quad \forall \alpha \in [0,1]$. Let $\Omega(0) = [m_1, m_2] \Rightarrow D(m_2) = 0$ outside $[m_1, m_2]$. Let $\Omega(1) = m$, with $m = d((x_1,y_1),(x_2,y_2))$. Since $k_1(\alpha)$ is increasing from m_1 to m and $k_2(\alpha)$ is decreasing from m_2 to m, it follows that D is increasing on the interval $[m_1, m]$, while it is decreasing on the interval $[m, m_2]$, and $D(m_2) = 1$ for $m_2 = m$. Consequently, $D(P_1, P_2)$ is a fuzzy (real) number and (ii) is proven.

□

A final remark is in order at this point. Let P_1, P_2 be fuzzy points at (x_1, y_1) and (x_2, y_2), respectively, and let $r \in \mathbb{R}^+$. Then, ${}^\alpha P_1 = \{(x_1, y_1)\}$ and ${}^\alpha P_2 = \{(x_2, y_2)\}$, $\forall \alpha \in [0, 1]$. Consequently, $D(P_1, P_2)(r) = \vee\{\alpha | \exists x \in \{(x_1, y_1)\}, \exists y \in \{(x_2, y_2)\} : r = d(x, y)\}$. Hence,

$$D(P_1, P_2)(r) = \begin{cases} 1, & \text{for} \quad r = d((x_1, y_1), (x_2, y_2)), \\ 0, & \text{otherwise.} \end{cases}$$

Therefore, as it is expected, the fuzzy distance D becomes the ordinary Euclidean distance d in $\mathbb{R}^2$ when P_1, P_2 are crisp.

Finally, an interesting notion is that of a *fuzzy metric*, that is, a mapping from pairs of fuzzy points into fuzzy numbers. The definition and the properties a fuzzy metric has to satisfy are very similar to that of the Euclidean metric (Definition 10.1.2) except that it refers to fuzzy points. For more information on the concept of fuzzy metric, the reader is referred, e.g., to [222] or [44].

11.2 Fuzzy Lines and Their Properties

We continue our study with the examination of *fuzzy lines*. There exist different possibilities to define a fuzzy line and six of them will be described here. For instance, for a set of given fuzzy numbers A, B, and C, a fuzzy line is the set of all fuzzy number pairs (X, Y) solving the linear equation

$$AX + BY = C.$$

However, by using standard fuzzy arithmetic it can be seen that this equation often has no solution (X, Y) [49].

Another way of defining a fuzzy line is, given any two fuzzy numbers A and B, by the set of all fuzzy numbers (X, Y) solving the linear equation

$$Y = AX + B.$$

Unfortunately, with this definition one cannot obtain any visualization (graph) of the fuzzy line considered.

A third possible way of defining a fuzzy line is provided as follows: Given any three fuzzy numbers A, B, and C with $A(1) = \{a\}$ and $B(1) = \{b\}$ and assuming that a, b are not both zero, let

$$\Omega_{11}(\alpha) = \{(x, y) | ax + by = c,\ a \in {}^\alpha A,\ b \in {}^\alpha B,\ c \in {}^\alpha C\}, \quad \forall \alpha \in [0, 1].$$

Also, let L_{11} be the fuzzy subset of $\mathbb{R}^2$ defined as

$$L_{11}(x,y) = \vee\{\alpha | (x,y) \in \Omega_{11}(\alpha)\}, \quad \forall (x,y) \in \mathbb{R}^2.$$

If a, b were both zero, then $A(1) = \{0\}$ and $B(1) = \{0\}$, and in this case, $\Omega_{11}(1)$ could be an empty set since the equation $0x + 0y = c$, $c \in C(1)$ has no solution when $c \neq 0$. Consequently, the assumption of a and b not being both zero is justified.

A fourth possibility for defining a fuzzy line is through the linear equation $y = ax + b$ and any two fuzzy numbers, say A and B. Then,

$$\Omega_{12}(\alpha) = \{(x,y) | y = ax + b, \ a \in {}^{\alpha}A, \ b \in {}^{\alpha}B\}, \quad \forall \alpha \in [0,1].$$

Then, L_{12} is a fuzzy subset of $\mathbb{R}^2$ defined by

$$L_{12}(x,y) = \vee\{\alpha | (x,y) \in \Omega_{12}(\alpha)\}, \quad \forall (x,y) \in \mathbb{R}^2.$$

A fifth possibility is provided by the use of the usual point-slope form. Suppose P is a fuzzy point in $\mathbb{R}^2$ and M is any fuzzy number. We have

$$\Omega_2(\alpha) = \{(x,y) | y - y_0 = a(x - x_0), \ (x_0, y_0) \in {}^{\alpha}P, \ a \in {}^{\alpha}M\}, \quad \forall \alpha \in [0,1].$$

Then, L_2 is a fuzzy subset of $\mathbb{R}^2$ defined by

$$L_2(x,y) = \vee\{\alpha | (x,y) \in \Omega_2(\alpha)\}, \quad \forall (x,y) \in \mathbb{R}^2.$$

A final possibility is given by the use of the two-point form. Suppose P_1, P_2 are any two fuzzy points in $\mathbb{R}^2$. We have

$$\Omega_3(\alpha) = \left\{ (x,y) \left| \frac{y - y_1}{x - x_1} = \frac{y_2 - y_1}{x_2 - x_1}, \ (x_1, y_1) \in c, \ (x_2, y_2) \in {}^{\alpha}P_2 \right. \right\}, \quad \forall \alpha \in [0,1].$$

Then, L_3 is a fuzzy subset of $\mathbb{R}^2$ defined by

$$L_3(x,y) = \vee\{\alpha | (x,y) \in \Omega_3(\alpha)\}, \quad \forall (x,y) \in \mathbb{R}^2.$$

In summary, we have four different ways for defining a fuzzy line, namely L_{11}, L_{12}, L_2, and L_3.

Naturally, α-cuts of fuzzy subsets of $\mathbb{R}^2$ will be crispy subsets of the plane. The following theorem holds for the α-cuts of fuzzy lines:

Theorem 11.2.1 *For the last four definitions of fuzzy lines L_{11}, L_{12}, L_2, and L_3 it holds*

$${}^{\alpha}L_{11} = \Omega_{11}(\alpha), \quad {}^{\alpha}L_{12} = \Omega_{12}(\alpha), \quad {}^{\alpha}L_2 = \Omega_2(\alpha), \quad {}^{\alpha}L_3 = \Omega_3(\alpha), \qquad \forall \alpha \in [0,1]$$

Proof: The proof runs along the same argumentation lines as in Theorem 11.1.1 and is left as an exercise for the reader. □

Let us present a few examples of fuzzy lines borrowed from [42] thereby stressing the fact that fuzzy lines can have a varying width, in other words they can be "thin" or "fat".

Example 11.2.1 Suppose we have the triangular fuzzy numbers (see Section 3.1.1) tfn$(-1, 0, 1)$, tfn$(-1, 1, 2)$, and tfn$(0, 1, 2)$. Then, the support of L_{11}, $L_{11}(0)$ is $\mathbb{R}^2$. Now, Theorem 11.2.1 holds and we speak of a *fat fuzzy line*. Furthermore, $L_{11}(1)$ is the crisp line with equation $y = 1$.

Example 11.2.2 Returning to the previous fourth possibility of defining a fuzzy line, suppose that L_{12} is defined by the linear equation $y = Ax + B$, where B is the triangular fuzzy number tfn$(0, 1, 2)$ and the crisp number A is 2. Then, the graph $z = \mu((x, y)|L_2)$ is produced if we put the triangular fuzzy number on the y-axis, the base put on the interval $[0, 2]$, and let the triangle made by the tfn$(0, 1, 2)$ "run" along the crisp line with equation $y = 2x + 1$. In this case, we speak of a *thin fuzzy line*.

Example 11.2.3 Going to the fifth possibility of defining a fuzzy line, let r be a crisp (real) number and $K(r, r)$ be a fuzzy point. Then, the set of all lines having slope r and passing through a point in ${}^{\alpha}K$ is given by ${}^{\alpha}L_2$, while ${}^{r}L_2$ is the crisp line described by the (bisector) equation $y = x$. ${}^{\alpha}L_2$ is considered as a thin fuzzy line when ${}^{0}K$ is "small".

Example 11.2.4 Suppose $P_1(0, 0)$ and $P_2(1, 1)$ are two fuzzy points with a right circular cone as a graph. The base of $P_1(0, 0)$ is $K_1 = \{(x, y)|x^2 + y^2 \leq 1/9\}$ and the vertex lies at $(0, 0)$. The base of $P_2(1, 1)$ is $K_2 = \{(x, y)|(x - 1)^2 + (y - 1)^2 \leq 1/9\}$ and the vertex lies at $(1, 1)$. Then, by the two-point form given previously, the support of L_3, ${}^{0}L_3$, is comprised of all lines going through a point in K_1 and a point in K_2. ${}^{1}L_3$ is the line with the bisector equation $y = x$. Thus, the graph $z = \mu((x, y)|L_3)$ is thin between the points K_1 and K_2, while it broadens getting thick along $y = x$ for $x > 1$ or $x < 0$.

In fact, based on what has been said so far, the following definition is fully justifiable:

Definition 11.2.1 A fuzzy line L contains a fuzzy point P if and only if $P \subset L$.

Some remarks are in order. The fuzzy line L_2 defined previously contains the fuzzy point P. If $P(x_1, y_1)$ is a fuzzy point at (x_1, y_1) that is contained in L_2, then we must have $(x_1, y_1) \in \Omega_2(1)$. Suppose that the α-cut ${}^{1}A = [a_1, a_2]$ is an interval.

$\Omega_2(1)$ is comprised of all lines passing through the point (x_0, y_0) and having slope $a : a_1 \leq a \leq a_2$. If A is a triangular fuzzy number, ${}^1A = \{a\}$, then $\Omega_2(1)$ is the crisp line with equation $y - y_0 = a(x - x_0)$.

Now, suppose $P_1(x_1, y_1)$ and $P_2(x_2, y_2)$ are two fuzzy points defining the line L_3. Then, by Definition 11.2.1, these points are contained in L_3 and ${}^1L_3 = \Omega_3(1)$ is the crisp line passing through the points (x_1, y_1) and (x_2, y_2). When some other fuzzy point P_3 is contained in L_3, then ${}^1P_3$ must belong to the line $\Omega_3(1)$.

Some important further properties of the previously defined fuzzy lines turn out to be very useful and can be summarized in the form of theorems as follows [42].

Theorem 11.2.2

$$L_{11} = L_{12}.$$

Proof: From Theorem 11.2.1 it suffices to prove that $\Omega_{11}(\alpha) = \Omega_{12}(\alpha)$, $\forall \alpha \in [0, 1]$. Let $(x, y) \in \Omega_{11}(\alpha)$, then $\exists\ a \subset {}^\alpha A,\ b \in {}^\alpha B,\ and\ c \in {}^\alpha C$ such that $ax + by = c$. But $y = a_1x + b_1y$ with $a_1 = -\frac{a}{b}$ and $b_1 = \frac{c}{b}$. Since $a_1 \in {}^\alpha A$ and $b_1 \in {}^\alpha B$, it follows that $(x, y) \in \Omega_{12}(\alpha)$ and thus $\Omega_{11}(\alpha)$ is a subset of $\Omega_{12}(\alpha)$. Now, let $(x, y) \in \Omega_{12}(\alpha)$, $\forall \alpha \in [0, 1]$. Then, $\exists\ a \in {}^\alpha A,\ and\ b \in {}^\alpha B$ such that $y = ax + b$. But $by + a_1x = c_1$, where $a_1 = ab$ and $c_1 = bc$. Since $a_1 \in {}^\alpha A$ and $c_1 \in {}^\alpha B$, it follows that $(x, y) \in \Omega_{11}(\alpha)$ and thus Ω_{12} is a subset of $\Omega_{11}(\alpha)$. Consequently, $\Omega_{11} = \Omega_{12}$, $\forall \alpha \in [0, 1]$, and hence $L_{11} = L_{12}$. □

Theorem 11.2.3

$$L_3 = L_2.$$

Proof: From Theorem 11.2.1 it suffices to prove that $\Omega_3(\alpha) = \Omega_2(\alpha)$, $\forall \alpha \in [0, 1]$. Let $(x, y) \in {}^\alpha L_3$. Then, $\exists\ (x_0, y_0) \in {}^\alpha P,\ a \in {}^\alpha M$ such that $y - y_0 = a(x - x_0)$. But $(x, y) \in \Omega_2(\alpha)$ and thus $\Omega_3(\alpha)$ is a subset of $\Omega_2(\alpha)$. As in the previous theorem, one can easily show that $\Omega_2(\alpha)$ is also a subset of $\Omega_3(\alpha)$. Consequently, $\Omega_3(\alpha) = \Omega_2(\alpha)$, $\forall \alpha \in [0, 1]$, and thus $L_3 = L_2$. □

Theorem 11.2.4

$$L_{12} = L_2.$$

Proof: Again, from Theorem 11.2.1 it suffices to show that $\Omega_{12}(\alpha) = \Omega_2(\alpha)$. The proof is omitted as it runs along the same argumentation lines as those of the previous two theorems, so it is left as an exercise for the reader. □

Finally, there are some properties of a fuzzy line L having either form L_{11}, L_{12}, L_2, or L_3, that hold in general and are worth to be given:

(i) The α-cuts of a fuzzy line L are closed, connected and arcwise connected, but they are not necessarily convex.
(ii) Since the α-cuts are closed, it follows that the fuzzy line L is upper semicontinuous.
(iii) A fuzzy line L is normalized or there is always at least one crisp line in ${}^1L$.

Next, the question arises when two fuzzy lines are parallel and when they intersect each other. Starting with parallelism, suppose we have the fuzzy lines L_a and L_b. Then, one can define a measure of parallelism between them as $\rho = 1 - \lambda$, with λ being the height of the intersection of L_a and L_b, as follows:

Definition 11.2.2

$$\lambda = \vee\{L_a(x, y) \vee L_b(x, y) | (x, y) \in \mathbb{R}^2\}.$$

From this definition, one can immediately infer that if the two fuzzy lines are parallel (i.e. $L_a \cap L_b = \emptyset$), then $\rho = 1$ (or $\lambda = 0$). Indeed, suppose that l_a and l_b are two crisp lines in ${}^1L_a$ and ${}^1L_b$, respectively. If these two crisp lines intersect, then $\rho = 0$ (or $\lambda = 1$). Thus, if L_a and L_b are both crisp lines, then $\rho = 1$ (or $\lambda = 0$) if and only if L_a and L_b are parallel. In the case where one of the lines, say L_a, is a crisp line while L_b is a fuzzy line, then L_a may intersect the crisp line in ${}^1L_b$ and thus $\rho = 0$ (or $\lambda = 1$).

Let us now examine the intersection of two fuzzy lines. We have the following definition:

Definition 11.2.3 Suppose L_a and L_b are two fuzzy lines and assume $\rho < 1$. Let $R = L_a \cap L_b$ be the (fuzzy) region of intersection of the two fuzzy lines. Then, $R(x, y) = L_a(x, y) \vee L_b$.

It is easily seen that $R = \emptyset$ when $\rho = 1$.

11.3 Fuzzy Circles

After having presented fuzzy points and lines, we continue with the study of fuzzy circles, whereby one must not forget that the central issue is again vagueness: the question is not about whether a certain figure describes a circle but to what extent (or degree) this figure represents a circle.

We know from plane analytic geometry that the equation of a circle with (x_0, y_0) the Cartesian coordinates of its center and r its radius, has the standard form $(x - x_0)^2 + (y - y_0)^2 = r^2$. This equation can be equivalently written as $x^2 + y^2 + ax + by = c$ defining a circle when $4c > a^2 + b^2$. Thus, a natural attempt to define a fuzzy circle would be to suppose fuzzy numbers A, B, and C and define a fuzzy circle as the set of all fuzzy number pairs (X, Y) that are solutions to the equation

$$X^2 + Y^2 + AX + BY = C,$$

where $4C > A^2 + B^2$.

However, using standard fuzzy arithmetic, this equation has no solution (X, Y) [49].

A second attempt would be to return to the standard equation form and write it for given fuzzy numbers A, B, and C. Then, a fuzzy circle would be the set of all fuzzy number pairs (X, Y) that are solutions to the equation

$$(X - A)^2 + (Y - B)^2 = C^2.$$

However, the problem of defining a fuzzy circle persists as this equation has just a few, if any, solutions (X, Y) [49].

A third approach to the problem is the following: Suppose A, B, and C are fuzzy numbers. Then, we have

$$\Omega(\alpha) = \{(x, y) | (x - a)^2 + (y - b)^2 = c^2,\ a \in {}^{\alpha}A,\ b \in {}^{\alpha}B,\ c \in {}^{\alpha}C\}, \quad \forall \alpha \in [0, 1]$$

and a *fuzzy circle* is defined as

$$C(x, y) = \vee\{\alpha | (x, y) \in \Omega(\alpha)\}, \quad \forall (x, y) \in \mathbb{R}^2.$$

The following theorem (see Ref. [43] or [222]) holds for α-cuts of fuzzy circles:

Theorem 11.3.1

$${}^{\alpha}C = \Omega(\alpha), \quad \forall \alpha \in [0, 1].$$

Proof: As a first step, we will show that the α-cuts are the same for $\alpha \in [0, 1]$. Suppose that $\eta \in \Omega(\alpha)$. Then, $C(\eta) \geq \alpha$ and $\Omega(\alpha) \subseteq {}^{\alpha}C$. Now, suppose $\eta \in {}^{\alpha}C$. Then, $C(\eta) \geq \alpha$. We set $C(\eta) = \theta$ and distinguish between the cases (i) $\theta > \alpha$ and (ii) $\theta = \alpha$.

(i) If $\theta > \alpha \Rightarrow \exists\ \lambda$ such that $\alpha < \lambda \leq \theta$, $\eta \in \Omega(\lambda)$. But $\Omega(\lambda) \subseteq \Omega(\alpha)$, therefore $\eta \in \Omega(\alpha)$.

(ii) Let $\theta > \alpha$ and $K = \{z | \eta \in \Omega(z)\}$. Then, $\vee K = \theta = \alpha = C(\eta)$. There exists a sequence s_n in K such that $\lim s_n \to \alpha$. So, for any $\varepsilon > 0$, $\exists\ N \in \mathbb{R}$ such that $\alpha - \varepsilon < s_n$, $\forall n \geq N$. From $\eta \in \Omega(s_n)\ \forall n$ we infer that $\eta \in \Omega(\alpha - \varepsilon)$, $\forall \varepsilon > 0$. If $\eta = (x, y) \Rightarrow (x - x_0)^2 + (y - y_0)^2 = r^2$ with x_0, y_0, r in the fuzzy subsets of $\mathbb{R}^n$, ${}^{\alpha-\epsilon}A$, ${}^{\alpha-\epsilon}B$, and ${}^{\alpha-\epsilon}R$, respectively. Consequently, ${}^{x_0}A \geq \alpha - \epsilon$, ${}^{y_0}B \geq \alpha - \epsilon$, and

${}^{r}R \geq \alpha - \epsilon$. Since $\varepsilon > 0$ is arbitrarily chosen, it follows that ${}^{x_0}A \geq \alpha$, ${}^{y_0}B \geq \alpha$, ${}^{r}R \geq \alpha$, and $x_0 \in {}^{\alpha}A$, $y_0 \in {}^{\alpha}B$, $r \in {}^{\alpha}R$. Thus, $\eta = (x, y) \in \Omega(\alpha)$ and ${}^{\alpha}C \subseteq \Omega(\alpha)$.

It follows that ${}^{0}C \subseteq \Omega(0)$, because ${}^{\alpha}C = \Omega(\alpha)$, $\forall \alpha \in [0, 1]$. □

Let us see an example adopted from [43].

Example 11.3.1 As in the case of fuzzy lines, vagueness may lead to *thick fuzzy circles*. Suppose we have three triangular fuzzy numbers, say A, B, and C, and equal to tfn(0, 1, 2). Then, the support of C, ${}^{0}C = \Omega(0)$, is the square $[-2, 4] \times [-2, 4]$ having rounded edges and all α-cuts of C, $\forall \alpha \in [0, 1]$, are rectangles with rounded corners, where ${}^{1}C$ is the crisp circle having equation $(x - 1)^2 + (y - 1)^2 = 1$. Thus, the graph of the membership function $\mu(x, y)$ is a pyramid having four sides, rounded edges, and a vertex lying at the point (1, 1).

A further example is the following:

Example 11.3.2 All α-cuts with $\alpha \in [0, 1)$ describe disks presented as *"regular" fuzzy circles*. Let the real numbers $A = B = 1$ and the triangular fuzzy number C given by tfn(1, 2, 5). Then, all the α-cuts with $\alpha \in [0, 1)$ are the disks with ${}^{0}C = \{(x, y) | 1 \leq (x - 1)^2 + (y - 1)^2 \leq 25\}$, while ${}^{1}C$ is the crisp circle having equation $(x - 1)^2 + (y - 1)^2 = 1$.

Let us now examine the notion of the *area* of a fuzzy circle. Suppose that A, B, and C are any fuzzy numbers (as those used previously in the third approach for the definition of a fuzzy circle). If we denote by $\Omega_a(\alpha) = \{f | f$ the area of $(x - x_0)^2 + (y - y_0)^2 = r^2$, $x_0 \in {}^{\alpha}A$, $y_0 \in {}^{\alpha}B$, $r \in {}^{\alpha}C\}$, $\forall \alpha \in [0, 1]$, then the area F of a fuzzy circle $C(x, y)$ is defined as:

Definition 11.3.1

$$F(f) = \vee\{\alpha | f \in \Omega(\alpha)\}.$$

At this point, some remarks are in order. If A, B, and $C \in \mathbb{R}$, then $C(x, y)$ becomes a crisp circle, while $F \in \mathbb{R}$. In the case where at least one of the A, B, C is a fuzzy (not real) number, F can become a crisp number only when $C \in \mathbb{R}$. For instance, if $C = r$, $r \in \mathbb{R}^+$, then $F = \pi r^2 \in \mathbb{R}$.

If $C \notin \mathbb{R}$, then we have the following theorem:

Theorem 11.3.2

(i) ${}^{\alpha}F = \Omega_a(\alpha)$, $\forall \alpha \in [0, 1]$*; and*
(ii) F *is a fuzzy number.*

Proof:

(i) The proof runs along the same argumentation lines as in Theorem 11.3.1 and it is left to the reader as an exercise.

(ii) The α-cuts of the fuzzy subsets A, B, and C are compact intervals, therefore $\Omega(\alpha)$ is a bounded and closed interval $\forall\alpha \in [0,1]$. Now, suppose that $\Omega_a(\alpha) = [p(\alpha), q(\alpha)], \forall\alpha \in [0,1]$. If the α-cuts are closed intervals, then the membership function is upper semicontinuous [49]. We have that ${}^{\alpha}F = \Omega_a(\alpha)$ is a closed interval, $\forall\alpha \in [0,1]$. Consequently, $F(f)$ is upper semicontinuous.
Let $\Omega(0) = [c,d]$. Then, $F(f) = 0$ outside the interval $[c,d]$. Now, if $\Omega(1) = [a,b]$, then $F(f) = 1$ in the interval $[a,b]$. But, since $\Omega_a(\alpha) = [p(\alpha), q(\alpha)] = {}^{\alpha}F$, $\forall\alpha \in [0,1]$, whereby $p(\alpha)$ increases from c to a and $q(\alpha)$ decreases from to d to b, we infer that $F(f)$ increases in the interval $[c,a]$, while it decreases in the interval $[b,d]$, and $F(f) = 1$ in the interval $[a,b]$. Therefore, F is a fuzzy number.

□

Next, we will consider the notion of a fuzzy circle's *circumference*. Suppose A, B, and C are any fuzzy numbers in the definition of the fuzzy circle $C(x,y)$ given previously. Then, we have the following definition:

Definition 11.3.2 ${}^{\alpha}\Omega_c = \{L|L$ is the circumference of $(x-x_0)^2 + (y-y_0)^2 = r^2$, $x_0 \in {}^{\alpha}A,\ y_0 \in {}^{\alpha}B,\ r \in {}^{\alpha}C\}$, $\forall\alpha \in [0,1]$, and $L'(L) = \vee\{\alpha|L \in {}^{\alpha}\Omega_c\}$ is the *circumference* L'.

At this point, a remark is in order: If C is a real number, then $L' \in \mathbb{R}$.

Finally, we come to the following theorem related to the circumference of a fuzzy circle:

Theorem 11.3.3

(i) ${}^{\alpha}L' = {}^{\alpha}\Omega_c$, $\forall\alpha \in [0,1]$; *and*

(ii) the circumference L' *is a fuzzy number.*

Proof: The proof runs along the same argumentation lines as in Theorem 11.3.2, and it is left to the reader as an exercise. □

Let us consider an example on the circumference of fuzzy circles.

Example 11.3.3 Suppose we have a *"regular" fuzzy circle* and consider the case $A = B = 1$ and the triangular fuzzy number C represented by tfn$(1,2,5)$. Then ${}^{0}C = \{(x,y)|1 \leq (x-1)^2 + (y-1)^2 \leq 25\}$, while ${}^{1}C$ is the crisp circle having equation $(x-1)^2+(y-1)^2=1$. So, the α-cuts correspond to the interval $[\alpha+1, 5-\alpha]$, $\forall \alpha \in [0,1]$. Hence, $F(\alpha)=[\pi(\alpha+1)^2, \pi(5-\alpha)^2]$ and $L'(\alpha)=[2\pi(\alpha+1), 2\pi(5-\alpha)]$, $\forall \alpha \in [0,1]$. Therefore, the fuzzy area F is a triangular-shaped fuzzy number having support $[\pi, 25\pi]$ and a vertex lying at the point $(4\pi, 0)$. Furthermore, its fuzzy circumference L' is the triangular fuzzy number $\text{tfn}(2\pi, 4\pi, 10\pi)$.

11.4 Regular Fuzzy Polygons

First, let us briefly recall what a *regular n-sided polygon* (from the Greek word πολυγώνιον denoting a figure with many angles) is in Euclidean plane geometry [80]: it is a closed and rotationally symmetric figure with n sides (an n-gon), with $n \geq 3$, that is *equiangular* and *equilateral*, while all its vertices are *concyclic*, that is, they lie on the circumscribed circle. Obviously, the polygons with the smallest number of sides are the equilateral triangles. In fact, every regular polygon has also an inscribed circle that is tangent to the midpoint of every side of the polygon. The circumscribed and the inscribed circles have a common center that is also the *center* of the polygon. All regular and *simple* (i.e. not self-intersecting) polygons are *convex*, that is, their interior is a convex set (in general, a set S in a vector space over $\mathbb{R}$ is called a convex set if the line segment joining any pair of points of S lies entirely in S).

The *area* of a regular n-sided polygon is given as

$$A = \frac{1}{2}nsa = \frac{1}{2}nR^2 \sin\left(\frac{2\pi}{n}\right),$$

where s is the polygon's side length, a is the *apothem* (the length of the line segment from the polygon's center perpendicular to any side of the polygon), and R is the radius of the circumscribed circle.

Let us now proceed to the introduction of fuzziness into the notion of regular polygons and see how the aforementioned concepts are changed.

Definition 11.4.1 Distinct points $P_1, P_2, \ldots, P_n \in \mathbb{R}^2$ are *convex independent* if and only if R_i, $i=1,2,\ldots,n$, does not belong to the convex hull (or closure) of the rest of the points P_j, $1 \leq j \leq n, j \neq i$.

Based on this definition, suppose that we have a collection of convex independent points $P_1, P_2, \ldots, P_n$ and let us assume that their numbering from P_1 to P_{n-1} to P_n runs counter-clockwise. We can connect adjacent points P_i with line segments l_{ab} of equal length, by drawing the line segment l_{12} from P_1 to P_2, l_{23} from P_2 to P_3, and so on, until we come to the last line segment l_{n1} from P_n to P_1. In this way, we can construct a regular n-sided polygon that consists of its vertices P_i and its sides (the line segments). As we pointed out above, the interior of such an n-sided regular polygon is convex.

Let us introduce *fuzzy line segments*. Suppose that P, Q are two distinct fuzzy points and $\Omega_l(\alpha)$ is the set of all line segments from a point in ${}^{\alpha}P$ to a point in ${}^{\alpha}Q$. Then, a fuzzy line segment L_{PQ} is defined as follows:

Definition 11.4.2

$$L_{PQ}(x,y) = \vee\{\alpha|(x,y) \in \ l \in \Omega_l(\alpha)\}.$$

One can proceed as in Theorem 11.3.1 to prove that

$${}^{\alpha}L_{PQ} = \Omega_l(\alpha), \quad \forall\alpha \in [0,1].$$

Suppose that we have $L_1, L_2, L_3, \ldots, L_n$ fuzzy line segments from the fuzzy points P_1 to P_2 to P_3, ..., to P_n to P_1, respectively. Then, a *regular n-sided fuzzy polygon* $\mathscr{P}$ is defined as follows:

Definition 11.4.3

$$\mathscr{P} = \bigcup_{i=1}^{n} L_i.$$

By use of this definition, we infer that

$$\mathscr{P}(x,y) = \vee\{L_i(x,y),\ i = 1,\ldots,n\}, \quad \forall(x,y) \in \mathbb{R}^2.$$

Now, if two fuzzy points, say P_1, P_2, are such that ${}^{0}P_2$ is a subset of ${}^{0}P_1$, then we have a *degenerate fuzzy polygon* in which the support ${}^{0}\mathscr{P}$ does not exhibit all the n (fuzzy) vertices.

The following example is borrowed from [43].

Example 11.4.1 Suppose that we have $n = 3$ and all the fuzzy points are right circular cones. P_1 has the base $B_1 = \{(x,y)|x^2 + y^2 \leq (0.1)^2\}$ and a vertex lying at the point $(0,0)$, while P_2 has the base $B_2 = \{(x,y)|(x-1)^2 + y^2 \leq (0.1)^2\}$ and a vertex lying at the point $(1,0)$, and P_3 has the base $B_3 = \{(x,y)|(x-1)^2 + (y-1/2)^2 \leq 1\}$ and a vertex lying at the point $(1,1/2)$. Then, ${}^{0}P_2 \subset {}^{0}P_3$, while ${}^{0}P_1$ and ${}^{0}P_3$ are disjoint. Hence, ${}^{0}L_2 \subset {}^{0}P_3$ and $\mathscr{P}$ is a degenerate fuzzy polygon. Furthermore, we have ${}^{0}L_1 \subset {}^{0}L_3$ and, therefore, the support of $\mathscr{P}$ is ${}^{0}L_3$.

We come to the following definition:

Definition 11.4.4 A regular n-sided fuzzy polygon is *nondegenerate* if ${}^{0}P_i$ is not a subset of ${}^{0}P_j$, $j \neq i$, $\forall i = 1,\ldots,n$. Further, the regular n-sided fuzzy polygon is called *strongly nondegenerate* if the ${}^{0}P_i$, $1 \leq i \leq n$, are pairwise disjoint.

Example 11.4.2 Suppose that we have $n = 4$ and P_1, P_2, P_3, P_4 four fuzzy points at $(0, 0)$, $(1, 0)$, $(1, 1)$, and $(0, 1)$, respectively. Let P_1 be a right circular cone with base $B_1 = \{(x, y)|x^2 + y^2 \leq (0.1)^2\}$ and its vertex lying at $(0, 0)$. The rest of the P_i, $i > 1$, will just be *rigid translations* of P_1 to their position in the plane. Then, $\mathscr{P}$ is *strongly nondegenerate* (it is a *fuzzy rectangle*).

What about α-cuts? The next theorem gives us the necessary information.

Theorem 11.4.1

$$^{\alpha}\mathscr{P} = \bigcup_{i=1}^{n} {}^{\alpha}L_i, \quad \forall \alpha \in [0, 1].$$

Proof: We have ${}^{\alpha}L_i = \Omega_i(\alpha)$ with $\Omega_i(\alpha)$ being the set of all line segments from ${}^{\alpha}P_i$ to ${}^{\alpha}P_{i+1}$, $\forall \alpha \in [0, 1]$, where we replace $i + 1$ with 1 when $i = n$. We must prove that

$$^{\alpha}\mathscr{P} = \bigcup_{i=1}^{n} \Omega_i(\alpha), \quad \forall \alpha \in [0, 1].$$

First, we will prove that the above expression holds for $\alpha \in (0, 1]$. Suppose that $u \in \bigcup_{i=1}^{n} \Omega_i(\alpha)$. Then, there exists an i (e.g. $i = 1$), so that $u \in \Omega_1(\alpha)$. So, $L_1(u) \geq \alpha$. Therefore, by $\mathscr{P}(x, y) = \vee\{L_i(x, y),\ i = 1, \dots, n\}$ we infer that $\mathscr{P}(u) \geq \alpha$ and $u \in {}^{\alpha}\mathscr{P}$. Hence, $\bigcup_{i=1}^{n} \Omega_i(\alpha) \subset {}^{\alpha}\mathscr{P}$.

Suppose that $u \in {}^{\alpha}\mathscr{P}$ and $\mathscr{P}(u) = m$. Then, we have that either $m = \alpha$ or $m > \alpha$. Let us begin with the case $m = \alpha$. By $\mathscr{P}(x, y) = \vee\{L_i(x, y),\ i = 1, \dots, n\}$. Then, there exists an i, e.g. $i = 1$, so that $L_1(u) = m$. Thus, $u \in \Omega_1(\alpha)$ and $u \in \bigcup_{i=1}^{n} \Omega_i(\alpha)$.

In the case $m > \alpha$, by $\mathscr{P}(x, y) = \vee\{L_i(x, y),\ i = 1, \dots, n\}$, there exists an i (e.g. $i = 1$), such that $L_1(u) = m$ and thus $u \in \Omega_1(m)$. But, since $m > \alpha$ we have that $u \in \Omega_1(\alpha)$. Consequently, $u \in \bigcup_{i=1}^{n} \Omega_i(\alpha)$. Thus ${}^{\alpha}\mathscr{P} \subset \bigcup_{i=1}^{n} \Omega_i(\alpha)$ and, since ${}^{\alpha}\mathscr{P} = \bigcup_{i=1}^{n} \Omega_i(\alpha)$ holds for $\forall \alpha \in (0, 1]$, it also holds for $\alpha = 0$. □

Next, we will examine the notion of *fuzzy area* of a regular n-sided fuzzy polygon. Let $\mathscr{P}$ be a strongly nondegenerate regular n-sided fuzzy polygon defined by fuzzy points P_i, $1 \leq i \leq n$ and define $\Omega_a(\alpha)$ as the set of all areas of n-sided polygons such that $u_i \in {}^{\alpha}P_i$, $1 \leq i \leq n$, $\forall \alpha \in [0, 1]$. Then, the fuzzy area F of $\mathscr{P}$ is defined as follows:

Definition 11.4.5

$$F(\theta) = \vee\{\alpha | \theta \in \Omega_a(\alpha)\}.$$

Similarly, one can define the *fuzzy perimeter* of a regular n-sided fuzzy polygon. Let $\mathscr{P}$ be a strongly nondegenerate regular n-sided fuzzy polygon defined by fuzzy points $P_i, 1 \le i \le n$ and define $\Omega_P(\alpha)$ as the set of all perimeters of n-sided polygons such that $u_i \in {}^{\alpha}P_i, 1 \le i \le n, \forall \alpha \in [0,1]$. Then, the fuzzy perimeter Π is defined as follows:

Definition 11.4.6

$$\Pi(L) = \vee\{\alpha | L \in \Omega_P(\alpha)\}.$$

Upon these two definitions, the following theorem on the fuzzy area and the fuzzy perimeter holds:

Theorem 11.4.2

$${}^{\alpha}F = \Omega_a(\alpha) \quad \textit{and} \quad {}^{\alpha}\Pi = \Omega_P(\alpha), \qquad \forall \alpha \in [0,1].$$

Proof: The proof is omitted as it runs along the same argumentation lines as those of Theorems 11.3.1 and 11.3.2, so it is left as an exercise for the reader. □

At this point, a remark is proper: when $\mathscr{P}$ is a *degenerate* regular n-sided fuzzy polygon, the relations ${}^{\alpha}F = \Omega_a(\alpha)$ and ${}^{\alpha}\Pi = \Omega_P(\alpha)$ are still valid; however, the area F and the perimeter Π may not be fuzzy numbers.

There are two interesting special cases that need some special attention: the *fuzzy triangle* ($n = 3$) and the *fuzzy rectangle* ($n = 4$). Let us suppose that a fuzzy triangle T is strongly nondegenerate, so that its fuzzy area and fuzzy perimeter are fuzzy numbers. Suppose P_1, P_2, and P_3 are fuzzy points defining the fuzzy triangle, and L_{12}, L_{23}, L_{31} are fuzzy line segments joining P_1 to P_2, P_2 to P_3, and P_3 to P_1, respectively. Let $\Omega_{\varphi}(\alpha)$ define the set of all angles (measured in radians) between l_{12} and l_{31}, so that l_{12} is a line segment from a point in ${}^{\alpha}P_1$ to a point in ${}^{\alpha}P_2$, l_{31} is a line segment from ${}^{\alpha}P_3$ to ${}^{\alpha}P_1$, and l_{12}, l_{31} start from the same point in ${}^{\alpha}P_1$, $\forall \alpha \in [0,1]$. Then, we have the following definition for the *fuzzy angle* Φ between L_{12} and L_{31}:

Definition 11.4.7

$$\Phi(\varphi) = \vee\{\alpha | \varphi \in \Omega_{\varphi}(\alpha)\},$$

and the following useful theorem:

Theorem 11.4.3

(i) ${}^{\alpha}\Phi = \Omega_{\varphi}(\alpha), \forall \alpha \in [0,1]$, *and*
(ii) Φ *is a fuzzy number.*

Proof: The proof runs along the same argumentation lines as those of Theorems 11.3.1 and 11.3.2, so it is left as an exercise for the reader. □

The dependence of Φ on the fuzzy points, i.e. $\Phi = \Phi(P_1, P_2, P_3)$, is justified as follows: Assume that P_i is a fuzzy point at $u_i \in \mathbb{R}^2$, $i = 1, 2, 3$. Then, P_i depends not only on the u_i but also on the "size" of the fuzzy points. In other words, substituting P_2 at u_2 with P'_2, having ${}^{\alpha}P_2 \subset {}^{\alpha}P'_2, \forall \alpha \in [0, 1]$ and with Φ' being the corresponding fuzzy angle, we obtain $\Phi \subset \Phi'$.

Now, suppose that we have the fuzzy points P_1, P_2, and P_3 at the positions $u_1 = (0, 0)$, $u_2 = (r, 0)$, and $u_3 = (s, 0)$, respectively, with $r, s \in \mathbb{R}^+$. Then, we have a fuzzy right triangle with vertices P_i, $1 \leq i \leq 3$, and Φ is the fuzzy angle between L_{12} and L_{31}.

Definition 11.4.8

$$\tan \Pi = \frac{D(P_2, P_3)}{D(P_1, P_2)},$$

where D is the *fuzzy distance* between the given fuzzy points and $\tan \Pi$ is a fuzzy number since (i) $D(P_1, P_2)$, $D(P_2, P_3)$, are both fuzzy numbers, and (ii) zero does not belong to the support of $D(P_1, P_2)$.

The trigonometric functions $\sin \Pi$ and $\cos \Pi$ can be defined in a similar way. At this point, one should stress that many identities from Euclidean trigonometry, for instance the well-known $\sin^2\varphi + \cos^2\varphi = 1$ or indeed the Pythagorean theorem, do not hold for fuzzy right triangles (for more on *fuzzy trigonometry* see, e.g. [44, Ch. 10]).

Let us now come to the case $n = 4$. Suppose that $u_1 = (x, y)$, $u_2 = (x + a, y)$, $u_3 = (x + a, y + b)$, and $u_4 = (x, y + b)$, $a, b \in \mathbb{R}^+$, are four points in the plane and P_i is a fuzzy point at the positions u_i, $i = 1, \ldots, 4$, respectively, and defining a fuzzy four-sided polygon. Assuming that the latter is strongly nondegenerate, this polygon is a *fuzzy rectangle* (naturally, for $a = b$ we get a *fuzzy square*).

The fuzzy area and the fuzzy perimeter of a fuzzy rectangle are fuzzy numbers, but they may not be obtained by the product and the sum of the side lengths, respectively, as in Euclidean geometry. Suppose that $A = D(P_1, P_2)$, $B = D(P_2, P_3)$, $C = D(P_3, P_4)$, and $E = D(P_4, P_1)$. Then, it is possible that $A \neq C$ and $B \neq E$ because the α-cuts of the fuzzy points may have different sizes, so let us assume that the fuzzy points are all the "same" but centered at different positions in the plane.

The following example from [43] illustrates the situation:

Example 11.4.3 Suppose that P_1, P_2, P_3, and P_4 are fuzzy points at the positions $u_1 = (0, 0)$, $u_2 = (0, 1)$, $u_3 = (1, 1)$, and $u_4 = (1, 0)$ in the plane, respectively, thus forming a fuzzy square. Let us assume that each P_i is a right circular cone with its base being a circle of radius equal to 0.1 and centered at u_i, and its vertex lying at u_i, $1 \leq i \leq 4$. With $A = D(P_1, P_2)$, $B = D(P_2, P_3)$, $C = D(P_3, P_4)$, and $E = D(P_4, P_1)$, we infer that ${}^0A = {}^0B = {}^0C = {}^0E = [0.8, 1.2]$. Therefore, ${}^0(AB) = [0.64, 1.44]$ and ${}^0\,(A + B + C + E) = [3.2, 4.8]$. If $\Omega_a(\alpha)$ is the set of all areas of a fuzzy rectangle with its vertices in ${}^\alpha P_i$, $1 \leq i \leq 4$, $\forall \alpha \in [0, 1]$, then we have that the fuzzy area F is such that $F(\theta) = \vee\{\alpha \in [0, 1] | \theta \in \Omega_a(\alpha)\}$. We can define the fuzzy parameter Π in a similar way. Consequently, the left end point of ${}^0F$ is greater than 0.64, and the left end point of ${}^0\Pi$ is greater than 3.2. Hence, we conclude that $F \neq AB$ and $\Pi \neq A + B + C + E$.

Before closing this section on fuzzy geometry, we would like to point out two intriguing topics: (i) the study of *fuzzy plane projective geometry*, and (ii) *random polygons*. The former is running on similar lines of argumentation as presented in this section. Here, some models of this kind of fuzzy geometry and some new notions of fuzzy points such as the so called vertical class of fuzzy points are introduced, while Desargues' theorem formulated in this fuzzy setting holds the central position just as it does in ordinary projective geometry. For an introduction to the topic, the reader is referred to [153].

Regarding the possibility of random polygons, we give only a comment in order to prevent the reader from running into possible misunderstandings in connection with fuzzy geometry. The existence of a random triangle is related to the question about the natural probability of taking at random three points on a plane and asking for the chance of their being the vertices of a triangle (or, generalizing the question for n vertices, of a random polygon both on a plane or in space). This rather old problem has been investigated by a number of authors and, despite its inherent randomness, it is not connected to the notion of fuzziness (for a review see, e.g. [57] or [260] and references therein).

11.5 Applications in Theoretical Physics

A final, very brief but hopefully informative, digression concerning the applications of fuzzy geometry in theoretical physics is deemed proper at this point. We start with the advanced notion of *fuzzy space-time* as it is considered one of the very interesting and "hot" topics in modern theoretical physics. The limitations on the measurability of distances on the grounds of Werner Karl Heisenberg's *uncertainty principle* in combination with the break-down of the classical space-time continuum and indeed of the geometric notion of the manifold at the Planck scale (with the Planck length of 10^{-35} m) have opened the possibility of the notion of quantum space-time, and a new branch of geometry called *quantum geometry* has emerged building the fundament of one of several theoretical approaches toward a theory of *quantum gravity*. In fact, under the aforementioned conditions, a precise localization of points is not possible on the classical space-time manifold (or on its spatial slices) which now becomes fuzzy, thus allowing the notion of space-time fuzziness and the definitions of fuzzy points presented in this chapter to come forward. Indeed, the ordinary commutative algebra of functions is now replaced by a noncommutative algebra. Simply put, the space-time coordinates or point observables do not commute anymore. They have been turned to operators, i.e. they are replaced by generators of a so called noncommutative Heisenberg algebra and, for a manifold parametrized by space-time coordinates x^μ, they satisfy commutator relations of the form $[x^\mu, x^\nu] = i\theta^{\mu\nu}$, where $\theta^{\mu\nu}$ is a real anti-symmetric tensor. These commutator relations lead to space-time uncertainty relations of the form $(\Delta x^\mu)(\Delta x^\nu) \geq \frac{1}{2}\theta^{\mu\nu}$. Thus, the notion of a point has actually lost its meaning, now points are ill-defined, they are "noncommutatively generalized points" and the associated geometry is *noncommutative* with the ordinary space (or space-time) geometry being fuzzy. In this context, the *fuzzy sphere* is one of the simplest examples of noncommutative geometry. For instance, the ordinary (commutative) two-dimensional sphere S^2 is a co-adjoint orbit which has a Poisson bracket and can be considered as a phase space. After the latter is "quantized," i.e. deformed, it is effectively divided into a number of so called *Bohr cells* each having (by use of the uncertainty relation) volume $2\pi\hbar$ and representing a field mode. The "quantum" space obtained in this way is a fuzzy sphere (for more implications and details see, e.g. [203]).

As it is evident, the central theme behind the fuzzy space emerging in the way described previously is a "discretization" of the manifold by means of quantization. Happily, things get back to "normal" when the effective Planck constant tends to zero. Naturally, noncommutativity has also entered other areas of theoretical physics such as, for example, quantum field theory and string theory (see, e.g. [105] or [280] for expository reviews, or the second volume of [149] for considerable details). In fact, much of the development of noncommutative geometry is inspired by the latter. Here, the need to express, e.g. differential operators in the noncommutative setting, leads the noncommutative geometric structure to be necessarily combined with an associated differential calculus over the algebra produced by the aforementioned generators (see, e.g. [205]). Finally, one should mention the attempts of Mayburov [211] for a possible fuzzy-geometric framework within which to reformulate the entire quantum mechanics formalism in strictly geometric terms, based on the assumption that the states of a massive particle correspond to fuzzy points, and to derive the evolution of a quantum system from geometric arguments.

There exists a rather large and on-going bibliography on the topics briefly touched upon previously. The interested reader is referred, for instance, to [13] on noncommutative space-times, or [19] on fuzzy physics and supersymmetry and the references therein, while the beginner may turn to [77] and [204] for two of the oldest but still valuable and indeed fascinating expositions given by two of the leading researchers on noncommutative geometry.

Exercises

11.1 Derive a thick fuzzy circle from the crisp equation $(x-2)^2 + (y-1)^2 = 4$.

11.2 Let the real numbers $A = B = 1$ and the triangular fuzzy number tfn(1, 2, 7). Give the set of disks ${}^0C$ and the crisp equation of the circle ${}^1C$.

11.3 Given a "regular" fuzzy circle with $A = B = 1$ and the triangular fuzzy number tfn(1, 2, 7), find the fuzzy circumference.

11.4 Describe mathematically (i) a fuzzy square obtained from a common square with sides of unit length in the Euclidean plane, and (ii) a regular fuzzy pentagon. Describe the fuzzy perimeter of the latter.

11.5 Choose three fuzzy points P_1, P_2, and P_3 so as to define a fuzzy triangle, and determine the fuzzy angle Φ between the fuzzy line segments L_{23} and L_{31}.

11.6 Compare the area A of a regular pentagon inscribed in a unit circle with the fuzzy area F of a regular fuzzy pentagon. What conclusions can be drawn?

12

Fuzzy Calculus

The importance of fuzzy calculus becomes strongly evident if one thinks of the role played by classical calculus concepts in the mathematical modeling of real-world phenomena. There, the complexity of nature forces one to make a lot of compromise, that is, to formulate assumptions, assertions, and premises not only in order to build and apply a mathematically tractable model, but also because of the imprecision or vagueness often characterizing the gathered experimental information or even the theoretical background needed to quantitatively describe a particular natural phenomenon. A possible way to deal with the aforesaid vagueness in mathematical modeling is offered by fuzzy calculus. In this chapter, we shall first define fuzzy functions and their properties and then examine their integration and differentiation. Thereafter, we consider a theory of fuzzy limits, and finally, we shall study fuzzy differential equations.

12.1 Fuzzy Functions

The possibility of fuzziness for a real-valued function of one independent real variable, which henceforth will be called *fuzzy function*, can rather naturally appear basically in three different ways (see, e.g., [190] or [110]):

(i) the nonfuzzy function considered has fuzzy constraints on its domain and codomain;
(ii) there is fuzziness in the independent variable, and the nonfuzzy function "transfers" this fuzziness to the codomain but without producing new fuzziness; and
(iii) a nonfuzzy function is properly extended so that it maps a fuzzy set (the function's domain) to a fuzzy set (the function's codomain).

It must be pointed out that in most "ordinary" applications, fuzzifying ordinary real-valued functions, that is, extensions of ordinary functions that belong to the

A Modern Introduction to Fuzzy Mathematics, First Edition.
Apostolos Syropoulos and Theophanes Grammenos.

third kind above (mainly based on Zadeh's extension principle, see Section 2.8) are used.

We proceed with the definition of a nonfuzzy function with constraints on the fuzzy domain and the fuzzy codomain.

Definition 12.1.1 Let a nonfuzzy function $f : X \rightarrow Y$, where X and Y are subsets of $\mathbb{R}$, and A and B are two fuzzy sets on X and Y, respectively. Then, A is the *fuzzy domain* and B is the *fuzzy codomain* of the function f if and only if the following condition holds:

$$A(x) \leq B(f(x)),$$

and the function $f(x)$ is a *nonfuzzy* function with constraints on the fuzzy domain A and the fuzzy codomain B.

Example 12.1.1 Consider two fuzzy sets

$$A = \{2/0.5, 4/0.7\} \quad \text{and} \quad B = \{4/0.7, 8/0.8\}$$

and the function $f(x) = 2x$. It is obvious that f satisfies $A(x) \leq B(f(x))$.

In a similar way, we can define the composition of such functions. Consider the functions that satisfy fuzzy constraints $f : A \rightarrow B$ and $g : B \rightarrow C$, where A, B, and C are fuzzy subsets of X, Y, and Z. Then, the composition $g \circ f$ is a fuzzy function with fuzzy constraints

$$g \circ f : A \rightarrow C$$

such that $A(x) \leq B(f(x))$ and $B(y) \leq C(g(y))$, where $x \in X$ and $y = f(x)$. In turn, this means that

$$A(x) \leq C(g(f(x))).$$

A function f is *fuzzy continuous* if and only if

$$R(f(x_1), f(x_2)) \geq S(x_1, x_2), \quad \text{for all } x_1, x_2 \in X,$$

where R, S are fuzzy proximity relations on X and Y, respectively (see Definition 4.5.4). Furthermore, the composition of fuzzy continuous functions is always a fuzzy continuous function [110]. It is pointed out that the continuity is introduced without employing the notion of limit as in crisp calculus.

Now, we come to the case of a nonfuzzy function $f : X \rightarrow Y$ that "transfers" the possible fuzziness of the independent variable to the codomain. Following the extension principle, an image $f(X)$ of the fuzzy subset X can be obtained by

$$f^{\rightarrow}(X)(y) = \begin{cases} \bigvee\{X(x) | x \in f^{-1}(y)\}, & \text{if } f^{-1}(y) \neq \emptyset, \\ 0, & \text{otherwise.} \end{cases}$$

Definition 12.1.2 A function of a crisp variable that maps a nonfuzzy domain X to a fuzzy codomain $\mathscr{P}(Y)$ is a *fuzzifying function* $f : X \to \mathscr{P}(Y)$.

There is a connection between fuzzy functions and fuzzy relations (see Definition 9.2.6). However, here this connection takes a different form. In particular, a fuzzifying function $f : X \to Y$ is associated with the notion of fuzzy relation R, in the sense that the former can be considered as a fuzzy relation such that

$$\forall(x,y) \in X \times Y,\ f(x)(y) = R(x,y).$$

One can define the *composition of fuzzifying functions* as the sup–min composition of the associated fuzzy relations. Thus, if $\tilde{h} : Y \to Z$ is a fuzzifying function, then

$$(h \circ f)(x)(z) = \bigwedge_{y \in Y} \bigvee (f(x)(y), h(y)(z)).$$

Next, we come to the concept of the fuzzy bunch of nonfuzzy functions:

Definition 12.1.3 A *fuzzy bunch* F of nonfuzzy functions from X to Y is a fuzzy set on Y^X, so that each function $f : X \to Y$ has a membership value $F(f)$ in F.

It must be stressed that a fuzzifying function is a fuzzy bunch in the following sense: for all $\alpha : 0 \leq \alpha \leq 1, f(x)(y) = \alpha$ defines one or more functions $f_a^i : X \to Y$ with the fuzzy bunch

$$F = \bigcup_i F^i \text{ and } F^i = \int_{a \in (0,1]} \alpha f_a^i.$$

However, a fuzzy bunch is not a fuzzifying function, because one can have two functions, say $f, h : X \to Y$ such that there are x with $f(x) = h(x) = y$, but $F(f) \neq F(h)$. This cannot happen in the case of a fuzzifying function since for each (x,y) one uniquely has $f(x)(y) = R(x,y)$, contrary to a fuzzy bunch where each (x,y) can have more membership values. Thus, a fuzzy bunch can be considered only as a multi-valued fuzzy relation [110].

The *composition of fuzzy bunches* is possible. If we have two fuzzy bunches, say $A : X \to Y$ and $B : Y \to Z$, then the composition $C = B \circ A$ is a fuzzy bunch from X to Z, such that

$$C(h) = \bigvee_{c=b \circ a} \min(A(a), B(b)), \text{ for all } h.$$

We turn now our attention to the concept of fuzzy extrema, starting with the important notion of a *maximizing* (respectively *minimizing*) set.

Definition 12.1.4 Let f be a function with real values in X and bounded by $\wedge(f)$ and $\vee(f)$. Then, the *maximizing set* M is a fuzzy set in X such that

$$M(x) = \frac{f(x) - \wedge(f)}{\vee(f) - \wedge(f)}, \quad \forall x \in X,$$

and it holds that

$$M(x_0) = 1, \quad \forall x_0 : f(x_0) = \vee(f),$$
$$M(x) = 0, \quad \forall x : f(x) = \wedge(f).$$

The *minimizing set* of f is defined as the maximizing set of $-f$.

Thus, the *fuzzy maximum* of the function f is defined as follows:

Definition 12.1.5 The *fuzzy maximum* of the function f, a fuzzy set of the function's codomain Y, is the image $f(M)$ of the maximizing set M under f such that

$$f(M)(y) = \bigvee_{x \in f^{-1}(y)} M(x), \quad \forall y \in Y.$$

Let us see an example of a maximizing set.

Example 12.1.2 Suppose that a function $f(x)$ is given such that $[\wedge(f), \vee(f)] = [20, 40]$, $\forall x \in [2, 10]$, $f(5) = 30$ and $f(8) = 36$. Then, the possibility of $x = 5$ and $x = 8$ to belong to the maximizing set M is obtained, respectively, as

$$\mu_M(5) = \frac{30 - 20}{40 - 20} = \frac{1}{2}, \quad \mu_M(8) = \frac{36 - 20}{40 - 20} = \frac{4}{5}.$$

In other words, the two values of the independent variable $x = 5$ and $x = 8$ give the maximum value of $f(x) = 40$ with possibilities $\frac{1}{2}$ and $\frac{4}{5}$, respectively.

Let us utilize the maximizing set to study the maximum value of a *nonfuzzy function* starting with the simpler case of a *nonfuzzy domain D*.

Suppose that the function f obtains its maximum value for x_o. Then,

$$\mu_M(x_0) = \bigvee_{x \in D} \mu_M(x) = \bigvee_{x \in X} \min[\mu_M(x), \mu_D(x)],$$

where $\mu_D(x)$ gives the possibility of x to belong to the domain D. Let us see a simple example [190].

Example 12.1.3 Suppose $f(x) = \sin x$, and, of course, $|\sin x| \leq 1$. Then,

$$\mu_M(x) = \frac{\sin x - \wedge(\sin x)}{\vee(\sin x) - \wedge(\sin x)} = \frac{1}{2}(\sin x + 1),$$

$$\mu_D(x) = \begin{cases} 1, & x \in [0, 2\pi], \\ 0, & \text{otherwise}. \end{cases}$$

Hence, the maximum value of $f(x)$ is given at x_0, where

$$\mu_M(x_0) = \bigvee \min[\mu_M(x), \mu_D(x)] = \bigvee_{x\in[0,2\pi]} \mu_M(x) = \begin{cases} 1, & x_0 = 0, \\ 1, & x_0 = 2\pi. \end{cases}$$

In other words, one gets the maximum value $f(x_0) = 1$ for $x_0 = 0$ and $x_0 = 2\pi$, as expected.

Now, we shall examine the more complicated case of the maximum value $f(x_0)$ of a *nonfuzzy function with a fuzzy domain D*. Here, we seek for an element of X that belongs as much as possible both to the fuzzy domain D and the maximizing set. The practical procedure contains two basic steps/conditions [190], (i) take $\mu_M(x)$ as a maximum, and (ii) take $\mu_D(x)$ as a maximum, that have to be satisfied by any element x_i. Thus, the possibility of any x_i leading to the maximum value of the function f is given by $\min[\mu_M(x_i), \mu_D(x_i)]$, while for the value x_0 for which the function f gets its maximum we have

$$\mu(x_0) = \bigvee_{x\in X} \min[\mu_M(x), \mu_D(x)], \tag{12.1}$$

where $\mu_M(x)$ is the membership function of the maximizing set, while $\mu_D(x)$ is the membership function of the fuzzy domain. From the comparison of x_0 with x_i in the graph of the function $f(x)$, we can conclude which of x_0 and x_i gives indeed the function its maximum. If, as an example, it is x_i for which the function f gets its maximum, then $f(x_i) > f(x_0)$ or $\mu_M(x_i) > \mu_M(x_0)$. However, $\mu_D(x_i) < \mu_D(x_0)$, so that we conclude that $f(x_0)$ is the maximum value. The following example for a nonfuzzy function with a fuzzy domain illuminates the situation [190].

Example 12.1.4 Suppose that $f(x) = 2 - x, x \in D$ and

$$\mu_D = \begin{cases} x^2, & x \in [0, 1], \\ 0, & \text{otherwise}. \end{cases}$$

Then, we have $\mu_M(x) = \frac{2-x-1}{2-1} = 1 - x$. From (12.1), we get x_0 when $\mu_M(x) = \mu_D(x)$, $\forall x \in [0, 1]$. Thus, $1 - x = x^2 \Rightarrow x = \frac{-1\pm\sqrt{5}}{2}$. Since $0 \le x \le 1$, we keep only $\frac{-1+\sqrt{5}}{2}$ which is the value of x_0 leading to the function's maximum value $f(x_0) = \frac{1}{2}(5 - \sqrt{5})$.

For the maximization of a *fuzzy function* $\tilde{f} : X \to \mathbb{R}$ *on a nonfuzzy domain D* of X, where X is assumed finite, a main idea is the use of the *extended operator* $\widetilde{\max}_{x\in D}\tilde{f}(x)$ in order to find the maximum value of the function, say, $\tilde{M}$. Due to the ambiguity regarding the choice of an element in D for the realization of $\widetilde{M}$ because the latter is, in general, not one of the $\widetilde{f}(x)$'s, it is necessary to find out which of the x's do contribute to $\widetilde{M}$'s membership function. In fact, if $|D| = n$, one has [110]

$$\begin{aligned}\mu_{\widetilde{M}(y)} &= \bigvee_{\substack{y_1 \dots y_n \\ y=\max(y_1 \dots y_n)}} \min \quad (\mu_{\widetilde{f}(x_1)}(y_1), \dots, \mu_{\widetilde{f}(x_n)}(y_n)) \\ &= \bigvee_{\substack{f \in \mathbb{R}^D \\ y=\max_{i=1,\dots n}(f(x_i))}} \min_{j=1,\dots n} \quad \mu_{\widetilde{f}(x_j)}(f(x_j))\end{aligned}$$

For possible alternatives to this procedure, which is rather difficult to implement, the interested reader is referred to [110].

12.2 Integrals of Fuzzy Functions

We begin our study with the *integration of a fuzzifying function over a nonfuzzy interval*. Suppose we have a real interval $(a, b) \subset \mathbb{R}$ and a fuzzifying function f : $[a, b] \to \mathbb{R}$ with the fuzzy value $\widetilde{f}(x)$ for $x \in [a, b]$, the fuzzy number $\widetilde{f}(x)$ being, as usually, a piecewise continuous convex normalized fuzzy set on $\mathbb{R}$. We assume that, for every $\alpha \in [0, 1]$, $\mu_{\widetilde{f}(x)}(y) = \alpha$ has only two continuous solutions $y = f_\alpha^-(x)$, $y = f_\alpha^+(x)$ for $\alpha \neq 1$, and only one continuous solution $y = f(x)$ for $\alpha = 1$. The properties of f_α^-, f_α^+ are such that [110]

$$f_{\alpha'}^+(x) \geq f_\alpha^+(x) \geq f(x) \geq f_\alpha^-(x) \geq f_{\alpha'}^-(x), \quad \forall \alpha, \alpha', \qquad \alpha' \leq \alpha$$

and f_α^-, f_α^+ are α-level curves of $\widetilde{f}(x)$. The integral of any continuous α-level curve of f over the real interval $[a, b]$ exists always. Thus, we define the *integral of the fuzzifying function over the nonfuzzy interval* $[a, b]$ as

Definition 12.2.1 $\widetilde{I}(a, b) = \int_a^b f_\alpha^-(x)dx + \int_a^b f_\alpha^+(x)dx$, $\alpha \in [0, 1]$ where f_α^-, f_α^+ are α-cut functions of $\widetilde{f}(x)$. So, for each α-cut function we have, respectively, the integrals $\widetilde{I}_\alpha^- = \int_a^b f_\alpha^-(x)dx$ and $\widetilde{I}_\alpha^+ = \int_a^b f_\alpha^+(x)dx$, with α giving the possibility of $\widetilde{I}_\alpha^-$ or $\widetilde{I}_\alpha^+$ to be a member of $\widetilde{I}(a, b)$.

Let us see a computing example to clarify the aforementioned definitions.

Example 12.2.1 Suppose we have to integrate over the interval $[2, 3]$ a fuzzy bunch of three functions $\widetilde{f}(x) = \{(f_1(x), \alpha_1), (f_2(x), \alpha_2), (f_3(x_3), \alpha_3)\}$, with $\alpha_1 = 0.4$, $\alpha_2 = 0.7$, and $\alpha_3 = 0.4$, and $f_1(x) = x$, $f_2(x) = x^2$, and $f_3(x) = x + 2$.

The integration for $\alpha_2 = 0.7$ gives $I_\alpha(2,3) = \int_2^3 x^2 dx = 7$, i.e. a result with possibility 0.7, and it follows that $\widetilde{I}_{0.7}(2,3) = \{(7, 0.7)\}$. For the integration at $\alpha = 0.4$, we have $f^+ = f_1(x) = x, f^- = f_3(x) = x + 2$. Subsequently, $I_\alpha^+(2,3) = \int_2^3 x dx = \frac{5}{2}$ and $I_\alpha^-(2,3) = \int_2^3 (x+2)dx = \frac{9}{2}$. In other words, we have the result $\frac{5}{2}$ with possibility 0.4 and the result $\frac{9}{2}$ again with possibility 0.4. Thus, $\widetilde{I}_{0.4}(2,3) = \left\{ (\frac{5}{2}, 0.4), (\frac{9}{2}, 0.4) \right\}$ and, in total, $\widetilde{I}(2,3) = \left\{ (7, 0.7), (\frac{5}{2}, 0.4), (\frac{9}{2}, 0.4) \right\}$.

We shall now treat the *integration of an L–R fuzzifying function over a nonfuzzy interval*. As we shall see, in this case it suffices to integrate the mean value and the left and right spread functions over the interval $[a, b]$, and the integration result is an L–R fuzzy number [110]. Before proceeding, we suggest the reader to return to Section 3.1.6 for a repetition of the definitions of L–R fuzzy numbers and L and R functions.

In the literature, an L–R fuzzy number is often denoted as $N(m, \alpha, \beta)_{LR}$, where m is the mean value of N, and α, β the left and right spread, respectively. If $\alpha, \beta = 0$, the fuzzy number N becomes a nonfuzzy number. In what follows, we use this notation.

Definition 12.2.2 A *fuzzifying function f is of L–R type* if and only if $(f(x), g(x), h(x))_{LR}$ is an L–R fuzzy number for every x in $[a, b]$.

Here, $f(x)$, $g(x)$, and $h(x)$ are assumed positive integrable functions on the interval $[a, b]$, while $f_\alpha^-(x) = f(x) - g(x)L^{-1}(\alpha)$ and $f_\alpha^+(x) = f(x) + h(x)R^{-1}(\alpha)$ are the α-level curves of $f(x)$. One can integrate the latter two functions and obtain

$$\int_a^b f_\alpha^-(x)dx = \int_a^b f(x)dx - L^{-1}(\alpha) \int_a^b g(x)dx = J$$

$$\int_a^b f_\alpha^+(x)dx = \int_a^b f(x)dx + R^{-1}(\alpha) \int_a^b h(x)dx = K$$

from which one gets

$$J = F(b) - F(a) - L^{-1}(a)[G(b) - G(a)],$$

$$K = F(b) - F(a) + R^{-1}(a)[H(b) - H(a)]$$

with F, G, and H, the antiderivatives of f, g, and h, respectively. Thus, we have

$$L\left(\frac{F(b) - F(a) - J}{G(b) - G(a)}\right) = \alpha, \quad \forall J \leq F(b) - F(a),$$

and $G(b) - G(a) \geq 0$, because $b \geq a$. With a similar argumentation, we have

$$R\left(\frac{-F(b) + F(a) + K}{H(b) - H(a)}\right) = \alpha$$

Hence, finally, we get the integration result

$$\widetilde{I}(a,b) = \left(\int_a^b f(x)dx, \int_a^b g(x)dx, \int_a^b h(x)dx \right)_{LR}$$

which is an L–R fuzzy number.

At this point, it must be stressed that, for a fuzzifying function of the L–R type, it can be shown that Zadeh's extension principle leads to a generalization of the ordinary Riemann sums, and the integrals presented previously are *generalizations of the ordinary Riemann integrals* (see [110] for an extended presentation).

Now, we come to the *integration of a nonfuzzy real-valued function over a fuzzy interval* (A, B), where A, B are fuzzy sets on $\mathbb{R}$. The integration is defined as follows:

Definition 12.2.3 $\mu_{I(A,B)}(z) = \bigvee_{x,y:z=\int_x^y f(u)du} \min(\mu_A(x), \mu_B(y)).$

For simplicity reasons, let us begin with the special case of only one of either A or B being nonfuzzy. So, suppose that the integration interval is $[a, B)$. Then,

$$\mu_{I(a,B)}(z) = \bigvee_{y:z=\int_a^y f(u)du} \mu_B(y) = \bigvee_{y:z=F(y)-F(a)} \mu_B(y),$$

with F being the antiderivative of f and $\widetilde{I}(a, B) = F(B) - F(a)$, where the minus sign denotes the *extended subtraction operation*. In the general case, when both ends A, B of the interval are fuzzy, Definition 12.2.3 becomes

$$\begin{aligned}
\mu_{I(A,B)}(z) &= \bigvee_{z=F(y)-F(x)} \min(\mu_A(x), \mu_B(y)) \\
&= \bigvee_{x\in\mathbb{R}} \min(\mu_A(x), \bigvee_{z=F(y)-F(x)} \mu_B(y)) \\
&= \bigvee_{x\in\mathbb{R}} \min(\mu_A(x), \mu_{I(x,B)}(z)).
\end{aligned}$$

In other words, the integration result $\widetilde{I}(A, B)$ is the fuzzy value of the extended fuzzifying function $y = F(B) - F(x)$, for $x = A$, or $\widetilde{I}(A, B) = F(B) - F(A) = \int_A^B f(x)dx$, where the minus sign denotes the *extended subtraction operation*. The calculation is modified when A, B are L–R fuzzy numbers and, in that case, $\widetilde{I}(A, B)$ is, in general, not an L–R fuzzy number (see [110] for more details on this matter and for an alternative approach as well).

Finally, we turn our attention to the problem of *integration of a fuzzifying function* $\widetilde{f}$ *over a fuzzy interval* (A, B). Following an analogous way as before, we begin with the definition of the integration based on Zadeh's extension principle [110]:

Definition 12.2.4 $\mu_{\widetilde{I}(A,B)}(z) = \bigvee_{\substack{l \in L \\ (x,y) \in \mathbb{R}^2,\ x \leq y \\ z = \int_x^y l(t)dt}} \min(\mu_A(x), \mu_B(y), \mu_{\widetilde{f}}(l)).$

Here, A and B are fuzzy numbers, $\widetilde{I}(A, B)$ is the fuzzy integral, i.e. the value of the extended $\widetilde{I}(x, y)$ when $x = A$ and $Y = B$, and L is the set of integrable functions $l : \mathbb{R} \to \mathbb{R}$. We can rewrite this expression in the form

$$\mu_{\widetilde{I}(A,B)}(z) = \bigvee_{x \leq y} \min\left(\mu_A(x), \mu_B(y), \bigvee_{\substack{l \in L \\ z = \int_x^y l(t)dt}} \mu_{\widetilde{f}}(l) \right)$$

$$= \bigvee_{x \leq y} \min(\mu_A(x), \mu_B(y), \mu_{\widetilde{I}(x,y)}(z)).$$

It must be pointed out that $\widetilde{I}(A, B) \neq \widetilde{F}(B) - \widetilde{F}(A)$, where $\widetilde{F}$ is the anti-derivative of $\widetilde{f}$ and the minus sign denotes the *extended subtraction operation*.

We wish to close this section by referring to some basic properties of fuzzy integrals beginning with *linearity*. Let the integrals $\int_a^b \widetilde{f}(x)dx$ and $\int_a^b \widetilde{g}(x)dx$ of the fuzzifying functions $\widetilde{f}$ and $\widetilde{g}$, respectively. Then

$$\int_a^b (\widetilde{f}(x) + \widetilde{g}(x))dx = \int_a^b \widetilde{f}(x)dx + \int_a^b \widetilde{g}(x)dx,$$

with the plus sign denoting the *extended addition operation*.

Now, let the integral $\int_a^B f(x)dx$ of the nonfuzzy function $f(x)$ with $[a, B)$ a fuzzy interval, where $a \in \mathbb{R}$ and B is a fuzzy number. Then

$$\int_a^{B \cup C} f(x)dx = \int_a^B f(x)dx \cup \int_a^C f(x)dx, \tag{12.2}$$

with C a fuzzy number. When $f(x) > 0$, then

$$\int_a^{B \cap C} f(x)dx = \int_a^B f(x)dx \cap \int_a^C f(x)dx. \tag{12.3}$$

In (12.2) and (12.3), the plus sign denotes the ordinary addition operation (for a proof of the three last expressions, see [110]).

There are special kinds of more sophisticated fuzzy integrals of fuzzy-valued functions, such as the *Aumann integral* and the *Henstock integral* that are also interrelated. The treatment of these integrals is beyond the scope of this book, but the interested reader is referred to, for example, [24] or, in particular, [101, 131, 301].

12.3 Derivatives of Fuzzy Functions

We begin with the case of a fuzzifying function $\widetilde{f} : \mathbb{R} \to \mathbb{R}$ with the image of any real x be a fuzzy number and each α-level curve f_α of $\widetilde{f}$ having a derivative at any point x_0 of the domain. Then, we define the (extended) derivative $\left.\frac{d\widetilde{f}}{dx}\right|_{x=x_0}$ through the membership function:

Definition 12.3.1 $\mu_{\left.\frac{d\widetilde{f}}{dx}\right|_{x=x_0}}(P) = \bigvee \mu(f_\alpha)$, with $\mu(f_\alpha) = \alpha$.

In fact, the derivative $\left.\frac{d\widetilde{f}}{dx}\right|_{x=x_0}$ gives an estimation of the degree of parallelism for the bunch of level curves at x_0. In other words, the level curves are the more parallel, the less fuzzy is $\left.\frac{d\widetilde{f}}{dx}\right|_{x=x_0}$ [110].

If we have an L–R fuzzifying function $\widetilde{f}(x) = (f(x), g(x), h(x))_{LR}$ that is a strictly convex fuzzy number, while $f(x)$, $g(x)$, and $h(x)$ are everywhere differentiable functions, then for $\alpha \neq 1$ there exist two α-level curves f_α^-, f_α^+ with

$$f_\alpha^-(x) = f(x) - L^{-1}(\alpha)g(x),$$

$$f_\alpha^+(x) = f(x) + R^{-1}(\alpha)h(x),$$

and, say, $f_1(x) = f(x)$.

Then, we have for the derivatives

$$\left.\frac{df_\alpha^-}{dx}\right|_{x=x_0} = \left.\frac{df}{dx}\right|_{x=x_0} - L^{-1}(\alpha)\left.\frac{dg}{dx}\right|_{x=x_0},$$

$$\left.\frac{df_\alpha^+}{dx}\right|_{x=x_0} = \left.\frac{df}{dx}\right|_{x=x_0} + R^{-1}(\alpha)\left.\frac{dh}{dx}\right|_{x=x_0}.$$

The sign of $\left.\frac{dg}{dx}\right|_{x=x_0}$ and $\left.\frac{dh}{dx}\right|_{x=x_0}$ determines the behavior of the α-level curves, so we shall discern between *four different cases*:

(a) **Both derivatives are strictly positive.** Then, $g(x)$ and $h(x)$ are monotonically increasing in a neighborhood of x_0 and, thus, as x increases the α-level curves recede from $f(x)$. So, for an α_1 such that $\alpha_1 < \alpha < 1$, it holds

$$\left.\frac{df_{\alpha_1}^+}{dx}\right|_{x=x_0} > \left.\frac{df_\alpha^+}{dx}\right|_{x=x_0} > \left.\frac{df}{dx}\right|_{x=x_0} > \left.\frac{df_\alpha^-}{dx}\right|_{x=x_0} > \left.\frac{df_{\alpha_1}^-}{dx}\right|_{x=x_0} \tag{12.4}$$

Here, there is at most one level curve which has a derivative at $x = x_0$, equal to a given slope s. From Definition 12.3.1, we get

$$\mu_{\left.\frac{d\tilde{f}}{dx}\right|_{x=x_0}}(s) = L\left(\frac{\left.\frac{df}{dx}\right|_{x=x_0} - s}{\left.\frac{dg}{dx}\right|_{x=x_0}}\right), \quad \text{when } s \le \left.\frac{df}{dx}\right|_{x=x_0},$$

$$\mu_{\left.\frac{d\tilde{f}}{dx}\right|_{x=x_0}}(s) = R\left(\frac{s - \left.\frac{df}{dx}\right|_{x=x_0}}{\left.\frac{dh}{dx}\right|_{x=x_0}}\right), \quad \text{when } s \ge \left.\frac{df}{dx}\right|_{x=x_0},$$

while $\left.\frac{d\tilde{f}}{dx}\right|_{x=x_0} = \left(\left.\frac{df}{dx}\right|_{x=x_0}, \left.\frac{dg}{dx}\right|_{x=x_0}, \left.\frac{dh}{dx}\right|_{x=x_0}\right)_{LR}$.

(b) **Both derivatives are strictly negative.** Then, $g(x)$ and $h(x)$ are monotonically decreasing in a neighborhood of x_0 and, thus, for an α_1 such that $\alpha_1 < \alpha < 1$, similarly to (12.4) it holds

$$\left.\frac{df^-_{\alpha_1}}{dx}\right|_{x=x_0} > \left.\frac{df^-_{\alpha}}{dx}\right|_{x=x_0} > \left.\frac{df}{dx}\right|_{x=x_0} > \left.\frac{df^+_{\alpha}}{dx}\right|_{x=x_0} > \left.\frac{df^+_{\alpha_1}}{dx}\right|_{x=x_0},$$

$$\left.\frac{d\tilde{f}}{dx}\right|_{x=x_0} = \left(\left.\frac{df}{dx}\right|_{x=x_0}, -\left.\frac{dh}{dx}\right|_{x=x_0}, -\left.\frac{dg}{dx}\right|_{x=x_0}\right)_{LR}.$$

(c) **One derivative is strictly positive while the other is strictly negative.** Let $\left.\frac{dg}{dx}\right|_{x=x_0} < 0$ and $\left.\frac{dh}{dx}\right|_{x=x_0} > 0$. Then,

$$\min\left(\left.\frac{df^+_{\alpha}}{dx}\right|_{x=x_0}, \left.\frac{df^-_{\alpha}}{dx}\right|_{x=x_0}\right) \ge \left.\frac{df}{dx}\right|_{x=x_0}.$$

As a result, $P < \left.\frac{df}{dx}\right|_{x=x_0} \Rightarrow \mu_{\left.\frac{d\tilde{f}}{dx}\right|_{x=x_0}}(P) = 0$.

While, if $P \ge \left.\frac{df}{dx}\right|_{x=x_0}$, then two level curves f^+_α, f^-_α can exist with $\left.\frac{df^+_\alpha}{dx}\right|_{x=x_0} = P$ and $\left.\frac{df^-_\alpha}{dx}\right|_{x=x_0} = P$. Consequently,

$$\left.\frac{d\tilde{f}}{dx}\right|_{x=x_0} = \left(\left.\frac{df}{dx}\right|_{x=x_0}, 0, -\left.\frac{dg}{dx}\right|_{x=x_0}\right)_L \bigcup \left(\left.\frac{df}{dx}\right|_{x=x_0}, 0, \left.\frac{dh}{dx}\right|_{x=x_0}\right)_R,$$

while the derivative $\left.\frac{d\tilde{f}}{dx}\right|_{x=x_0}$ can be either of L or of R type.

(d) **This is the opposite of case (c).** Let $\left.\frac{dg}{dx}\right|_{x=x_0} > 0$ and $\left.\frac{dh}{dx}\right|_{x=x_0} < 0$. Then, following argumentation lines similar to the previous case, we are led to

$$\left.\frac{d\tilde{f}}{dx}\right|_{x=x_0} = \left(\left.\frac{df}{dx}\right|_{x=x_0}, \left.\frac{dg}{dx}\right|_{x=x_0}, 0\right)_L \bigcup \left(\left.\frac{df}{dx}\right|_{x=x_0}, -\left.\frac{dh}{dx}\right|_{x=x_0}, 0\right)_R.$$

A comment is in order regarding the validity of the *sum rule of differentiation*. Namely, if $\frac{dg}{dx}\big|_{x=x_0}, \frac{dh}{dx}\big|_{x=x_0}$ are of the same sign, then for two fuzzifying functions $\widetilde{f}_1, \widetilde{f}_2$ we have

$$\frac{d}{dx}(\widetilde{f}_1 + \widetilde{f}_2)\Big|_{x=x_0} = \frac{d\widetilde{f}_1}{dx}\Big|_{x=x_0} + \frac{d\widetilde{f}_2}{dx}\Big|_{x=x_0},$$

where the plus sign is denoting the *extended addition operation*.

For a study of the integral and the derivative in the special case where the integrable function depends on a fuzzy parameter, the interested reader is referred to [67]. Furthermore, concerning the differentiability of fuzzy-valued functions, one should also mention the rather difficult notion of the *Hukuhara derivative* introduced in [244] and based on the assumption that the differentiable function has an increasing length of its support interval.

12.4 Fuzzy Limits of Sequences and Functions

Among several approaches to the concept of the *fuzzy limit of a function* we shall focus on one that, in our opinion, stands out for its elegance and rigor as well: The approach of Mark Burgin [53] based on the notion of the *fuzzy limits of sequences* and developed in the frame of *neoclassical analysis* [52], that is, the study of ordinary (real) functions on the basis of fuzzy set theory, and set-valued analysis. Thus, before we consider the fuzzy limit of a function, we begin by studying the essential properties of the fuzzy limit of a sequence.

Definition 12.4.1 Let $r \geq 0$ be a real number and $s = \{a_i \in \mathbb{R},\ i \in a_n\}$ a sequence of real numbers, where a_n is the sequence of all numbers belonging to $\mathbb{N}$. Then, the number a is called *r-limit of the sequence s*, symbolically $a = r\text{-}\lim_{i\to\infty} a_i$ or $a = r\text{-}\lim s$, if $\rho(a, a_i) < r + \varepsilon$ for almost all a_i and $\forall \varepsilon \in \mathbb{R}^+$, where we define $\rho(a, a_i) = |a - a_i|$.

Equivalently, $\exists n : \forall i > n \Rightarrow \rho(a, a_i) < r + \varepsilon$, that is, the sequence is r-converging to a. For $r = 0$, the r-limit becomes the ordinary limit of the sequence. Indeed, the following lemma holds:

Lemma 12.4.1 *$a = r\text{-}lim\ s \Rightarrow a = t\text{-}lim\ s,\ \forall t > r$.*

The following illuminating example is from [54].

Example 12.4.1 Let the sequence $s = \{\frac{1}{i},\ i \in a_n\}$. Then, the numbers 1 and -1 are 1-limits of s, while $\frac{1}{2}$ is a $\frac{1}{2}$-limit of s, but 1 is *not* a $\frac{1}{2}$-limit of s.

We come to the definition of the fuzzy limit of a sequence:

Definition 12.4.2 A number a is the *fuzzy limit of sequence s* if a is the r-limit of s.

Thus, the sequence s *converges fuzzily* if it has a fuzzy limit. It must be pointed out that a sequence that does not have an ordinary limit may, nevertheless, have a fuzzy limit. Further,

Lemma 12.4.2 $a = q\text{-}lim\ s \Rightarrow a = t\text{-}lim\ k$, *with k any subsequence of s.*

Lemma 12.4.3 *For any convergent subsequence k of a sequence s,* $a = r\text{-}lim\ k \Rightarrow a = r\text{-}lim\ s$ *(see [54] for a proof).*

Infinite fuzzy limits for real number sequences are also possible:

Definition 12.4.3 $\infty = r\text{-lim } s$ if for almost all a_i we have $a_i > r$. Similarly, $-\infty = r\text{-lim } s$ if for almost all a_i we have $a_i < -r$.

A notion exhibiting significant applications in the theory of timed automata (i.e. models for analyzing the behavior of real-time computing systems over time, see e.g., [8]) is the notion of a *weak fuzzy limit*:

Definition 12.4.4 Consider the sequence $s = \{a_i \in \mathbb{R},\ i \in a_n\}$. Then, the number d is the *weak fuzzy limit* of s, symbolically $d = r\text{-wlim } s$, if there exists a subsequence k of the sequence s such that $a = \lim k$.

We now proceed to the concept of the *fuzzy limit of a set of sequences.*

Definition 12.4.5 The number a is an r-limit of a set of real number sequences $E = \{s_i,\ i \in a_n\}$, symbolically $a = r\text{-lim } E$, if a is an r-limit of each sequence $l_i \in E$.

In fact, if there exists an $a = r\text{-lim } E$ with $r = 0$ for the set E, then this limit is unique and each $l_i \in E$ converges to that a. On the other hand, sequences from a given set E may have different limits but only one common fuzzy limit. The following example from [54] illustrates this situation.

Example 12.4.2 Let the set of sequences

$$E = \left\{ \left\{ \frac{1}{2^n},\ n \in \mathbb{N} \right\}, \left\{ 1 + \frac{1}{3^n},\ n \in \mathbb{N} \right\}, \left\{ 2 + \frac{1}{5^n},\ n \in \mathbb{N} \right\} \right\}.$$

Then, E has an $a = 1$-limit with $a = 1$, but it does not have an ordinary limit since $\lim \frac{1}{2^n} \to 0$, $\lim \left(1 + \frac{1}{3^n}\right) \to 1$, and $\lim \left(2 + \frac{1}{5^n}\right) \to 2$.

What changes when the set of sequences E is *finite* or *countable*? The following theorem [54] gives an answer.

Theorem 12.4.1 *If the set of sequences E is* finite *or* countable, *then there exists a sequence* $s = \{a_i \in \mathbb{R},\ i \in a_n\}$ *such that for any real number a and any nonnegative real number r it holds* $a = r\text{-lim } E$ *if and only if* $a = r\text{-lim } s$.

We shall close with some basic definitions and properties of the *fuzzy limit of sequences in metric spaces*. Suppose we have a *metric space M* possessing a metric ρ, r is a nonnegative real number, and $s = \{a_i | a_i \in M,\ i \in a_n\}$ is a sequence in M.

Definition 12.4.6 $a = r\text{-lim } s,\ a \in M$, if $\rho(a, a_i) < r + k$ for any positive real number k, and $s = \{a_i | a_i \in M,\ i \in a_n\}$.

Lemma 12.4.4 $a = \lim s$ *if and only if* $a = 0\text{-lim } s$.

Definition 12.4.7 $a \in M$ is a *fuzzy limit* of the sequence $s = \{a_i | a_i \in M,\ i \in a_n\}$ if $\mu(x = \lim s) > 0$, where μ is the *upper measure of convergence* of s to a point $x \in M$. The sequence $s = \{a_i | a_i \in M,\ i \in a_n\}$ *converges fuzzily* if it has a fuzzy limit.

Here, the upper measure of convergence of the sequence s to a point $x \in M$ defines the set called *normal fuzzy set* which contains the fuzzy limits of $s = \{a_i | a_i \in M,\ i \in a_n\}$.

Now, let $b \in M$. Then, we have the following theorem:

Theorem 12.4.2 *If a is a fuzzy limit of* $s = \{a_i | a_i \in M,\ i \in a_n\}$ *and* $\rho(a, b)$ *is finite, then b is a fuzzy limit of* $s = \{a_i | a_i \in M,\ i \in a_n\}$.

The following theorem reminds us of the analogous important theorem of real analysis.

Theorem 12.4.3 *A sequence* $s = \{a_i | a_i \in M,\ i \in a_n\}$ converges fuzzily *if and only if it is bounded.*

For a proof, see [53].

Now, based on the previous definitions and results about fuzzy limits and weak fuzzy limits of sequences, let us turn our attention to *fuzzy limits* and *weak fuzzy limits of functions*.

Definition 12.4.8 Let a function $f : \mathbb{R} \to \mathbb{R}$. The real number b is an r-limit of f at a point $x_0 \in \mathbb{R}$, symbolically $b = r\text{-lim}_{x \to x_0} f(x)$, if $a = \lim S \Rightarrow b = r\text{-lim}_{i \to \infty} f(a_i)$

for any sequence $S = \{a_i \in X,\ i \in a_n,\ a_i \neq a\}$, where $X \subseteq \mathbb{R}$ is the domain of the function f.

Lemma 12.4.5 *If $X = \mathbb{R}$, then $b = 0\text{-}lim_{x\to x_0} f(x)$ if and only if $b = \lim_{x\to x_0} f(x)$ in the classical sense.*

If the domain X is a discrete set, then any function with this domain has a limit at any point of X and this limit equals the value of the function at this point. Hence, the function is *continuous*.

We come to the definition of the fuzzy limit of a function:

Definition 12.4.9 A real number a is a *fuzzy limit of a function* $f(x)$ at a point $x_0 \in \mathbb{R}$ if it is an r-limit of $f(x)$ at the point x_0 for some nonnegative real number r. Thus, the function $f(x)$ converges fuzzily at the point x_0 if it has a fuzzy limit at this point.

There exist functions not having a limit at some point in the classical sense, but having many fuzzy limits. The following example from [54] illustrates such a situation.

Example 12.4.3 Let the function $f(x)$ be defined as

$$f_n(x) = \begin{cases} 1 + \dfrac{1}{n}, & x = 1 - \dfrac{1}{n}, \quad n \in \mathbb{N} \\ 2 - \dfrac{1}{n}, & x = 1 + \dfrac{1}{2n}, \quad n \in \mathbb{N} \\ 1 + (-1)^n, & x = 1 + \dfrac{1}{2n+1}, \quad n \in \mathbb{N} \\ x, & \text{otherwise.} \end{cases}$$

We see that, in the classical sense, $f(x)$ does not possess a limit at point 1 but it does have several fuzzy limits at that point, for example, the number 1 is a 1-limit of $f(x)$, and the numbers 0, 0.5, 1.5, 1.7, and 2 are 2-limits of $f(x)$.

If a function $f(x)$ has fuzzy limits at a point x_0, it follows that if $f(x)$ is locally bounded at x_0 then all real numbers become its fuzzy limits at this point. However, as it is difficult to consider the set of all fuzzy limits of a locally bounded function at a given point, Burgin's theory introduces a classification of these fuzzy limits by using the notions of the measure of convergence of a function $f(x)$ and the inferred normal fuzzy set, that is, the set of the fuzzy limits of $f(x)$ at x_0. Both notions are defined analogously to the case of sequences (see Definition 12.4.7). Now, similarly to Definition 12.4.4, one defines the concept of the *weak r-limit of a function* $f(x)$:

Definition 12.4.10 A real number d is the *weak r-limit of a function* $f(x)$ at $x_0 \in \mathbb{R}$, symbolically $d = r\text{-wlim}_{x\to x_0} f(x)$, if there exists a sequence $S = \{a_i \in X,\ i \in a_n,\ a_i \neq a\}$ such that $a = \lim S$ and $d = r\text{-wlim}_{i\to\infty} f(a_i)$.

Based on this definition, one is led to the concept of the *weak limit of a function*:

Definition 12.4.11 A real number e is the *weak limit of a function* $f(x)$ at x_0, symbolically $e = \text{wlim}_{x\to x_0} f(x)$, if $a = \lim S \Rightarrow e = \lim_{i\to\infty} f(a_i)$ for any sequence $S = \{a_i \in X,\ i \in a_n,\ a_i \neq a\}$, where $X \subseteq \mathbb{R}$ is the domain of the function $f(x)$.

Lemma 12.4.6 *If* $X \equiv \mathbb{R}$, *as in our case, then* $e = wlim_{x\to x_0} f(x)$ *if and only if* $e = 0\text{-}wlim_{x\to x_0} f(x)$.

Hence, the *weak fuzzy limit of a function* is defined as follows:

Definition 12.4.12 A real number e is a weak fuzzy limit of the function $f(x)$ at $x_0 \in \mathbb{R}$, if it is a weak r-limit of $f(x)$ at x_0 for some non-negative real value of r.

How is the notion of *boundedness* associated with *fuzzy convergence*? First, let us give the definition of a bounded function.

Definition 12.4.13 A real function $f : \mathbb{R} \to \mathbb{R}$ is *bounded* at a point $x_0 \in \mathbb{R}$ if there exists a number r and a neighborhood N of x_0 such that $\rho(f(x_0), f(x)) < r$, $\forall x \in N$.

Then, the following theorem holds:

Theorem 12.4.4 *A real function* $f : \mathbb{R} \to \mathbb{R}$ converges fuzzily *at a point* $x_0 \in \mathbb{R}$ *if and only if* $f(x)$ *is bounded at* x_0.

The proof can be found in [52]. For a treatment of real functions defined on discrete sets as well as the more complicated notions of *fuzzy–fuzzy limit*, *fuzzy–fuzzy convergence*, and *fuzzy–fuzzy continuity*, the concepts presented previously must be properly extended. The interested reader is referred to [54].

12.4.1 Fuzzy Ordinary Differential Equations

One of the basic questions encountered in the process of classical mathematical modeling in various areas of science, engineering, or economics concerns the nature of the coefficients, parameters, and the initial and/or boundary conditions of an ordinary (or partial) differential equation, i.e. whether they are constant or variable. Things become complicated and the question changes into whether coefficients, dependent variables, and/or associated conditions are sharp (crisp) or nonsharp when they are experimentally determined and characterized by imprecision and uncertainty, by vagueness. Naturally, at this point, the fuzzy setting comes into play and various definitions of fuzzy derivatives and functions form the frame and lead to different (analytical and numerical) methods and solutions of fuzzy initial and/or boundary value problems [64, 145]. Historically, however, the term "fuzzy differential equation" was first used in 1980 by Abraham Kandel and William J. Byatt [172], followed by a study of fuzzy differential equations presented by Kaleva [170] and based on the notion of the Hukuhara derivative that we have mentioned only in passing at the end of Section 12.3. In Kaleva's approach, fuzzy differential equations have solutions with an increasing length of support and lead to a, temporally, increasing imprecise behavior of fuzzy dynamical systems rendering impossible any periodic solution. Another approach is based on the notion of (fuzzy) differential inclusion (see, e.g., [188]). Here, we shall focus on the approach based on the extension principle and proposed in [46], and we will concentrate on the practical side of solving a *fuzzy ordinary differential equation of second order*.

Let us consider the classical *initial value problem* for a linear and nonhomogeneous, second order ordinary differential equation with constant coefficients:

$$\begin{aligned} &y'' + a_1 y' + a_2 y = f(x), \qquad x \in [0, +\infty), \\ &\quad y(0) = b_0, \quad y'(0) = b_1, \quad b_0, b_1 \in \mathbb{R}, \end{aligned} \tag{12.5}$$

where the nonhomogeneous term $f(x)$ is a continuous function in $[0, +\infty)$. We turn the problem into a fuzzy problem by fuzzifying the initial conditions through replacing b_0, b_1 with the triangular fuzzy numbers $\overline{b}_0$, $\overline{b}_1$; thus, we have a differential equation with *crisp coefficients* but *fuzzy initial conditions* [50]. Then, the solution of the initial value problem (12.5) is a triangular fuzzy number $\overline{y} = [y_1(x,\alpha), y_2(x,\alpha)]$, $\alpha \in [0,1]$, with $y_1(x,\alpha)$, $y_2(x,\alpha)$ differentiable functions at least up to second order. The substitution of $\overline{y}$ into the problem (12.5) yields

$$\begin{aligned} &[y_1''(x,\alpha), y_2''(x,\alpha)] + a_1[y_1'(x,\alpha) + y_2'(x,\alpha)] + a_2[y_1(x,\alpha) + y_2(x,\alpha)] \\ &\quad = [f(x), f(x)]. \end{aligned}$$

Through interval arithmetic, two fuzzy initial value problems are obtained for y_1, y_2 with the corresponding initial conditions

$$\begin{aligned} y_1(0,\alpha) &= \overline{b}_{01}(\alpha), \quad y_1'(0,\alpha) = \overline{b}_{11}(\alpha), \\ y_2(0,\alpha) &= \overline{b}_{02}(\alpha), \quad y_2'(0,\alpha) = \overline{b}_{12}(\alpha), \end{aligned}$$

respectively. Indeed, we can write

$$\overline{b}_0 = [b_{01}(\alpha), b_{02}(\alpha)], \quad \overline{b}_1 = [b_{11}(\alpha), b_{12}(\alpha)]. \tag{12.6}$$

If we assume that y_1, y_2 are also differentiable w.r.t α, then in order for $\overline{y}$ to be a fuzzy solution of the aforementioned *fuzzy initial value problem*, we must also have [50]

$$\frac{\partial}{\partial\alpha}[y_1(x,\alpha)] > 0, \quad \frac{\partial}{\partial\alpha}[y_2(x,\alpha)] < 0, \tag{12.7}$$

$$y_1(x,1) = y_2(x,1), \quad \forall x \in [0,\infty) \tag{12.8}$$

Let us consider an example.

Example 12.4.4 Let the initial value problem

$$y'' + 5y' + 4y = 0, \tag{12.9}$$

$$y(0) = 1, \quad y'(0) = 0. \tag{12.10}$$

Classically, the solution of this problem is given by $y = \frac{4}{3}e^{-x} - \frac{1}{3}e^{-4x}$. Now, by fuzzifying the initial conditions (12.10) we obtain the triangular fuzzy numbers

$$\overline{b}_0 = \text{tfn}(0,1,2), \quad \overline{b}_1 = \text{tfn}(-1,0,1),$$

while the differential equation (12.9) leads to the couple of equations

$$y_1'(x,\alpha) + 5y_1'(x,\alpha) + 4y_1(x,\alpha) = 0,$$
$$y_2'(x,\alpha) + 5y_2'(x,\alpha) + 4y_2(x,\alpha) = 0,$$

with the classical solutions

$$y_1 = C_1e^{-x} + C_2e^{-4x}, \tag{12.11a}$$
$$y_2 = C_3e^{-x} + C_4e^{-4x}, \tag{12.11b}$$

respectively. In order to determine the four unknown constants as functions of the parameter α, we need to solve the following two algebraic systems:

$$C_1 + C_2 = \alpha,$$
$$-C_1 - 4C_2 = -1 + \alpha,$$

and

$$C_3 + C_4 = 2 - \alpha,$$
$$-C_3 - 4C_4 = 1 - \alpha.$$

Thus, we obtain

$$C_1 = -\frac{1}{3}(1-5\alpha), \quad C_2 = \frac{1}{3}(1-2\alpha), \quad C_3 = (3 - \frac{5}{3}\alpha), \quad C_4 = -1 + \frac{2}{3}\alpha,$$

and the classical solutions (12.11) become, respectively

$$y_1(x,\alpha) = -\frac{1}{3}(1-5\alpha)\, e^{-x} + \frac{1}{3}(1-2\alpha)\, e^{-4x}, \tag{12.12a}$$
$$y_2(x,\alpha) = \left(3 - \frac{5}{3}\alpha\right) e^{-x} + \left(-1 + \frac{2}{3}\alpha\right) e^{-4x}. \tag{12.12b}$$

Finally, one can readily see that conditions (12.7) and (12.8) hold for every $x \in [0, \infty)$ and, hence, $\bar{y} = [y_1(x, \alpha), y_2(x, \alpha)]$ exists as a solution of the fuzzy initial value problem considered, while, as one can see from (12.12), it vanishes and the fuzziness disappears as $x \to \infty$.

Let us now examine what happens if we solve the problem (12.5) classically and then *fuzzify the solution obtained.* Suppose that we have found a *fundamental set* of solutions $\{y_1(x), y_2(x)\}$ for the corresponding homogeneous equation and a *particular solution* $y_p(x)$ of the full equation. Then, by the superposition principle, the classical *general solution* of (12.5) is

$$y(x) = C_1 y_1(x) + C_2 y_2(x) + y_p(x) \tag{12.13}$$

Applying the initial conditions $y(0) = b_0$ and $y'(0) = b_1$, we get

$$\begin{aligned} b_0 &= C_1 y_1(0) + C_2 y_2(0) + y_p(0), \\ b_1 &= C_1 y_1'(0) + C_2 y_2'(0) + y_p'(0), \end{aligned}$$

and solving by Cramer's rule the system for the integration constants C_1 and C_2, we obtain

$$\begin{aligned} C_1 &= f_1(b_0, b_1) + g_1, \\ C_2 &= f_2(b_0, b_1) + g_2, \end{aligned}$$

where

$$f_1(b_0, b_1) = \frac{b_0 y_2'(0) - b_1 y_2(0)}{D}, \quad f_2(b_0, b_1) = \frac{b_1 y_1(0) - b_0 y_1'(0)}{D},$$

with the determinants

$$D = \begin{vmatrix} y_1(0) & y_2(0) \\ y_1'(0) & y_2'(0) \end{vmatrix}, \quad D_1 = \begin{vmatrix} y_2(0) & y_p(0) \\ y_2'(0) & y_p'(0) \end{vmatrix}, \quad D_2 = \begin{vmatrix} y_p(0) & y_1(0) \\ y_p'(0) & y_1'(0) \end{vmatrix},$$

while

$$g_1 = \frac{D_1}{D}, \quad g_2 = \frac{D_2}{D},$$

and $D \neq 0$ is the Wronski determinant of the homogeneous solution at $x = 0$.

Hence, the classical general solution (12.13) becomes

$$y(x) = f_1(b_0, b_1) y_1(x) + f_2(b_0, b_1) y_2(x) + f_3(x), \tag{12.14}$$

with

$$f_3(x) = g_1 y_1(x) + g_2 y_2(x) + y_p(x),$$

the latter being independent of the initial conditions.

Let us now apply Zadeh's extension principle. To this purpose, we have to replace in (12.14) b_0, b_1 with the triangular fuzzy numbers $\overline{b}_0, \overline{b}_1$. Then, the solution of the differential equation will be a triangular fuzzy number $\overline{y}_{\text{ext}} = [y_{\text{ext1}}(x, a), y_{\text{ext2}}(x, a)]$ with $y_{\text{ext1}}, y_{\text{ext2}}$ differentiable at least up to second order w.r.t. x, and

$$y_{\text{ext1}} = \min\{f_1(b_0, b_1)y_1(x) + f_2(b_0, b_1)y_2(x) + f_3(x) | b_0 \in \overline{b}_0[\alpha], b_1 \in \overline{b}_1[\alpha]\},$$
$$y_{\text{ext2}} = \max\{f_1(b_0, b_1)y_1(x) + f_2(b_0, b_1)y_2(x) + f_3(x) | b_0 \in \overline{b}_0[\alpha], b_1 \in \overline{b}_1[\alpha]\},$$

$\forall a \in [0, 1]$, $\forall x \in [0, \infty)$, while also satisfying the initial conditions $y_{\text{ext1}}(0, \alpha) = b_{01}(\alpha)$, $y'_{\text{ext1}}(0, \alpha) = b_{11}(\alpha)$, $y_{\text{ext2}}(0, \alpha) = b_{02}(\alpha)$, $y'_{\text{ext2}}(0, a) = b_{12}(\alpha)$. Furthermore, y_{ext1} and y_{ext2} will be solutions of the initial value problem (12.5) if the interval condition holds [50] that can be generally defined as follows:

Definition 12.4.14 **(Interval Condition)** $\forall I_n \exists\, b_i^*, b_i^{**} \in \overline{b}_i[\alpha], i = 0, 1$:

$$y_{\text{ext1}} = f_1(b_0^*, b_1^*)y_1(x) + f_2(b_0^*, b_1^*)y_2(x) + f_3(x)$$
$$y_{\text{ext2}} = f_1(b_0^{**}, b_1^{**})y_1(x) + f_2(b_0^{**}, b_1^{**})y_2(x) + f_3(x),$$

where $I_n = [\varepsilon_{n-1}, \varepsilon_n]$ and $0 = \varepsilon_0 < \varepsilon_1 < \ldots < \varepsilon_{n-1} < \varepsilon_n < \ldots$, with $I = [0, \infty)$, $\forall x \in I_n$, $\forall n$.

To put it differently, f_1, and f_2 are independent of x in each interval I_n, $\forall a \in [0, 1]$. In other words, f_1 and f_2 are constant in each interval I_n for each fixed value of α and $y_{\text{ext1}}, y_{\text{ext2}}$ are solutions of the differential equation. In fact, the following two theorems help clarify the situation:

Theorem 12.4.5 *The interval condition is valid and $\overline{y}_{ext}(x)$ is a solution to the fuzzy initial value problem.*

Theorem 12.4.6 *Assuming that the coefficients a_1, a_2 of the differential equation (12.5) are positive, it holds that $\overline{y}(x) \leq \overline{y}_{ext}(x) \leq \overline{y}_1(x)$.*

The solution $\overline{y}_1(x)$ also satisfies the initial value problem and is obtained if one substitutes b_0 and b_1 in (12.14) with $\overline{b}_0 = [b_{01}(\alpha), b_{02}(\alpha)]$ and $\overline{b}_1 = [b_{11}(\alpha), b_{12}(\alpha)]$ from (12.6), and then applies interval arithmetic.

For a proof of Theorem 12.4.5, see [47], while for a proof of Theorem 12.4.6, see [50].

The following example from classical mechanics is adapted from [50] and illustrates the situation in the form of an "algorithm."

Example 12.4.5 Let a one-dimensional spring-mass system with a unit mass and a spring constant $k = 4$ N/m (in SI units) perform forced oscillations due to an external periodic force $F(t)$, and suppose there is no damping. Then, the equation of motion of the vibrating system is given as

$$\ddot{x}(t) + \omega_0^2 x(t) = F(t), \text{ or}$$
$$\ddot{x}(t) + 4x(t) = F(t)$$

where $\omega_0 = \sqrt{k/m} = 2$ is the system's *natural frequency*, and the dot denotes differentiation w.r.t. the time variable t. Now, let us assume that the external periodic force has the form

$$F(t) = 2\cos(\omega t), \quad \omega \in (0, 2),$$

with ω its frequency and 2 its amplitude, and let us further impose the initial conditions $x(0) = b_0$, $\dot{x}(0) = b_1$ expressing the position and velocity of the system at $t = 0$. Then, the solution of this initial value problem is given by

$$x(t) = b_0 \cos(2t) + \frac{b_1}{2} \sin(2t) + x_p(t),$$
$$x_p(t) = \frac{2}{4 - \omega^2}[\cos(\omega t) - \cos(\omega_0 t)].$$

At this point, a comment on the value of the external frequency $\omega \to 2$ is in order. In this case, the external frequency equals the system's natural frequency $\omega \to \omega_0$ and, thus, becomes the *resonant frequency* leading the system to *resonance*, that is, to oscillations with maximum amplitude.

Let us now introduce to the spring-mass system *fuzzy initial conditions* through the triangular fuzzy numbers $\overline{b}_0 = \text{tfn}(0, 2, 4)$ and $\overline{b}_1 = \text{tfn}(-2, 0, 2)$. Then, we have

$$\overline{b}_0[\alpha] = [2\alpha, 4 - 2\alpha], \quad \overline{b}_1[\alpha] = [2(-1 + \alpha), 2(1 - \alpha)].$$

Further, we have the validity of the interval condition given by Definition 12.4.14 and the intervals $I_n = [\varepsilon_{n-1}, \varepsilon_n]$ with $\varepsilon_0 = 0$, $\varepsilon_1 = \pi/4$, $\varepsilon_2 = 2\pi/4 = \pi/2$, and so forth, while $\frac{\partial x}{\partial b_0} > 0$ in the intervals $I_1, I_4, I_5, \ldots$, but $\frac{\partial x}{\partial b_0} < 0$ in the other intervals, and similarly $\frac{\partial x}{\partial b_1} > 0$ in the intervals $I_1, I_2, I_5, I_6, \ldots$, but $\frac{\partial x}{\partial b_1} < 0$ in the other intervals. Now, the α-cuts for $\overline{x}_{\text{ext}}(t) = [x_{\text{ext1}}(t, \alpha), x_{\text{ext2}}(t, \alpha)]$ can be determined through the following steps, starting with $x_{\text{ext1}}(t, \alpha)$:

(i) in the interval I_1, we use $b_{01}(\alpha)$ instead of b_0, and $b_{11}(\alpha)$ instead of b_1,
(ii) in the interval I_2, we use $b_{01}(\alpha)$ instead of b_0, and $b_{11}(\alpha)$ instead of b_1,
(iii) in the interval I_3, we use $b_{02}(\alpha)$ instead of b_0, and $b_{12}(\alpha)$ instead of b_1,
(iv) in the interval I_4, we use $b_{01}(\alpha)$ instead of b_0, and $b_{12}(\alpha)$ instead of b_1.

For $x_{\text{ext2}}(t, \alpha)$, we must repeat the four steps above, but interchange $b_{01}(\alpha)$ with $b_{02}(\alpha)$ and $b_{11}(\alpha)$ with $b_{12}(\alpha)$. As it is expected, the fuzziness vanishes for $t \to \infty$.

Turning to the classical solution $\overline{x}(t)$, the initial values are changed to the triangular fuzzy numbers $\overline{b}_0 = \text{tfn}(0, 1, 2)$ and $\overline{b}_1 = \text{tfn}(-1, 0, 1)$. So,

$\bar{x}(t)[\alpha] = [x_1(t,\alpha), x_2(t,\alpha)]$ with $x_1(t,\alpha)$, $x_2(t,\alpha)$ to be determined in a way similar to Example 12.4.4. The corresponding four integration constants have to be calculated, while the conditions (12.7) must be examined for their validity. We have $\frac{\partial x_1}{\partial \alpha} = \cos(2t) + \frac{\sin(2t)}{2}$, which can take negative values for $0 \leq t < \infty$, hence $\bar{x}(t)$ does not exist $\forall t \geq 0$.

As a general conclusion, it is possible that the classical solution may not exist, but the extended solution exists always. For a treatment of the interesting special combinations for the coefficients a_1 and a_2 of the differential equation (12.5), where a_1 is fuzzy and $a_2 = 0$ or $a_1 = 0$ and a_2 is fuzzy, and the corresponding difficulties that emerge, the reader is referred to [50].

12.4.2 Fuzzy Partial Differential Equations

In this section, we shall consider fuzzy linear partial differential equations with two independent variables in *boundary value problems*, adopting the strategy of [50] and [45] and based on the use of *triangular fuzzy numbers*, as we did in the previous section. We distinguish between two kinds of solutions: (i) the classical solution $\bar{z}$ obtained by solving the fuzzified partial differential equation, and (ii) the solution $\bar{z}_{ext}$ obtained by solving the nonfuzzified partial differential equation and then applying Zadeh's extension principle. Furthermore, the accompanying boundary conditions and the nonhomogeneous term (if there is one) of the equation will be also fuzzified.

First, we present some preliminaries in the study of partial differential equations assuming that their solutions are given in *closed form*. The general form of a classical linear and second-order partial differential equation in two independent variables, say x and y, is given as

$$F(x, y, z(x, y), \partial_x z, \partial_y z, \partial_{xx} z, \partial_{xy} z, \partial_{yy} z) = 0, \tag{12.15}$$

with $z(x, y)$ a continuous function. Let the differential operator $D(\partial_x, \partial_y)$ be a polynomial in ∂_x and ∂_y

$$D(\partial_x, \partial_y) = a_{11}\partial_{xx} + a_{12}\partial_{xy} + a_{22}\partial_{yy}, \tag{12.16}$$

with real constant coefficients a_{11}, a_{12}, and a_{22}. Thus, (12.15) can be written in the form

$$D(\partial_x, \partial_y) = f(x, y, K), \tag{12.17}$$

where $f(x, y, K)$ is a continuous function and K is a vector comprised of constants k. Further, (12.17) has associated *boundary conditions* on z and/or its derivatives and depending on x, y, and possibly some constants c comprising the vector C, appropriate for the problem to be *well-posed*. So, the solution of this boundary value problem is of the form $z = g(x, y, K, C)$. Since, some or all of the constants in K and C may not be known with precision as being observational or experimental results, triangular fuzzy numbers $\overline{K}, \overline{C}$ are chosen to represent them, while by applying the extension principle, the function $f(x, y, K)$ leads to the fuzzy $\overline{f}(x, y, \overline{K})$. Recalling the previous section, we have

$$\overline{f}(x, y, \overline{K})[\alpha] = [f_1(x, y, \alpha), f_2(x, y, \alpha)], \ 0 \le a \le 1,$$
$$f_1(x, y, \alpha) = \min\{f(x, y, k) | k \in \overline{K}[\alpha]\},$$
$$f_2(x, y, \alpha) = \max\{f(x, y, k) | k \in \overline{K}[\alpha]\}.$$

The fuzzification of $z = g(x, y, K, C)$ yields $\overline{z} = \overline{g}(x, y, \overline{K}, \overline{C})$ with

$$\overline{g}(x, y, \overline{K}, \overline{C})[\alpha] = [g_1(x, y, \alpha), g_2(x, y, \alpha)],$$
$$g_1(x, y, \alpha) = \min\{g(x, y, k, c) | k \in \overline{K}[\alpha], \ c \in C[\alpha]\},$$
$$g_2(x, y, \alpha) = \max\{g(x, y, k, c) | k \in \overline{K}[\alpha], \ c \in C[\alpha]\},$$

and the *fuzzified partial differential equation* now reads

$$D(\partial_x, \partial_y)\overline{z}(x, y) = \overline{f}(x, y, \overline{K}).$$

Finally, one can, again by use of the extension principle, fuzzify the associated, homogeneous or nonhomogeneous, boundary conditions.

Let us now present the practical steps for solving a fuzzy partial differential equation

$$D(\partial_x, \partial_y)[z_1(x, y, \alpha), z_2(x, y, \alpha)] = [f_1(x, y, \alpha), f_2(x, y, \alpha)] \tag{12.18}$$

accompanied by fuzzy boundary conditions. In other words, we need to solve the two equations comprising (12.18) seeking for the solutions z_1 and z_2 comprising $\overline{z}(x, y)[\alpha] = [z_1(x, y, \alpha), z_2(x, y, \alpha)]$ for the entire domain of the independent variables x, y and for each value of $\alpha \in [0, 1]$, and assuming that $z_1(x, y, \alpha)$, $z_2(x, y, \alpha)$ have continuous first and second partial derivatives. Suppose that the fuzzy boundary conditions are $\overline{z}(0, y) = \overline{C}_1$ and $\overline{z}(L, y) = \overline{C}_2$, where 0 and L are the boundary values of x, and $\overline{C}_1[\alpha] = [c_{11}(\alpha), c_{12}(\alpha)]$, $\overline{C}_2[\alpha] = [c_{21}(\alpha), c_{22}(\alpha)]$. Then, the fuzzy boundary conditions can be written as

$$y_1(0, y, \alpha) = c_{11}, \quad y_2(0, y, \alpha) = c_{12}, \quad y_1(L, y, \alpha) = c_{21}, \quad y_2(L, y, \alpha) = c_{22}$$

Furthermore, for the triangular fuzzy number $\overline{z}(x, y)[\alpha]$ as solution it is also necessary that, assuming that $z_1(x, y, \alpha)$, $z_2(x, y, \alpha)$ are differentiable with respect to α, it holds

$$\frac{\partial z_1}{\partial \alpha} > 0, \quad \frac{\partial z_2}{\partial \alpha} < 0, \quad \forall \alpha \in (0, 1),$$
$$z_1(x, y, 1) = z_2(x, y, 1), \quad \forall x, y.$$

Let us see an example before we continue with the extended solution $\overline{z}_{\text{ext}}$.

Example 12.4.6 Let the equation $\partial^2_{xy} z = k_1 y e^y + k_2 x$, $k_1, k_2 \in [0, \infty)$, with the boundary conditions $z(x, 0) = c_1$ and $\partial_y z(0, y) = c_2 y$, where $c_1, c_2 \in$

$[0, \infty)$. Thus, we have $f(x, y, k) = k_1 ye^y + k_2 x$ with $\partial_{k_1} f > 0$, $\partial_{k_2} f > 0$, so that $f_1(x, y, k) = k_{11}(\alpha)ye^y + k_{21}(\alpha)x$, $f_2(x, y, k) = k_{12}(\alpha)ye^y + k_{22}(\alpha)x$. Therefore, we have $\overline{K}_1(\alpha) = [k_{11}(\alpha), k_{12}(\alpha)]$ and $\overline{K}_2(\alpha) = [k_{21}(\alpha), k_{22}(\alpha)]$. Hence, the problem splits into

$$\partial^2_{xy} z_1(x, y, \alpha) = f_1(x, y, \alpha), \quad z_1(x, 0, \alpha) = c_{11}(\alpha), \quad \partial_y z_1(0, y, \alpha) = c_{21}(\alpha)y,$$

and

$$\partial^2_{xy} z_2(x, y, \alpha) = f_2(x, y, \alpha), \quad z_2(x, 0, \alpha) = c_{12}(\alpha)y, \quad \partial_y z_2(0, y, \alpha) = c_{22}(\alpha)y.$$

The solutions are, respectively,

$$z_1(x, y, \alpha) = k_{11}(\alpha)x(y-1)e^y + k_{21}(\alpha)\tfrac{x^2 y}{2} + c_{11}(\alpha) + c_{21}(\alpha)\tfrac{y^2}{2} + k_{11}(\alpha)x,$$
$$z_2(x, y, \alpha) = k_{12}(\alpha)x(y-1)e^y + k_{22}(\alpha)\tfrac{x^2 y}{2} + c_{12}(\alpha) + c_{22}(\alpha)\tfrac{y^2}{2} + k_{22}(\alpha)x,$$

and

$$z_1(x, y, 1) = z_2(x, y, 1), \qquad \frac{\partial z_1}{\partial \alpha} > 0, \quad \frac{\partial z_2}{\partial \alpha} < 0.$$

Thus, the triangular fuzzy number $\overline{z}(x, y)$ is the solution given by

$$\overline{z}(x, y) = \overline{C}_1 + \overline{C}_2 \frac{y^2}{2} + \overline{K}_1 x(y-1)e^y + \overline{K}_2 x\left(\frac{xy}{2} + 1\right).$$

Now, let us turn our attention to the study of a fuzzy solution $\overline{z}_{ext}(x, y)$ based on the extension principle. Based on the discussion in the previous section, we have

$$\overline{z}_{\text{ext}}(x, y)[\alpha] = [z_{\text{ext1}}(x, y, \alpha), z_{\text{ext2}}(x, y, \alpha)],$$

leading to the two partial differential equations

$$D(\partial_x, \partial_y)\overline{z}_{\text{ext1}} = f_1(x, y, \alpha),$$
$$D(\partial_x, \partial_y)\overline{z}_{\text{ext2}} = f_2(x, y, \alpha),$$

with the associated fuzzy boundary conditions. The following example is adopted from [50].

Example 12.4.7 Let the linear, nonhomogeneous partial differential equation of second order with constant coefficients

$$(\partial^2_{xy} - \partial_x)z(x, y) = k, \quad k \geq 0,$$

with the boundary conditions $z(0, y) = c_1$ and $\partial_x z(x, 0) = c_2 x^2$, where $c_1, c_2 > 0$. The solution of this boundary value problem is given by

$$z(x, y) = c_1 + \frac{c_2}{3}x^3 e^y + kx(e^y - 1).$$

Hence, we have

$$g_1(x, y, \alpha) = c_{11}(\alpha) + \tfrac{c_{21}(\alpha)}{3}x^3 e^y + k_1(\alpha)x(e^y - 1),$$
$$g_2(x, y, \alpha) = c_{12}(\alpha) + \tfrac{c_{22}(\alpha)}{3}x^3 e^y + k_2(\alpha)x(e^y - 1),$$

and $\overline{K}[\alpha] = [k_1(\alpha), k_2(\alpha)]$, so that we have

$$(\partial_{xy}^2 - \partial_x)z_1(x, y, \alpha) = k_1(\alpha),$$
$$(\partial_{xy}^2 - \partial_x)z_2(x, y, \alpha) = k_2(\alpha).$$

Correspondingly, the given boundary conditions become

$$z_1(0, y, \alpha) = c_{11}(\alpha),$$
$$z_2(0, y, \alpha) = c_{12}(\alpha),$$
$$\partial_x z_1(x, 0, \alpha) = c_{21}(\alpha)x^2,$$
$$\partial_x z_2(x, 0, \alpha) = c_{22}(\alpha)x^2.$$

Hence, with $\overline{C}_1[\alpha] = [c_{11}(\alpha), c_{12}(\alpha)]$ and $\overline{C}_2[\alpha] = [c_{21}(\alpha), c_{22}(\alpha)]$, the final solution is given as

$$\overline{z}_{\text{ext}}(x, y) = \overline{C}_1 + \frac{\overline{C}_2}{3}x^3e^y + \overline{K}x(e^y - 1).$$

When both $\overline{z}(x, y)$ and $\overline{z}_{\text{ext}}(x, y)$ are solutions of a given boundary value problem, then they are equal if all the constants in the problem's differential operator $D(\partial_x, \partial_y) = a_{11}\partial_{xx} + a_{12}\partial_{xy} + a_{22}\partial_{yy}$ introduced in (12.16) are positive (see, e.g., [50, chapter 9] for a relevant theorem and its proof).

A final remark is deemed proper concerning the theory of *fuzzy Fourier series and/or transforms*, since the classical Fourier series expansion is necessary for the analytical solution of initial-boundary value problems of equations such as the wave equation and the heat equation, appearing in numerous applications of physics and engineering (particularly in signal analysis, where randomness and vagueness are in the foreground). Indeed, there exist several attempts (but still not a fully developed and unique approach) of solving the fuzzy versions of these equations based on various ideas, for example, the idea of finding a representation of an interval-valued fuzzy set in terms of its level sets aiming at the study of Fourier series for periodic fuzzy-valued functions [168], the application of a generalized Hukuhara derivative [148], the introduction of pentagonal fuzzy numbers [235], fuzzy discrete signal analysis [55], or fuzzy time series [288], just to name but a few.

Exercises

12.1 Let a function $f(x)$ such that $[\bigwedge(f), \bigvee(f)] = [7, 15]$, $\forall x \in [1, 10]$, $f(3) = 9$, $f(9) = 14$. What is the possibility of $x = 3$ and $x = 9$ to belong to the maximizing set M?

12.2 Suppose $f(x) = \cos x, 0 \leq x \leq 2\pi$. Find the possibility $\mu_M(x)$ of x to belong to the maximizing set M, and $\mu_D(x)$ to belong to the domain D.

12.3 Let $f(x) = 3x + 2$ with $\mu_D = \begin{cases} x^2, & x \in [0, 1] \\ 0, & \text{otherwise} \end{cases}$. Find x_0 for which $\mu_M(x) = \mu_D(x)$, and the function's maximum value $f(x_0)$.

12.4 Let the fuzzy bunch $\widetilde{f}(x) = \{(x^3, \alpha_1), (x^4, \alpha_2), (5x^5 + 1, \alpha_3)\}$, with $\alpha_1 = 0.5$, $\alpha_2 = 0.6$, and $\alpha_3 = 0.8$. Integrate the given fuzzy bunch over the interval $[4, 6]$.

12.5 Let the initial value problem $y'' - 6y' + 5y = 0, y(0) = 0, y'(0) = 1$. Fuzzify the initial conditions and then find the fuzzy solution $\overline{y}$.

12.6 Let the two-dimensional boundary value problem $(\partial^2_{xy} - \partial_y)u(x, y) = 1$, $u(0, y) = 2, \partial_y u(x, 0) = y^2$. Find $\overline{u}_{\text{ext}}(x, y)$.

A

Fuzzy Approximation

Classical approximation theory of real-valued continuous functions by algebraic or trigonometric polynomials has been a subject of research for more than two centuries (see, e.g. [72, 164, 194, 286] for a rigorous as well as instructive presentation). Here, we are going to give a brief account of some basic results of approximation theory put in a fuzzy setting.

A.1 Weierstrass and Stone–Weierstrass Approximation Theorems

Let us recall two of the most fundamental questions in classical approximation theory:

(i) Can every continuous function, $f(x) \in C[a, b]$, $[a, b] \subset \mathbb{R}$, be arbitrarily well approximated by algebraic polynomials?
(ii) Can every continuous periodic function with period $T = 2\pi$ be arbitrarily well approximated by trigonometric polynomials?

Both questions, that are actually intrinsically interrelated, were answered affirmatively by Karl Theodor Wilhelm Weierstrass in 1885 through perhaps the most significant result in approximation theory known as *Weierstrass approximation theorem* which, in its basic form, reads:

Theorem A.1.1 *Let a continuous real-valued function $f(x)$ be defined on an interval $[a, b] \subset \mathbb{R}$. Then, there exists a real algebraic polynomial $p(x)$ such that*

$$|f(x) - p(x)| < \varepsilon, \quad \forall \varepsilon > 0, \ \forall x \in [a, b].$$

A Modern Introduction to Fuzzy Mathematics, First Edition.
Apostolos Syropoulos and Theophanes Grammenos.

Equivalently, the *supremum norm*[1] can be used, so that Theorem A.1.1 can be formulated in the following equivalent form:

The algebraic polynomials $p(x)$ are *dense* in $C[a, b]$ with respect to the supremum norm, and every continuous function $f \in C[a, b]$ can be arbitrarily well approximated with respect to the supremum norm, that is, there is an algebraic polynomial $p(x)$ such that

$$||f(x) - p(x)||_\infty < \varepsilon, \quad \forall \varepsilon > 0, \ \forall x \in [a, b]. \tag{A.1}$$

A constructive proof (by use of Korovkin sequences and Bernstein polynomials) of Theorem A.1.1 can be found, for example, in [72, 165]. Now, Theorem A.1.1 leads to the corollary that the linear space $C[a, b]$ is separable, i.e. the real polynomials are dense and each one of them can be approximated by a polynomial with rational coefficients. Further, it can be proved that the aforementioned theorem can be "transferred" to trigonometric polynomials as well, whereby the space $C[a, b]$ is replaced by the linear space of continuous 2π-periodic functions and the interval $[a, b]$ is replaced by $[0, 2\pi]$ (see, e.g. [165] for a proof).

The first generalization of the Weierstrass approximation theorem was formulated in 1937 by Marshall Harvey Stone [267], who replaced the interval $[a, b]$ by a compact Hausdorff space and the algebra of real polynomials by a more general (sub)algebra. The *Stone–Weierstrass theorem*, as it is called, states the following:

Theorem A.1.2 *Let X a compact Hausdorff space and A a subalgebra $C(X, \mathbb{R})$ that contains a nonzero constant function. Then, the subalgebra A is dense in $C(X, \mathbb{R})$ if and only if it separates points.*

For a proof see, for example [88, 268].

A.2 Weierstrass and Stone–Weierstrass Fuzzy Analogs

Before we present the fuzzy analogs for the two theorems presented in the previous section, let us introduce some basic concepts and notions. We follow [9, 130], written by perhaps the leading researchers on fuzzy approximation theory. We begin with a possible definition of a fuzzy function that is more appropriate for our purpose, in relation to the *fuzzified graph* of an ordinary function [129]:

Definition A.2.1 Let an ordinary real-valued function $f : [a, b] \to \mathbb{R}$. A fuzzy function is the fuzzified graph of the function f, i.e. the pair (f, Λ_f), where Λ_f :

1 A supremum norm is defined as $||f||_\infty = \max_{x \in S} |f(x)|, \forall f \in C(S)$, where $S \subset \mathbb{R}^d, d \in \mathbb{N}$ is a compact set and $C(S)$ is the normed space of continuous functions f. In fact, $C(S)$ equipped with the supremum norm is complete, that is, it is a Banach space.

$[a,b] \times \mathbb{R} \to [0,1]$ gives the degree of membership of a point $(x,y) \in [a,b] \times \mathbb{R}$ to the graph $\mathrm{Gr}(f) = \{(x,y) \in [a,b] \times \mathbb{R} \to [0,1] | y = f(x)\}$. Hence,

$$\Lambda_f = \begin{cases} F(x,y), & (x,y) \in \mathrm{Gr}(f), \\ 0, & \text{otherwise}, \end{cases}$$

with $F(x,y) \in (0,1]$.

Consequently, if $F(x,y)$ is an algebraic polynomial P, then the pair (P, Λ_P) is a *fuzzy algebraic polynomial* from $[a,b]$ to $\mathbb{R}$, and

$$(P, \Lambda_P) = \begin{cases} F(x,y), & (x,y) \in \mathrm{Gr}(P), \\ 0, & \text{otherwise}, \end{cases}$$

where now $F(x,y)$ is a bivariate algebraic polynomial in x and y.

Similarly, if $F(x,y)$ is a trigonometric polynomial T, then the pair (T, Λ_T) is a real-valued *fuzzy trigonometric polynomial*, and

$$(T, \Lambda_T) = \begin{cases} F(x,y), & (x,y) \in \mathrm{Gr}(T), \\ 0, & \text{otherwise}, \end{cases}$$

where $F(x,y)$ is a trigonometric polynomial with respect to x and an algebraic polynomial with respect to y.

The following two definitions are more familiar and consistent with Definition A.2.1:

Definition A.2.2 A *fuzzy algebraic polynomial* of degree n is expressed as

$$P_n(x) = \sum_{k=0}^{n} x^k a_k, \quad k \in \mathbb{N}_0,$$

where $x \in \mathbb{R}$, while a_k are real fuzzy numbers.

Definition A.2.3 A *fuzzy trigonometric polynomial* of degree n is expressed as

$$T_n(x) = \sum_{k=0}^{n} [\cos(kx) a_k + \sin(kx) b_k], \qquad k \in \mathbb{N}_0,$$

where $x \in \mathbb{R}$, while the coefficients a_k and b_k are real fuzzy numbers.

We will further need the notion of *distance between two fuzzy functions*. Recall that we have already given the Hausdorff distance in Definition 10.2.3 and have discussed the distance between two fuzzy points in Section 11.1. The distance $d((f, \Lambda_f), (g, \Lambda_g))$ between two fuzzy functions (f, Λ_f) and (g, Λ_g) as given in Definition A.2.1, is the Hausdorff distance d_H between the sets

$$A(\Lambda_f) = \{(x,y,z) \in [a,b] \times \mathbb{R} \times (0,1] | y = f(x),\ 0 < z = F(x,y),\ (x,y) \in \mathrm{Gr}(f)\}$$

and

$$B(\Lambda_g) = \{(x, y, z) \in [a, b] \times \mathbb{R} \times (0, 1] | y = g(x),\ 0 < z = F(x, y),\ (x, y) \in \mathrm{Gr}(g)\}.$$

Hence, we have

Definition A.2.4 $d\,((f, \Lambda_f), (g, \Lambda_g)) = d_H\,(A(\Lambda_f), B(\Lambda_g))$.

In fact, if $A(\Lambda_f)$ and $B(\Lambda_g)$ are compact sets, then $d\,((f, \Lambda_f), (g, \Lambda_g))$ is a metric [131].

Now we have everything needed to come to the *fuzzy analog of the Weierstrass theorem* (of which a non-constructive proof is given in [129]):

Theorem A.2.1 *For any fuzzy function (f, Λ_f) continuous on $[a, b]$ and on $Gr(f)$, there exists a sequence $(P_n, \Lambda_{P_n})_n$ of fuzzy algebraic polynomials $P_n : [a, b] \to \mathbb{R}$ such that $\lim_{n\to\infty} d\,((f, \Lambda_f), (P_n, \Lambda_{P_n})) = 0$.*

Let us turn to the fuzzy analog of the Stone–Weierstrass theorem. To this purpose, the interval $[a, b]$ is replaced by (M, d), where M is a *compact metric space* (see Definition 10.1.2 for a metric space) and d is a distance. Consider two *subalgebras* A_1 and A_2 with $A_1(M) \subset C(M, \mathbb{R}), A_2(M) \subset C(\mathbb{R}, \mathbb{R})$. Then, the *fuzzy analog of the Stone–Weierstrass theorem* states the following [130]:

Theorem A.2.2 *Let the compact metric space (M, d) and the subalgebras $A_1(M) \subset C(M, \mathbb{R})$ and $A_2(M) \subset C(\mathbb{R}, \mathbb{R})$, with A_1 containing the constants on the compact metric space M and A_2 containing the constants on $\mathbb{R}$. Further, A_1 separates the points of M and A_2 separates the points of $\mathbb{R}$. Then, for any fuzzy function (f, Λ_f) with f, Λ_f, and F continuous on M, there exists a sequence $(f_n, \Lambda_{f_n})_n$ of (A_1, A_2)-fuzzy functions from M to $\mathbb{R}$, such that $\lim_{n\to\infty} d((f, \Lambda_f), (f_n, \Lambda_{f_n})) = 0$.*

Naturally, if M is replaced by the real interval $[a, b]$, the aforementioned fuzzy analog of the Stone–Weierstrass theorem gives its place to the fuzzy analog of the Weierstrass theorem, Theorem A.2.1.

In this brief presentation, we have left out some further important issues of fuzzy approximation theory such as quantitative estimates of the approximation error, fuzzy interpolation by use of fuzzy Lagrange interpolating polynomials and splines, and fuzzy Taylor expansion, to name but a few still significant topics. For a thorough presentation, the reader is referred to [9, 130].

B

Chaos and Vagueness

There are "simple" systems whose parts are governed by well-understood physical laws, yet these systems behave unpredictably. If vagueness is indeed a fundamental property of this world, then what is its role in this unpredictable behavior?

B.1 Chaos Theory in a Nutshell

It is a fact that weather forecasting is not easy. It is also known that we cannot predict how the weather will be in 10 or 20 days from now. What is not widely known is why this happens. Moreover, it is not known whether we can make our predictions more accurate and if we can predict the weather for long periods.

Edward Norton Lorenz [195] was a mathematician and meteorologist who wanted to use computers to predict the weather. He created a simple model, but soon he realized that when he slightly changed the input parameters, he was getting completely different results. For example, he noticed that, when his computer simulation was fed with decimal numbers having three decimal digits instead of decimal numbers with six decimal digits, the results were not the same as he would expect. One does not need to reimplement his simulation, which involved 12 ordinary differential equations, instead, one can observe the same effect using a much simpler mathematical expression. For example, one can use an iteration of the quadratic expression $p + rp(1 - p)$ with initial value $p_0 = 0.1$ and $r = 3$. If, at some point, the iteration stops and p is truncated to three decimal digits and the operation resumes with this value and goes to the end without any other interruption, the result will be quite different from what we will get if no interruption happened. This effect, that is known in the literature as the *sensitive dependence on initial conditions* effect, is a manifestation of a *chaotic system*. The theory that studies such phenomena is known as *chaos theory* (see Ref. [239] for an accessible overview).

A Modern Introduction to Fuzzy Mathematics, First Edition.
Apostolos Syropoulos and Theophanes Grammenos.

Recently, in a study [34], it was *discovered* that there are systematic distortions in the statistical properties of chaotic dynamical systems when represented and simulated on digital computers using standard IEEE floating-point numbers (i.e. the most common representation for real numbers on computers today). However, this is not a limitation of modern computers or modern programming languages, but a limitation imposed by people who implement their simulations using this specific approach. In fact, modern programming languages such as Java and Perl support arbitrary precision decimal numbers that can be used to compute anything to any accuracy. Of course, the calculations are much slower, but this is the price one has to pay when not using IEEE floating-point numbers.

Chaos theory is usually studied alongside *fractal* geometry (again, see Ref. [239] for an accessible overview of fractal geometry). Fractals are geometric figures, where each part of such a figure looks like the whole figure. This property is known as *self-similarity*. In general, objects encountered in Nature do not have the precise properties of the objects of Euclidean geometry. Nevertheless, they can be "approached" by fractals. In a layman's terms, the word *dimension* is associated with the numbers 1, 2, or 3 if we are talking about points, plane objects, or solid objects, respectively. However, each fractal object has its own fractal dimension, which is a rational or, more generally, a real number. Most of the time, when we say fractal dimension, we actually mean *Hausdorff dimension*.[1]

Definition B.1.1 Suppose that (X, d) is a metric space. Then, for any subset $E \subset X$,

$$\operatorname{diam}(E) = \bigvee \{d(x, y) \mid x, y \in E\}$$

is the *diameter* of E.

Definition B.1.2 For any $E \subset X$, any $\delta \in (0, \infty]$ and any $\alpha \in [0, \infty)$, we consider the *outer measure*[2]

$$\mathscr{H}_\delta^\alpha = \bigwedge \left\{ \sum_{i=1}^{\infty} (\operatorname{diam}(E_i))^\alpha \;\middle|\; E \subset \bigcup_i E_i \text{ and } \operatorname{diam}(E_i) < \delta \right\}.$$

1 The definitions that follow are from *Hausdorff dimension. Encyclopedia of Mathematics*. URL: http://www.encyclopediaofmath.org/index.php?title=Hausdorff_dimension&oldid=35823.

2 A measure μ is *countably subadditive* if

$$\mu\left(\bigcup_{i=1}^{\infty} E_i\right) \le \sum_{i=1}^{\infty} \mu(E_i)$$

for any disjoint class $\{E_1, E_2, \dots, E_n\}$ of sets in $\mathbf{C}$ whose union is also in $\mathbf{C}$. A countably subadditive measure is an outer measure if its domain is the power set of $\mathbf{C}$, whereby it holds that $\mu(\emptyset) = 0$, and $A \subseteq B$ implies $\mu(A) \le \mu(B)$.

The map $\delta \mapsto \mathscr{H}^{\alpha}_{\delta}(E)$ is monotonically nonincreasing and thus, we can define the *Hausdorff α-dimensional measure* of E as

$$\mathscr{H}^{\alpha}(E) = \lim_{\delta\to 0} \mathscr{H}^{\alpha}_{\delta}(E).$$

Theorem B.1.1 *For* $0 \leq s < t < \infty$ *and* $A \subset X$, *we have*

- $\mathscr{H}^{s}(A) < \infty$ *implies* $\mathscr{H}^{t}(A) = 0$;
- $\mathscr{H}^{t}(A) > 0$ *implies* $\mathscr{H}^{s}(A) = \infty$.

The Hausdorff dimension $\dim_{\mathrm{H}}(A)$ of a subset $A \subset X$ is then defined as

$$\begin{aligned}\dim_{\mathrm{H}}(A) &= \bigvee \{s | \mathscr{H}^{s}(A) > 0\} = \bigvee \{s | \mathscr{H}^{s}(A) = \infty\} \\ &= \bigwedge \{t | \mathscr{H}^{t}(A) = 0\} = \bigwedge \{t | \mathscr{H}^{t}(A) < \infty\}.\end{aligned}$$

In order to give a formal definition of fractals, we need to be familiar with the notion of *topological* dimension (also known as *Lebesgue covering* dimension) [118].

Definition B.1.3 A cover $\mathscr{B}$ is a *refinement* of another cover $\mathscr{A}$ of the same space, in other words $\mathscr{B}$ *refines* $\mathscr{A}$, if for every $B \in \mathscr{B}$ there exists an $A \in \mathscr{A}$ such that $B \subset A$. Obviously, every subcover $\mathscr{A}_0$ of $\mathscr{A}$ is a refinement of $\mathscr{A}$.

Definition B.1.4 For every topological space (X, τ), the *topological* dimension, denoted by $\dim_T(X)$, is an integer greater than or equal to -1 or the "infinite number ∞"; the definition of the dimension function $\dim_{\mathrm{T}}$ consists of the following conditions:

(i) $\dim_{\mathrm{T}}(X) \leq n$, where $n = -1, 0, 1, 2, \ldots$, if every finite cover of the space X has a finite refinement of order less than or equal to n;
(ii) $\dim_{\mathrm{T}}(X) = n$ if $\dim_{\mathrm{T}}(X) \leq n$ and $\dim X > n - 1$;
(iii) $\dim_{\mathrm{T}}(X) = \infty$ if $\dim_{\mathrm{T}}(X) > n$ for $n = -1, 0, 1, 2, \ldots$

Using these notions, Benoit B. Mandelbrot [209], the father of fractal geometry, gave the following definition:

Definition B.1.5 A fractal is, by definition, a set of which the Hausdorff dimension strictly exceeds the topological dimension.

In addition, every set with an noninteger $\dim_{\mathrm{H}}$ is a fractal.

Unfortunately, the definition of the Hausdorff dimension is such that it is causing difficulty to calculate on computers. Thus, we typically use the *capacity* dimension that can be computed with the *box-counting* algorithm. The description of the algorithm that follows is based on its presentation in [128].

Suppose the space is such that one can define sensible distances and its integer dimension is d. In this space, we are able to cover a fractal with d-dimensional boxes (i.e. squares, cubes, etc.). Their sizes are regulated by one dimensionless parameter ℓ. This is always possible by choosing nondimensionless basic units so that ℓ becomes just a scaling factor. Next, we count the number $N(\ell)$ of boxes that cover some part of the fractal. $N(\ell)$ is connected to the capacity or box-counting dimension, d_c, by

$$N(\ell) = \text{constant} \times \ell^{-d_c}.$$

By plotting $\ln(N(\ell))$ against $\ln(\ell)$ for many values of ℓ, we get a straight line whose slope is the fractal dimension. When the resulting curve is not a straight line, we employ the least square method to compute an approximation of the fractal dimension.

B.2 Fuzzy Chaos

A form of *fuzzy chaos* was proposed by Oscar Castillo and Patricia Melin [60]. Their starting point is dynamical systems that can be expressed as a nonlinear differential equation:

$$\frac{dy}{dt} = f(t, y), \quad \text{where } y(0) = y_0.$$

Or as a nonlinear difference equation:

$$y_{t+1} = f(y_t, \ldots), \quad \text{where } y(0) = y_0.$$

In order to define and then to study fuzzy chaos, the authors considered a dynamical system on the real line given as follows:

$$y_t = f(y_{t-1}, \theta).$$

The idea is to associate chaotic behavior with the number of period doublings (or *bifurcations*, that is, a separation of a structure into two branches) that occur when the parameter θ is varied.

Definition B.2.1 A one-dimensional system shows fuzzy chaos, when the number of period doublings is considered to be large.

An alternative definition that relies on fractal dimensions follows:

Definition B.2.2 A one-dimensional dynamical system shows fuzzy chaos, when the value of the fractal dimension is large (close to a numeric value of 2 for the plane).

As was explained earlier, a numerical simulation of a dynamical system is realized by a feedback process, where the output produced by a function at some stage i is given as input to the same function at stage $i + 1$. The output generated by such a simulation is called a *time series*: $y_1, y_2, \ldots, y_n$. Next, we need to make a behavior identification for the dynamic system. For example, we can define the typical dynamic behaviors of a three-dimensional system with the following fuzzy rules:

If fractal dimension is low,	Then behavior is cycle of period 2.
If fractal dimension is small,	Then behavior is cycle of period 4.
If fractal dimension is regular,	Then behavior is cycle period 8.
If fractal dimension is medium,	Then behavior is cycle period 16.
If fractal dimension is high,	Then behavior is high order cycles.
If fractal dimension is large,	Then behavior is fuzzy chaos.

Here the fractal dimension can be calculated using the box-counting algorithm and the time series. In addition, the fractal dimension and the behavior are linguistic variables and one needs to define their membership functions.

Now, let us consider a dynamic system that is described by the following mathematical model:

$$
\begin{aligned}
X' &= \sigma(Y - X)\\
Y' &= rX - Y - XZ\\
Z' &= XY - bZ
\end{aligned}
$$

Here $X, Y, Z, \sigma, r, b \in \mathbb{R}$ and σ, r, and b are positive numbers. However, there is a problem here: How do we choose these parameters? Fortunately, there is an algorithm that can be used to choose the *best* set of parameters. This algorithm is presented in Figure B.1. Also, the genetic algorithm mentioned in Figure B.1 is presented in Figure B.2.

Step 1	Read the mathematical model M.
Step 2	Analyze the model M so to "understand" its complexity.
Step 3	Generate a set of admissible parameters using the understanding of the model.
Step 4	Choose the *best* set of parameter values using a specific genetic algorithm.
Step 5	Perform the simulations by solving numerically the equations of the mathematical model. At this time, the different types of dynamical behaviors are identified using a fuzzy rule base.

Figure B.1 Algorithm for choosing the best set of parameter values.

Step 1	Initialize a population with randomly generated individuals (parameters) and evaluate the fitness value of each individual.
Step 2	(a) Select two members from the population with probabilities proportional to their fitness values. (b) Apply crossover with a probability equal to the crossover rate. (c) Apply mutation with a probability equal to the mutation rate. (d) Repeat (a)–(d) until enough members are generated to form the next generation.
Step 3	Repeat steps 2 and 3 until the stopping criterion is met.

Figure B.2 Genetic algorithm for choosing parameter values.

It is possible to build a set of fuzzy rules for dynamic behavior identification based on the analytical properties of the mathematical models and using known results from dynamical systems theory. In particular, we should be able to get this dynamic behavior identification from the time series that we get from the simulation of a dynamical system. More specifically, we can calculate the Lyapunov exponents[3] (see Ref. [299] for the description of an algorithm that solves this problem) of the dynamical system and also the fractal dimension of the time series. Having this information, we can easily identify the corresponding behaviors of the system.

B.3 Fuzzy Fractals

Fractal images have entered the pop culture mainly because they are stunning and relatively easy to create. The algorithms used to draw Julia sets (named after Gaston Maurice Julia), the Mandelbrot set, and images generated by iterated function systems (IFSs), which were conceived in their present form by John E. Hutchinson [164] (the term IFS has been introduced in [21]), are the tools that most people use to draw these impressive images.

Carlos A. Cabrelli et al. [56] presented a variant of the IFSs that can be used to draw "fuzzy" fractals. An IFS consists of a set of *contraction maps*[4] $w_i : X \to X$, $i = 1, 2, \ldots, N$ and associated probabilities p_i, $i = 1, 2, \ldots, N$, $\sum_{i=1}^{N} p_i = 1$, where X is a compact metric space. Now, we can use the *chaos game* and an IFS to actually draw fractals. The chaos game is a method that can be used to generate the *attractor*

3 Lyapunov exponents are a measure of a system's sensitivity to changes to its initial conditions.
4 A contraction map on a metric space (X, d) is a function $f : X \to X$ with the property that there is some nonnegative real number $0 \leq k < 1$ such that, for all $x, y \in X$, $d(f(x), f(y)) \leq k\, d(x, y)$.

(usually a fractal) of an IFS. We start with a random point p_0 that is supposed to belong to the fractal, then we randomly apply any of the contraction maps w_i and get a new point $q_1 = w_i(q_0)$. We continue this process and get a sequence of points $q_k = w_j(q_{k-1})$ that form a fractal.

A grayscale digitized picture is actually a table (p_{ij}), where p_{ij} corresponds to the (i,j) screen pixel ($i \geq 0$ and $j \geq 0$). Each p_{ij} represents the amount of light that is emitted by the pixel (i,j). We can assume that $p_{ij} \in [0,1]$, where 0 denotes no emission (i.e. black) and 1 denotes maximum luminosity (i.e. white).

Definition B.3.1 The function $h : (p_{ij}) \to [0,1]$, defined by the pixel luminosity distribution of an image, is called *image function*.

Thus, a grayscale digitized picture is described by an image function. Similarly, the image generated by the chaos game can be represented by a function $u : X \to [0,1]$, where for each point x, the value $u(x)$ represents a normalized *gray-level value*.

Definition B.3.2 An *iterated fuzzy sets system* (abbreviated as IFZS) consists of a collection $w_i : X \to X$ of contraction maps and an associated collection $\phi_i : [0,1] \to [0,1]$ of gray-level maps having the following properties:

(i) each ϕ_i is nondecreasing and right continuous on $[0,1)$;
(ii) $\phi_i(0) = 0$; and
(iii) for at least one $j \in \{1,2,\ldots,n\}$, $\phi_j(1) = 1$.

The collection of maps (w_i, ϕ_i) is used to define an operator $T : \mathcal{F}^*(X) \to \mathcal{F}^*(X)$, where $\mathcal{F}^*(X)$ is a subclass of the class $\mathcal{F}(X)$, which is contractive with respect to a metric d_∞ on $\mathcal{F}^*(X)$. This metric is induced by the Hausdorff distance on the nonempty closed subsets of X. Starting with an (arbitrary) initial fuzzy set $u_0 \in \mathcal{F}^*(X)$, the sequence $u_n \in \mathcal{F}^*(X)$ produced by the iteration $u_{n_1} = Tu_n$ converges in the d_∞ metric to a unique and invariant fuzzy set $u^* \in \mathcal{F}^*(X)$, that is, $Tu^* = u^*$. The unique, invariant fuzzy set $u^*(x)$ is an attractor for the IFZS.

Iterated fuzzy function systems have been used to study fuzzy turbulence [206]. In addition, IFZS have been used for the study of fuzzy hyperfractals [10]. Also, a similar path was used to "fuzzify" the Mandelbrot set and Julia sets [318].

Works Cited

1 Abramsky, S. (1994). Proofs as processes. *Theoretical Computer Science* 135 (1: 5–9. [cited on page(s) 188]

2 Adlassnig, K.-P. (1986). Fuzzy set theory in medical diagnosis. *IEEE Transactions on Systems, Man, and Cybernetics* 16 (2): 260–265. [cited on page(s) 185, 186]

3 Akbari, M. and Sadegh, M.K. (2012). Estimators based on fuzzy random variables and their mathematical properties. *Iranian Journal of Fuzzy Systems* 9 (1): 79–95. [cited on page(s) 141]

4 Akiba, K. (2014). Boolean-valued sets as vague sets. In: *Vague Objects and Vague Identity* (ed. K. Akiba and A. Abasnezhad) [5], 175–195. Dordrecht: Springer. [cited on page(s) 34]

5 Akiba, K. and Abasnezhad, A. (eds.) (2014). *Vague Objects and Vague Identity, Logic, Epistemology, and the Unity of Science, No. 33*. Dordrecht, The Netherlands: Springer. [cited on page(s) 327, 332]

6 Akram, M. (2018). *Fuzzy Lie Algebras*. Singapore: Springer Nature Singapore. [cited on page(s) 231]

7 Alex, R. (2007). Fuzzy point estimation and its application on fuzzy supply chain analysis. *Fuzzy Sets and Systems* 158 (14): 1571–1587. [cited on page(s) 139]

8 Alur, R. and Dill, D.L. (1994). A theory of timed automata. *Theoretical Computer Science* 126 (2): 183–235. [cited on page(s) 301]

9 Anastasiou, G.A. (2010). *Fuzzy Mathematics: Approximation Theory, Studies in Fuzziness and Soft Computing, No. 251*. Berlin: Springer-Verlag. [cited on page(s) 316, 318]

10 Andres, J. and Rypka, M. (2016). Fuzzy fractals and hyperfractals. *Fuzzy Sets and Systems* 300: 40–56. Theme: Fuzzy Metric Spaces and Topology. [cited on page(s) 325]

11 Anthony, J. and Sherwood, H. (1979). Fuzzy groups redefined. *Journal of Mathematical Analysis and Applications* 69 (1): 124–130. [cited on page(s) 220]

A Modern Introduction to Fuzzy Mathematics, First Edition.
Apostolos Syropoulos and Theophanes Grammenos.

12 Aristotle (1994). *Όργανον 1: Κατηγορίαι - Περὶ ἑρμηνείας*. Αθήνα: Εκδόσεις Κάκτος. [cited on page(s) 161]

13 Aschieri, P., Dimitrijevic, M., Kulish, P. et al. (2009). *Noncommutative Spacetimes: Symmetries in Noncommutative Geometry and Field Theory*. Berlin: Springer-Verlag. [cited on page(s) 286]

14 Atanassov, K.T. (1986). Intuitionistic fuzzy sets. *Fuzzy Sets and Systems* 20 (1): 87–96. [cited on page(s) 27]

15 Atanassov, K.T. (2012). *On Intuitionistic Fuzzy Sets Theory, Studies in Fuzziness and Soft Computing, No. 283*. Berlin: Springer-Verlag. [cited on page(s) 27, 28]

16 Ayres, F. Jr. and Jaisingh, L.R. (2004). *Schaum's Outline of Theory and Problems of Abstract Algebra*, 2e. New York: McGraw-Hill. [cited on page(s) 215]

17 Baaz, M. and Metcalfe, G. (2007). Proof theory for first order Łukasiewicz logic. In: *Automated Reasoning with Analytic Tableaux and Related Methods* (ed. N. Olivetti), 28–42. Berlin: Springer-Verlag. [cited on page(s) 177]

18 Baaz, M., Preining, N., and Zach, R. (2007). First-order Gödel logics. *Annals of Pure and Applied Logic* 147 (1): 23–47. [cited on page(s) 176]

19 Balachandran, A., Kürkçüoglu, S., and Vaidya, S. (2007). *Lectures on Fuzzy and Fuzzy SUSY Physics*. Singapore: World Scientific Publishing Co. [cited on page(s) 286]

20 Ban, A. (2008). Approximation of fuzzy numbers by trapezoidal fuzzy numbers preserving the expected interval. *Fuzzy Sets and Systems* 159 (11): 1327–1344. [cited on page(s) 41]

21 Barnsley, M.F. and Demko, S. (1985). Iterated function systems and the global construction of fractals. *Proceedings of the Royal Society of London. Series A: Mathematical and Physical Sciences* 399 (1817): 243–275. [cited on page(s) 324]

22 Barr, M. (1979). **-Autonomous Categories, Lecture Notes in Mathematics, No. 752*. Berlin: Springer-Verlag. [cited on page(s) 100]

23 Barr, M. (1996). Fuzzy models of linear logic. *Mathematical Structures in Computer Science* 6 (3): 301–312. [cited on page(s) 99]

24 Bede, B. (2013). *Mathematics of Fuzzy Sets and Fuzzy Logic, Studies in Fuzziness and Soft Computing, No. 295*. Berlin: Springer-Verlag. [cited on page(s) 44, 81, 297]

25 Bell, J.L. (2005). *Set Theory: Boolean-Valued Models and Independence Proofs: Oxford Logic Guides*, 3e, vol. 47. Oxford: Oxford University Press. [cited on page(s) 35]

26 Bělohlávek, R., Dauben, J.W., and Klir, G.J. (2017). *Fuzzy Logic and Mathematics: A Historical Perspective*. New York: Oxford University Press. [cited on page(s) 166]

27 Benovsky, J. (2015). Vague objects with sharp boundaries. *Ratio: An International Journal of Analytic Philosophy* 28 (1): 29–39. [cited on page(s) 1]

28 Berge, C. (1976). *Graphs and Hypergraphs*, 2e. Amsterdam: North-Holland Publishing Company. [cited on page(s) 90]

29 Bhattacharya, P. and Suraweera, F. (1991). An algorithm to compute the supremum of max-min powers and a property of fuzzy graphs. *Pattern Recognition Letters* 12 (7): 413–420. [cited on page(s) 83, 92]

30 Birkhoff, G. (1967). *Lattice Theory*, 3e, Colloquium Publications, No. 25. Providence, RI: American Mathematical Society. [cited on page(s) 11]

31 Biswas, R. (1989). Fuzzy fields and fuzzy linear spaces redefined. *Fuzzy Sets and Systems* 33 (2): 257–259. [cited on page(s) 229]

32 Bochvar, D.A. (1938). Об одном трехзначном исчислении и его применении к анализу парадоксов классического расширенного функционального исчисления. *Математический сборник* 4 (46), 2: 287–308. English translation: [33]. [cited on page(s) 163]

33 Bochvar, D.A. and Bergmann, M. (1981). On a three-valued logical calculus and its application to the analysis of the paradoxes of the classical extended functional calculus. *History and Philosophy of Logic* 2 (1–2): 87–112. [cited on page(s) 329]

34 Boghosian, B.M., Coveney, P.V., and Wang, H. (2019). A new pathology in the simulation of chaotic dynamical systems on digital computers. *Advanced Theory and Simulations* 1900125. https://doi.org/10.1002/adts.201900125. [cited on page(s) 320]

35 Boole, G. (2009). *The Mathematical Analysis of Logic: Being an Essay Towards a Calculus of Deductive Reasoning, Cambridge Library Collection - Mathematics*. Cambridge: Cambridge University Press. [cited on page(s) 11]

36 Borkowski, L. (ed.) (1970). *Jan Łukasiewicz: Selected Works*. Amsterdam: North-Holland Publishing Company. [cited on page(s) 339]

37 Bouchon-Meunier, B., Coletti, G., and Marsala, C. (2002). Conditional possibility and necessity. In: *Technologies for Constructing Intelligent Systems 2: Tools* (ed. B. Bouchon-Meunier, J. Gutierrez-Rios, L. Magdalena, and R.R. Yager), 59–71. Heidelberg: Physica-Verlag HD. [cited on page(s) 117, 118]

38 Bredon, G.E. (1993). *Topology and Geometry, Graduate Texts in Mathematics, No. 139*. New York: Springer-Verlag. [cited on page(s) 235, 238]

39 Brown, J.G. (1971). A note on fuzzy sets. *Information and Control* 18 (1): 32–39. [cited on page(s) 35]

40 Buckley, J.J. (1989). Fuzzy complex numbers. *Fuzzy Sets and Systems* 33 (3): 333–345. [cited on page(s) 40]

41 Buckley, J.J. (2006). *Fuzzy Probability and Statistics, Studies in Fuzziness and Soft Computing*, vol. 196. Berlin: Springer-Verlag. [cited on page(s) 153]

42 Buckley, J.J. and Eslami, E. (1997a). Fuzzy plane geometry I: points and lines. *Fuzzy Sets and Systems* 86 (2): 179–187. [cited on page(s) 269, 273, 275]

43 Buckley, J.J. and Eslami, E. (1997b). Fuzzy plane geometry II: circles and polygons. *Fuzzy Sets and Systems* 87 (1): 79–85. [cited on page(s) 269, 277, 278, 281, 284]

44 Buckley, J.J. and Eslami, E. (2002). *An Introduction to Fuzzy Logic and Fuzzy Sets, Advances in Soft Computing, No. 13*. Berlin: Springer-Verlag. [cited on page(s) 55, 60, 272, 284]

45 Buckley, J.J. and Feuring, T. (1999). Introduction to fuzzy partial differential equations. *Fuzzy Sets and Systems* 105 (2): 241–248. [cited on page(s) 310]

46 Buckley, J.J. and Feuring, T. (2000). Fuzzy differential equations. *Fuzzy Sets and Systems* 110 (1): 43–54. [cited on page(s) 304]

47 Buckley, J.J. and Feuring, T. (2001). Fuzzy initial value problem for Nth-order linear differential equations. *Fuzzy Sets and Systems* 121 (2): 247–255. [cited on page(s) 308]

48 Buckley, J.J. and Jowers, L.J. (2008). *Monte Carlo Methods in Fuzzy Optimization*. Berlin: Springer-Verlag. [cited on page(s) 55]

49 Buckley, J. and Qu, Y. (1991). Solving systems of linear fuzzy equations. *Fuzzy Sets and Systems* 43 (1): 33–43. [cited on page(s) 271, 272, 277, 279]

50 Buckley, J.J., Eslami, E., and Feuring, T. (2002). *Fuzzy Mathematics in Economics and Engineering, Studies in Fuzziness and Soft Computing*, vol. 91. Berlin: Springer-Verlag. [cited on page(s) 305, 308, 310, 312]

51 Bueno, O. and Colyvan, M. (2012). Just what is vagueness? *Ratio: An International Journal of Analytic Philosophy* 25: 19–33. [cited on page(s) 4]

52 Burgin, M. (1995). Neoclassical analysis: fuzzy continuity and convergence. *Fuzzy Sets and Systems* 75 (3): 291–299. [cited on page(s) 300, 304]

53 Burgin, M. (2000). Theory of fuzzy limits. *Fuzzy Sets and Systems* 115 (3): 433–443. [cited on page(s) 300, 302]

54 Burgin, M. (2006). Fuzzy limits of functions, arXiv:math/0612676v1 [math.CA]. [cited on page(s) 300, 301, 302, 303, 304]

55 Butkiewicz, B.S. (2007). An approach to theory of fuzzy discrete signals. In: *Foundations of Fuzzy Logic and Soft Computing* (ed. P. Melin, O. Castillo, L.T. Aguilar et al.), 646–655. Berlin: Springer-Verlag. [cited on page(s) 313]

56 Cabrelli, C.A., Forte, B., Molter, U.M., and Vrscay, E.R. (1992). Iterated fuzzy set systems: a new approach to the inverse problem for fractals and other sets. *Journal of Mathematical Analysis and Applications* 171 (1): 79–100. [cited on page(s) 324]

57 Cantarella, J., Needham, T., Shonkwiler, C., and Stewart, G. (2019). Random triangles and polygons in the plane. *The American Mathematical Monthly* 126 (2): 113–134. [cited on page(s) 285]

58 Carroll, K.K. (1975). Experimental evidence of dietary factors and hormone-dependent cancers. *Cancer Research* 35 (11, Part 2): 3384–3386. [cited on page(s) 123]

59 Casals, M.R., Gil, M.A., and Gil, P. (1986). On the use of Zadeh's probabilistic definition for testing statistical hypotheses from fuzzy information. *Fuzzy Sets and Systems* 20 (2): 175–190. [cited on page(s) 141, 146]

60 Castillo, O. and Melin, P. (2001). A new theory of fuzzy chaos and its application for simulation and control of robotic dynamic systems. In: *10th IEEE International Conference on Fuzzy Systems*. (Cat. No.01CH37297), vol. 1, 151–154. [cited on page(s) 322]

61 Chachi, J. and Taheri, S.M. (2011). Fuzzy confidence intervals for mean of Gaussian fuzzy random variables. *Expert Systems with Applications* 38 (5): 5240–5244. [cited on page(s) 146, 148]

62 Chachi, J., Taheri, S.M., and Viertl, R. (2016). Testing statistical hypotheses based on fuzzy confidence intervals. *Austrian Journal of Statistics* 41 (4): 267–286. [cited on page(s) 146, 147]

63 Chakrabarty, K. (2001). On fuzzy lattice. In: *Rough Sets and Current Trends in Computing* (ed. W. Ziarko and Y. Yao), 238–242. Berlin: Springer-Verlag. [cited on page(s) 88]

64 Chakraverty, S., Tapaswini, S., and Behera, D. (2016). *Fuzzy Differential Equations and Applications for Engineers and Scientists*. Boca Raton, FL: CRC Press. [cited on page(s) 304]

65 Chalco-Cano, Y., Báez-Sánchez, A., Román-Flores, H., and Rojas-Medar, M. (2011). On the approximation of compact fuzzy sets. *Computers & Mathematics with Applications* 61 (2): 412–420. [cited on page(s) 105]

66 Chang, C.L. (1968). Fuzzy topological spaces. *Journal of Mathematical Analysis and Applications* 24 (1): 182–190. [cited on page(s) 244]

67 Chang, S.S.L. and Zadeh, L.A. (1972). On fuzzy mapping and control. *IEEE Transactions on Systems, Man, and Cybernetics* SMC-2 (1): 30–34. [cited on page(s) 300]

68 Chapin, E.W. Jr. (1974). Set-valued set theory: Part one. *Notre Dame Journal of Formal Logic* 15 (4): 619–634. [cited on page(s) 36]

69 Chapin, E.W. Jr. (1975). Set-valued set theory: Part two. *Notre Dame Journal of Formal Logic* 16 (2): 255–267. [cited on page(s) 36]

70 Chen, S.-H. (1985). Operation on fuzzy numbers with function principle. *Tamkang Journal of Management Science* 6 (1): 13–25. [cited on page(s) 49]

71 Chen, S.-J. and Chen, S.-M. (2003). Fuzzy risk analysis based on similarity measures of generalized fuzzy numbers. *IEEE Transactions on Fuzzy Systems* 11 (1): 45–56. [cited on page(s) 46]

72 Cheney, E.W. (1966). *Introduction to Approximation Theory*. New York: McGraw-Hill Book Co. [cited on page(s) 315, 316]

73 Cheng, C.-B. (2004). Group opinion aggregation based on a grading process: a method for constructing triangular fuzzy numbers. *Computers & Mathematics with Applications* 48 (10): 1619–1632. [cited on page(s) 60]

74 Chiu, C.-H. and Wang, W.-J. (2002). A simple computation of min and max operations for fuzzy numbers. *Fuzzy Sets and Systems* 126 (2): 273–276. [cited on page(s) 51]

75 Cho, Y.J., Rassias, T.M., and Saadati, R. (2018). *Fuzzy Operator Theory in Mathematical Analysis*. Cham, Switzerland: Springer Nature Switzerland AG. [cited on page(s) 229, 231]

76 Coletti, G. and Vantaggi, B. (2007). Comparative models ruled by possibility and necessity: a conditional world. *International Journal of Approximate Reasoning* 45 (2): 341–363. Eighth European Conference on Symbolic and Quantitative Approaches to Reasoning with Uncertainty (ECSQARU 2005). [cited on page(s) 117]

77 Connes, A. (1994). *Noncommutative Geometry*. San Diego, CA: Academic Press. Freely available from http://www.alainconnes.org. [cited on page(s) 286]

78 Coppi, R., Gil, M.A., and Kiers, H.A. (2006). The fuzzy approach to statistical analysis. *Computational Statistics and Data Analysis* 51 (1): 1–14. [cited on page(s) 123, 132]

79 Corral, N. and Gil, M.A. (1988). A note on interval estimation with fuzzy data. *Fuzzy Sets and Systems* 28 (2): 209–215. [cited on page(s) 142, 144]

80 Coxeter, H.S.M. (1969). *Introduction to Geometry*, 2e. New York: Wiley. [cited on page(s) 280]

81 Craine, W.L. (1993). Fuzzy hypergraphs and fuzzy intersection graphs. PhD thesis. University of Idaho. [cited on page(s) 96]

82 Cường, B.C. (2014). Picture fuzzy sets. *Journal of Computer Science and Cybernetics* 30 (4). Freely available from https://doi.org/10.15625/1813-9663/30/4/5032. [cited on page(s) 30]

83 Darby, G. (2014). Vague objects in quantum mechanics? In: *Vague Objects and Vague Identity* (ed. K. Akiba and A. Abasnezhad [5], 69–108. Dordrecht, The Netherlands: Springer. [cited on page(s) 4]

84 Das, S., Guha, D., and Dutta, B. (2016). Medical diagnosis with the aid of using fuzzy logic and intuitionistic fuzzy logic. *Applied Intelligence* 45 (3): 850–867. [cited on page(s) 187]

85 Dat, L.Q., Chou, S.-Y., Dung, C.C., and Yu, V.F. (2013). Improved arithmetic operations on generalized fuzzy numbers. In: *2013 International Conference on Fuzzy Theory and Its Applications (iFUZZY)*, 407–414. IEEE. [cited on page(s) 50]

86 de Andrés Sánchez, J. and Gómez, A.T. (2004). Estimating a fuzzy term structure of interest rates using fuzzy regression techniques. *European Journal of Operational Research* 154 (3): 804–818. [cited on page(s) 153]

87 De Baets, B. and De Meyer, H. (2003). Transitive approximation of fuzzy relations by alternating closures and openings. *Soft Computing* 7 (4): 210–219. [cited on page(s) 82]

88 De Branges, L. (1959). The Stone-Weierstrass Theorem. *Proceedings of the American Mathematical Society* 10 (5): 822–824. [cited on page(s) 316]

89 de Paiva, V. (1989). A dialectica-like model of linear logic. In: *Proceedings of Category Theory and Computer Science, Manchester, UK, September 1989, Lecture Notes in Computer Science, No. 389* (ed. D. Pitt, D. Rydeheard, P. Dybjer et al.), 341–356. Springer-Verlag. [cited on page(s) 264]

90 de Paiva, V. (1991). The Dialectica Categories. Tech. Rep. 213. Computer Laboratory, University of Cambridge, UK, [cited on page(s) 100]

91 de Paiva, V. (2000). Dialectica and Chu constructions: cousins? Talk presented at the Workshop on *Chu Spaces: Theory and Applications*, Santa Barbara, California, USA. [cited on page(s) 100, 102]

92 de Paiva, V. (2002). Lineales: algebraic models of linear logic from a categorical perspective. In: *Words, Proofs and Diagrams* (ed. D. Barker-Plummer, D.I. Beaver, J. van Benthem, and P.S. di Luzio), 123–142. Stanford, CA: CLSI Publications. [cited on page(s) 102]

93 de Paiva, V. (2006). Dialectica and Chu constructions: cousins? *Theory and Applications of Categories* 17 (7): 127–152. [cited on page(s) 263]

94 Demirci, M. (1999). Vague groups. *Journal of Mathematical Analysis and Applications* 230 (1): 142–156. [cited on page(s) 220]

95 Denœux, T. (2011). Maximum likelihood estimation from fuzzy data using the EM algorithm. *Fuzzy Sets and Systems* 183 (1): 72–91. [cited on page(s) 141]

96 Deschrijver, G. and Kerre, E.E. (2003). On the relationship between some extensions of fuzzy set theory. *Fuzzy Sets and Systems* 133 (2): 227–235. [cited on page(s) 231]

97 Deschrijver, G. and Král, P. (2007). On the cardinalities of interval-valued fuzzy sets. *Fuzzy Sets and Systems* 158: 1728–1750. [cited on page(s) 22]

98 Diamond, P. (1988). Fuzzy least squares. *Information Sciences* 46 (3): 141–157. [cited on page(s) 152]

99 Diamond, P. and Kloeden, P. (1990). Metric spaces of fuzzy sets. *Fuzzy Sets and Systems* 35 (2): 241–249. [cited on page(s) 242]

100 Diamond, P. and Kloeden, P. (1994). *Metric spaces of fuzzy sets: Theory and Applications*. Singapore: World Scientific. [cited on page(s) 242]

101 Diamond, P. and Kloeden, P. (2000). Metric topology of fuzzy numbers and fuzzy analysis. In: *Fundamentals of Fuzzy Sets* (ed. D. Dubois and H. Prade), 583–641. Boston, MA: Springer US. [cited on page(s) 297]

102 Dixit, V., Kumar, R., and Ajmal, N. (1992). On fuzzy rings. *Fuzzy Sets and Systems* 49 (2): 205–213. [cited on page(s) 233]

103 Dongale, T.D., Ghatage, S.R., and Mudholkar, R.R. (2013). Application philosophy of fuzzy regression. *International Journal of Soft Computing and Engineering* 2 (6): 170–172. [cited on page(s) 151]

104 Doostfatemeh, M. and Kremer, S.C. (2005). New directions in fuzzy automata. *International Journal of Approximate Reasoning* 38 (2): 175–214. [cited on page(s) 205]

105 Douglas, M.R. and Nekrasov, N.A. (2001). Noncommutative field theory. *Reviews of Modern Physics* 73 (4): 977–1029. [cited on page(s) 286]

106 Draper, N.R. and Smith, H. (1998). *Applied Regression Analysis*, 3e. New York: Wiley. [cited on page(s) 148]

107 Drossos, C.A., Markakis, G., and Theodoropoulos, P.L. (2000). $\mathbb{B}$-fuzzy stochastics. In: *Asymptotics in Statistics and Probability: Papers in Honor of George Gregory Roussas* (ed. M.L. Puri), 155–170. Utrecht, Holland: VSP International Science Publishers. [cited on page(s) 35]

108 Dubois, D. (2012). Have fuzzy sets anything to do with vagueness? In: *Understanding Vagueness: Logical, Philosophical, and Linguistic Perspectives* (ed. P. Cintula, C.G. Fermuller, L. Godo, and P. Hájek), 317–346. London: College Publications. [cited on page(s) 6]

109 Dubois, D. and Prade, H. (1979). Fuzzy real algebra: some results. *Fuzzy Sets and Systems* 2 (4): 327–348. [cited on page(s) 44]

110 Dubois, D. and Prade, H. (1980). *Sets and Systems—Theory and Applications, Mathematics in Science and Engineering, No. 144*. New York: Academic Press. [cited on page(s) 289, 290, 291, 293, 294, 295, 296, 297, 298]

111 Dubois, D. and Prade, H. (1988). *Possibility Theory: An Approach to Computerized Processing of Uncertainty*. New York: Plenum Press. [cited on page(s) 115]

112 Dubois, D. and Prade, H. (2009). Possibility theory. In: *Encyclopedia of Complexity and Systems Science* (ed. R.A. Meyers), 6927–6939. New York: Springer. [cited on page(s) 122]

113 Dubois, D., Prade, H., and Sandri, S. (1993). On possibility/probability transformations. In: *Fuzzy Logic: State of the Art* (ed. R. Lowen and M. Roubens), 103–112. Dordrecht: Springer Netherlands. [cited on page(s) 120]

114 Einstein, A. (1921). Geometrie und Erfahrung. In: *Geometrie und Erfahrung: Erweiterte Fassung des Festvortrages gehalten an der Preussischen Akademie der Wissenschaften zu Berlin am 27. Januar 1921*, 2–20. Heidelberg: Springer-Verlag. [cited on page(s) 269]

115 Einstein, A. (2015). *Sidelights on Relativity*. London: Forgotten Books. [cited on page(s) 269]

116 Eklund, P., Galán, M., Medina, J. et al. (2002). Set functors, L-fuzzy set categories, and generalized terms. *Computers & Mathematics with Applications* 43 (6): 693–705. [cited on page(s) 99]

117 Encyclopedia of Mathematics. Semi-continuous function. http://www.encyclopediaofmath.org/index.php?title=Semicontinuous_function&oldid=40145. [cited on page(s) 242]

118 Engelking, R. (1978). *Dimension Theory*. Amsterdam: North-Holland. [cited on page(s) 321]

119 Evans, G. (1978). Can there be vague objects? *Analysis* 38 (4): 208. [cited on page(s) 3]

120 Feferman, S., Dawson, J.W., Kleene, S.C. et al. (eds.) (1986). *Kurt Gödel: Collected Works Volume I: Publications 1929–1936*. Oxford: Oxford University Press. [cited on page(s) 337]

121 Feferman, S., Dawson, J.W., Kleene, S.C. et al. (eds.) (1990). *Kurt Gödel: Collected Works Volume II: Publications 1938–1974*. Oxford: Oxford University Press. [cited on page(s) 337]

122 Feng, Y. (2000). Gaussian fuzzy random variables. *Fuzzy Sets and Systems* 111 (3): 325–330. [cited on page(s) 136]

123 Fiadeiro, J.L. (2004). *Categories for Software Engineering*. New York: Springer-Verlag. [cited on page(s) 97]

124 Flori, C. (2013). *A First Course in Topos Quantum Theory, Lecture Notes in Physics*, vol. 868. Berlin: Springer-Verlag. [cited on page(s) 97]

125 Forrest, P. (2016). The identity of indiscernibles. In: *The Stanford Encyclopedia of Philosophy* (ed. E.N. Zalta), winter 2016 ed. Metaphysics Research Lab, Stanford University. [cited on page(s) 3]

126 French, S. and Krause, D. (2003). Quantum vagueness. *Erkenntnis* 59: 97–124. [cited on page(s) 4]

127 French, S. and Krause, D. (2006). *Indentity in Physics: A Historical, Philosophical and Formal Analysis*. Oxford: Oxford University Press. [cited on page(s) 4]

128 Frøyland, J. (1992). *Introduction to Chaos and Coherence*. Bristol: Institute of Physics Publishing. [cited on page(s) 321]

129 Gal, G.S. (1993). A fuzzy variant of the Weierstrass' approximation theorem. *Journal of Fuzzy Mathematics* 1 (4): 865–872. [cited on page(s) 316, 318]

130 Gal, G.S. (1995). Fuzzy variant of the Stone-Weierstrass approximation theorem. *Mathematica (Cluj)* 37 (60): 103–108. [cited on page(s) 316, 318]

131 Gal, S.G. (2000). Approximation theory in fuzzy setting. In: *Handbook of Analytic Computational Methods in Applied Mathematics* (ed. G. Anastassiou). Boca Raton, FL: Chapman and Hall/CRC. [cited on page(s) 297, 318]

132 Galindo, J., Urrutia, A., and Piattini, M. (2006). *Fuzzy Databases: Modeling, Design and Implementation*. Hershey, PA: IGI Global. [cited on page(s) 107]

133 Gallager, R.G. (2013). *Stochastic Processes: Theory for Applications*. Cambridge: Cambridge University Press. [cited on page(s) 130]

134 George, A. and Veeramani, P. (1994). On some results in fuzzy metric spaces. *Fuzzy Sets and Systems* 64 (3): 395–399. [cited on page(s) 241]

135 Ghosh, D. and Chakraborty, D. (2012). Analytical fuzzy plane geometry I. *Fuzzy Sets and Systems* 209: 66–83. Theme: Fuzzy numbers and Analysis. [cited on page(s) 270]

136 Girard, J.-Y. (1987). Linear logic. *Theoretical Computer Science* 50: 1–102. [cited on page(s) 99, 188]

137 Girard, J.-Y. (1995). Linear logic: its syntax and semantics. In: *Advances in Linear Logic, London Mathematical Society Lecture Note Series*, vol. 222 (ed. J.-Y. Girard, Y. Lafont, and L. Regnier), 1–42. Cambridge University Press. [cited on page(s) 188]

138 Girard, J.-Y. (2003). From foundations to ludics. *Bulletin of Symbolic Logic* 9 (2): 131–168. [cited on page(s) 187]

139 Girard, J.-Y. (2007). Truth, modality and intersubjectivity. *Mathematical Structures in Computer Science* 17 (6): 1153–1167. [cited on page(s) 188]

140 Girard, J.-Y. (2011). *The Blind Spot: Lectures on Logic*. Zürich: European Mathematical Society Publishing House. [cited on page(s) 187]

141 Glöckner, I. (2006). *Fuzzy Quantifiers: A Computational Theory, Studies in Fuzziness and Soft Computing, No. 193*. Berlin: Springer-Verlag. [cited on page(s) 177]

142 Goguen, J.A. (1967). L-fuzzy sets. *Journal of Mathematical Analysis and Applications* 18 (1): 145–174. [cited on page(s) 26]

143 Goguen, J.A. (1969). The logic of inexact concepts. *Synthese* 19 (3–4): 325–373. [cited on page(s) 166]

144 Goldblatt, R. (1998). *Lectures on the Hyperreals: An Introduction to Nonstandard Analysis, Graduate Texts in Mathematics, No. 188*. Berlin: Springer-Verlag. [cited on page(s) 29]

145 Gomes, L.T., de Barros, L.C., and Bede, B. (2015). *Fuzzy Differential Equations in Various Approaches*. Heidelberg: Springer Cham. [cited on page(s) 304]

146 Gottwald, S. (2006). Universes of fuzzy sets and axiomatizations of fuzzy set theory. Part II: Category theoretic approaches. *Studia Logica* 84: 23–50. [cited on page(s) 99]

147 Goudarzi, M. and Vaezpour, S.M. (2009). On the definition of fuzzy Hilbert spaces and its application. *The Journal of Nonlinear Science and Applications* 2 (1): 46–59. [cited on page(s) 261]

148 Gouyandeh, Z., Allahviranloo, T., Abbasbandy, S., and Armand, A. (2017). A fuzzy solution of heat equation under generalized Hukuhara differentiability by fuzzy Fourier transform. *Fuzzy Sets and Systems* 309: 81–97. [cited on page(s) 313]

149 Grensing, G. (2013). *Structural Aspects of Quantum Field Theory and Noncommutative Geometry*. Singapore: World Scientific. [cited on page(s) 286]

150 Grosan, C. and Abraham, A. (2011). *Intelligent Systems: A Modern Approach, Intelligent Systems Reference Library*, vol. 17. Berlin: Springer-Verlag. [cited on page(s) 183]

151 Grzegorzewski, P. and Hryniewicz, O. (1997). Testing statistical hypotheses in fuzzy environment. *Mathware & Soft Computing* 4: 203–217. [cited on page(s) 146]

152 Gupta, M.M. and Qi, J. (1991). Theory of T-norms and fuzzy inference methods. *Fuzzy Sets and Systems* 40 (3): 431–450. [cited on page(s) 24]

153 Gupta, K. and Ray, S. (1993). Fuzzy plane projective geometry. *Fuzzy Sets and Systems* 54 (2): 191–206. [cited on page(s) 285]

154 Guts, A.K. (1991). A Topos-theoretic approach to the foundations of relativity theory. *Doklady Akademii Nauk SSSR* 318 (6): 1294–1297. English translation in *Soviet Mathematics Doklady* 43 (3): 904–907, 1991. [cited on page(s) 97]

155 Gödel, K. (1932). Zum intuitionistischen Aussagenkalkül. *Anzeiger der Akademie der Wissenschaften in Wien* 69: 65–66. English translation with comments in [[120], p. 222–225]. [cited on page(s) 163]

156 Gödel, K. (1958). Über eine bisher noch nicht benützte Erweiterung des finiten Standpunktes. *Dialectica* 12 (3–4): 280–287. English translation with comments in [[121], p. 217–251]. [cited on page(s) 99]

157 Hájek, P. (1998). *Metamathematics of Fuzzy Logic, Trends in Logic, No. 4.* Dordrecht, The Netherlands: Springer Netherlands. [cited on page(s) 155, 167, 184]

158 Halmos, P.R. (1974). *Naive Set Theory.* New York: Springer-Verlag. [cited on page(s) 36]

159 Hanss, M. (2005). *Applied Fuzzy Arithmetic: An Introduction with Engineering Applications.* Berlin: Springer-Verlag. [cited on page(s) 40, 44, 63]

160 Hanss, M. and Nehls, O. (2000). Simulation of the human glucose metabolism using fuzzy arithmetic. In: *PeachFuzz 2000: 19th International Conference of the North American Fuzzy Information Processing Society* (ed. T. Whalen), 201–205. IEEE. [cited on page(s) 63]

161 Hogg, R.V., Tanis, E., and Zimmerman, D. (2013). *Probability and Statistical Inference*, 9e. Harlow, Essex: Pearson Education Limited. [cited on page(s) 136, 142]

162 Holzmann, C.A., Perez, C.A., and Rosselot, E. (1988). A fuzzy model for medical diagnosis. *Medical Progress Through Technology* 13 (4): 171–178. [cited on page(s) 186]

163 Hong, D.H. and Kim, K.T. (2006). An easy computation of MIN and MAX operations for fuzzy numbers. *Journal of Applied Mathematics and Computing* 21 (1–2): 555–561. [cited on page(s) 53]

164 Hutchinson, J.E. (1981). Fractals and self similarity. *Indiana University Mathematics Journal* 30 (5): 713–747. [cited on page(s) 315, 324]

165 Iske, A. (2018). *Approximation.* Berlin: Springer Spektrum. [cited on page(s) 316]

166 Jech, T. (2003). *Set Theory: The Third Millennium Edition, revised and expanded.* Berlin: Springer-Verlag. [cited on page(s) 36]

167 Joshi, K.D. (1983). *Introduction to General Topology.* New Delhi: Wiley Eastern Limited. [cited on page(s) 235]

168 Kadak, U. and Başar, F. (2014). On Fourier series of fuzzy-valued functions. *The Scientific World Journal 2014* Article ID 782652. [cited on page(s) 313]

169 Kahraman, C. and Kabak, O. (eds.) (2016). *Fuzzy Statistical Decision-Making: Theory and Applications, Studies in Fuzziness and Soft Computing, No. 343*. Basel, Switzerland: Springer. [cited on page(s) 142]

170 Kaleva, O. (1987). Fuzzy differential equations. *Fuzzy Sets and Systems* 24 (3): 301–317. [cited on page(s) 304]

171 Kaleva, O. and Seikkala, S. (1984). On fuzzy metric spaces. *Fuzzy Sets and Systems* 12 (3): 215–229. [cited on page(s) 243]

172 Kandel, A. and Byatt, W.J. (1980). Fuzzy processes. *Fuzzy Sets and Systems* 4 (2): 117–152. [cited on page(s) 304]

173 Katsaras, A.K. and Liu, D.B. (1977). Fuzzy vector spaces and fuzzy topological vector spaces. *Journal of Mathematical Analysis and Applications* 58 (1): 135–146. [cited on page(s) 229]

174 Kaufmann, A. (1973). *Eléments Théoriques de Base*, vol. 1 of *Introduction à la Théorie des Sous-Ensembles Flous à l'usage des ingénieurs*. Masson, Paris. [cited on page(s) 89]

175 Kleene, S.C. (1938). On notation for ordinal numbers. *Journal of Symbolic Logic* 3 (4): 150–155. [cited on page(s) 162]

176 Klement, E.P., Puri, M.L., and Ralescu, D.A. (1986). Limit theorems for fuzzy random variables. *Proceedings of the Royal Society of London A* 407 (1832): 171–182. [cited on page(s) 132]

177 Klir, G.J. and Parviz, B. (1992). Probability-possibility transformations: a comparison. *International Journal of General Systems* 21 (3): 291–310. [cited on page(s) 120]

178 Klir, G.J. and Yuan, B. (1995). *Fuzzy Sets and Fuzzy Logic: Theory and Applications*. Prentice Hall (Sd). [cited on page(s) 20, 21]

179 Kolmogoroff, A. (1933). *Grundbegriffe der Wahrscheinlichkeitsrechnung*. Berlin: Verlag von Julius Springer. [cited on page(s) 5]

180 Kosko, B. (1990). Fuzziness vs. Probability. *International Journal of General Systems* 17 (2): 211–240. [cited on page(s) 266]

181 Kramosil, I. and Michálek, J. (1975). Fuzzy metrics and statistical metric spaces. *Kybernetika* 11 (5): 336–344. [cited on page(s) 240]

182 Kreinovich, V. (2013). Membership functions or α-cuts? Algorithmic (constructivist) analysis justifies an interval approach. *Applied Mathematical Sciences* 7 (5): 217–228. [cited on page(s) 20]

183 Kruse, R. and Meyer, K.D. (1987). *Statistics with Vague Data, Theory and Decision Library B*, 2e, vol. 6. Dordrecht, The Netherlands: Kluwer. [cited on page(s) 132, 134, 135, 136]

184 Kuchta, D. (2012). Application of fuzzy numbers to the estimation of an ongoing project's completion time. *Operations Research and Decisions* 22 (4): 87–103. [cited on page(s) 66]

185 Kuncheva, L.I. and Steimann, F. (1999). Fuzzy diagnosis. *Artificial Intelligence in Medicine* 16 (2): 121–128. [cited on page(s) 108]

186 Kwakernaak, H. (1978). Fuzzy random variables—I. Definitions and theorems. *Information Sciences* 15 (1): 1–29. [cited on page(s) 132]

187 Kwakernaak, H. (1979). Fuzzy random variables—II. Algorithms and examples for the discrete case. *Information Sciences* 17 (3): 253–278. [cited on page(s) 132]

188 Lakshmikantham, V. and Mohapatra, R.N. (2003). *Theory of Fuzzy Differential Equations and Inclusions*. Boca Raton, FL: CRC Press. [cited on page(s) 304]

189 Lawvere, F.W. and Schanuel, S.H. (2009). *Conceptual Mathematics: A First Introduction to Categories*, 2e. Cambridge: Cambridge University Press. [cited on page(s) 97]

190 Lee, K.H. (2005). *First Course on Fuzzy Theory and Applications, Advances in Soft Computing*, vol. 27. Berlin: Springer-Verlag. [cited on page(s) 91, 289, 292, 293]

191 Lehmann, E.L. and Casella, G. (1998). *Theory of Point Estimation*, 2e. New York: Springer-Verlag. [cited on page(s) 136]

192 Lehmann, E.L. and Romano, J.P. (2005). *Testing Statistical Hypotheses*, 3e. New York: Springer-Verlag. [cited on page(s) 144]

193 Liu, W.-J. (1982). Fuzzy invariant subgroups and fuzzy ideals. *Fuzzy Sets and Systems* 8 (2): 133–139. [cited on page(s) 227]

194 Lorentz, G.G. (1966). *Approximation of Functions*. New York: Holt, Rinehart and Winston Inc. [cited on page(s) 315]

195 Lorenz, E.N. (1963). Deterministic nonperiodic flow. *Journal of the Atmospheric Sciences* 20 (2): 130–141. [cited on page(s) 319]

196 Lowe, E.J. (1994). Vague identity and quantum indeterminacy. *Analysis* 54 (2): 110–114. [cited on page(s) 3]

197 Lowen, R. (1976). Fuzzy topological spaces and fuzzy compactness. *Journal of Mathematical Analysis and Applications* 56 (3): 621–633. [cited on page(s) 245, 246, 247]

198 Łukasiewicz, J. (1920). O logice trójwartościowej. *Ruch Filozoficzny* 5: 170–171. English translation in [[36], p. 87–88]. [cited on page(s) 161]

199 Łukasiewicz, J. (1930). Philosophische Bemerkungen zu mehrwertigen Systemen des Aussagenkalküls. *Comptes rendus des séance de la Société des Sciences et des Lettres de Varsovie* 23: 51–77. Classe III (English translation in [[36], p. 153–178]. [cited on page(s) 161]

200 Łukasiewicz, J. and Tarski, A. (1930). Untersuchungen über den Aussagenkalkül. *Comptes rendus des séance de la Société des Sciences et des Lettres de Varsovie* 23: 30–50. Classe III (English translation in [[284], p. 153–178]. [cited on page(s) 165]

201 Mac Lane, S. (1986). *Mathematics: Form and Function*. New York: Springer-Verlag. [cited on page(s) 7, 97]

202 Mac Lane, S. (1998). *Category Theory for the Working Mathematician*, 2e. New York: Springer-Verlag. [cited on page(s) 97, 104]

203 Madore, J. (1992). The fuzzy sphere. *Classical and Quantum Gravity* 9 (1): 69–89. [cited on page(s) 285]

204 Madore, J. (1997). Fuzzy space-time. *Canadian Journal of Physics* 75 (6): 385–399. [cited on page(s) 286]

205 Madore, J. (2000). *An Introduction to Noncommutative Differential Geometry and its Physical Applications*, 2e. Cambridge: Cambridge University Press. [cited on page(s) 286]

206 Majumdar, K.K. (2004). Fuzzy fractals and fuzzy turbulence. *IEEE Transactions on Systems, Man, and Cybernetics Part B: Cybernetics* 34 (1): 746–751. [cited on page(s) 325]

207 Malpass, A. and Marfori, M.A. (eds.) (2017). *The History of Philosophical and Formal Logic: From Aristotle to Tarski*. London: Bloomsbury Academic. [cited on page(s) 155]

208 Mandelbrot, B. (1967). How long is the coast of Britain? Statistical self-similarity and fractional dimension. *Science* 156 (3775): 636–638. [cited on page(s) 2]

209 Mandelbrot, B.B. (1982). *The Fractal Geometry of Nature*. San Francisco, CA: W. H. Freeman and Co.. [cited on page(s) 321]

210 Matheron, G. (1975). *Random Sets and Integral Geometry*. New York: Wiley. [cited on page(s) 133]

211 Mayburov, S.N. (2018). Commutative fuzzy geometry and geometric quantum mechanics. *Journal of Physics: Conference Series* 1051: 012022. [cited on page(s) 286]

212 McCain, R.A. (1983). Fuzzy confidence intervals. *Fuzzy Sets and Systems* 10 (1): 281–290. [cited on page(s) 142]

213 McNaughton, R. (1951). A theorem about infinite-valued sentential logic. *Journal of Symbolic Logic* 16 (1): 1–13. [cited on page(s) 165]

214 Meeden, G. and Noorbaloochi, S. (2013). Hypotheses testing as a fuzzy set estimation problem. *Communications in Statistics—Theory and Methods* 42 (10): 1806–1820. [cited on page(s) 146]

215 Meeker, W.Q., Hahn, G.J., and Escobar, L.A. (2017). *Statistical Intervals: A Guide for Practitioners and Researchers*, 2e. New York: Wiley. [cited on page(s) 141]

216 Mendel, J.M. and John, R.I.B. (2002). Type-2 fuzzy sets made simple. *IEEE Transactions on Fuzzy Systems* 10 (2): 117–127. [cited on page(s) 22]

217 Mendelson, E. (1997). *Introduction to Mathematical Logic*, 4e. London: Chapman & Hall. [cited on page(s) 155, 156]

218 Metcalfe, G., Olivetti, N., and Gabbay, D. (2009). *Proof Theory for Fuzzy Logics, Applied Logic Series, No. 36*. Dordrecht, The Netherlands: Springer Netherlands. [cited on page(s) 155, 159]

219 Molodtsov, D.A. (1999). Soft set theory—first results. *Computers & Mathematics with Applications* 37 (4): 19–31. [cited on page(s) 31, 237]

220 Mordeson, J.N. and Malik, D.S. (2002). *Fuzzy Automata and Languages: Theory and Applications*. Boca Raton, FL: Chapman & Hall/CRC. [cited on page(s) 191]

221 Mordeson, J.N. and Nair, P.S. (2000). *Fuzzy Graphs and Fuzzy Hypergraphs, Studies in Fuzziness and Soft Computing*, vol. 46. Heidelberg: Physica-Verlag HD. [cited on page(s) 89, 96]

222 Mordeson, J.N. and Nair, P.S. (2001). *Fuzzy Mathematics—An Introduction for Scientists and Engineers*, 2e. Heidelberg: Physica Verlag. [cited on page(s) 272, 277]

223 Mordeson, J.N., Bhutani, K.R., and Rosenfeld, A. (2005). *Fuzzy Group Theory, Studies in Fuzziness and Soft Computing*, vol. 182. Berlin: Springer. [cited on page(s) 219]

224 Morreau, M. (2002). What vague objects are like. *The Journal of Philosophy* 99 (7): 333–361. [cited on page(s) 2]

225 Mostowski, A. (1957). On a generalization of quantifiers. *Fundamenta Mathematicae* 44 (1): 12–36. Freely available from http://eudml.org/doc/213418. [cited on page(s) 160]

226 Muller, E.N. (1972). A test of a partial theory of potential for political violence. *American Political Science Review* 66 (3): 928–959. [cited on page(s) 124]

227 Neyman, J. (1937). Outline of a theory of statistical estimation based on the classical theory of probability. *Philosophical Transactions of the Royal Society of London. Series A: Mathematical and Physical Sciences* 236 (767): 333–380. [cited on page(s) 141]

228 Nobuhara, H., Bede, B., and Hirota, K. (2006). On various eigen fuzzy sets and their application to image reconstruction. *Information Sciences* 176 (20): 2988–3010. [cited on page(s) 81]

229 Ovchinnikov, S. (1991). Similarity relations, fuzzy partitions, and fuzzy orderings. *Fuzzy Sets and Systems* 40 (1): 107–126. [cited on page(s) 83, 86]

230 Palaniappan, N. (2002). *Fuzzy Topology*. Pangbourne: Alpha Science International. [cited on page(s) 244]

231 Papadopoulos, B.K. and Syropoulos, A. (2000). Fuzzy sets amd fuzzy relational structures as Chu spaces. *International Journal of Uncertainty, Fuzziness and Knowledge-Based Systems* 8 (4): 471–479. [cited on page(s) 99]

232 Papadopoulos, B.K. and Syropoulos, A. (2005). Categorical relationships between Goguen sets and "two-sided" categorical models of linear logic. *Fuzzy Sets and Systems* 149: 501–508. [cited on page(s) 99]

233 Papoulis, A. and Pillai, S.U. (2002). *Probability, Random Variables, and Stochastic Processes*, 4e. Boston, MA: McGraw-Hill. [cited on page(s) 130, 132]

234 Parsaye, K. and Chignell, M. (1988). *Expert Systems for Experts*. New York: Wiley. [cited on page(s) 182]

235 Pathinathan, T. and Dolorosa, E.A. (2019). Symmetric periodic Fourier series using pentagonal fuzzy number. *Journal of Computer and Mathematical Sciences* 10 (3): 510–518. [cited on page(s) 313]

236 Pavlačka, O. and Talašová, J. (2010). Fuzzy vectors as a tool for modeling uncertain multidimensional quantities. *Fuzzy Sets and Systems* 161 (11): 1585–1603. Theme: Decision Systems. [cited on page(s) 105]

237 Pawlak, Z. (1982). Rough sets. *International Journal of Computer and Information Sciences* 11 (5): 341–356. [cited on page(s) 7]

238 Pedrycz, W. and Gomide, F. (2007). *Fuzzy Systems Engineering: Toward Human-Centric Computing*. Wiley. [cited on page(s) 82]

239 Peitgen, H.-O., Jürgens, H., and Saupe, D. (2004). *Chaos and Fractals: New Frontiers of Science*. New York: Springer. [cited on page(s) 319, 320]

240 Pelletier, F.J. (1984). The not-so-strange modal logic of indeterminacy. *Logique et Analyse* 27 (108): 415–422. [cited on page(s) 3]

241 Perezgonzalez, J.D. (2015). Fisher, Neyman-Pearson or NHST? A tutorial for teaching data testing. *Frontiers in Psychology* 6: 223. [cited on page(s) 122]

242 Post, E.L. (1921). Introduction to a general theory of elementary propositions. *American Journal of Mathematics* 43 (3): 163–185. [cited on page(s) 162]

243 Pultr, A. (1984). Fuzziness and fuzzy equality. In: *Aspects of Vagueness* (ed. H.J. Skala, S. Termini, and E. Trillas), 119–135. Dordrecht, Holland: D. Reidel Publishing Company. [cited on page(s) 36]

244 Puri, M.L. and Ralescu, D.A. (1983). Differentials of fuzzy functions. *Journal of Mathematical Analysis and Applications* 91 (2): 552–558. [cited on page(s) 300]

245 Puri, M.L. and Ralescu, D.A. (1986). Fuzzy random variables. *Journal of Mathematical Analysis and Applications* 114 (2): 409–422. [cited on page(s) 132]

246 Ramadan, A. (1992). Smooth topological spaces. *Fuzzy Sets and Systems* 48 (3): 371–375. [cited on page(s) 259]

247 Ronzitti, G. (ed.) (2011). *Vagueness: A Guide, Logic, Epistemology, and the Unity of Science, No. 19*. Dordrecht, The Netherlands: Springer. [cited on page(s) 2]

248 Rosen, K.H. (2012). *Discrete Mathematics and Its Applications*, 7e. New York: McGraw-Hill. [cited on page(s) 71, 89]

249 Rosenfeld, A. (1971). Fuzzy groups. *Journal of Mathematical Analysis and Applications* 35 (3): 512–517. [cited on page(s) 219]

250 Rosenfeld, A. (1975). Fuzzy graphs. In: *Fuzzy Sets and their Applications to Cognitive and Decision Processes* (ed. L.A. Zadeh, K.-S. Fu, K. Tanaka, and M. Shimura), 77–95. New York: Academic Press. [cited on page(s) 89]

251 Rosenfeld, A. (1998). Fuzzy geometry: an updated overview. *Information Sciences* 110 (3): 127–133. [cited on page(s) 269]

252 Saadati, R. and Vaezpour, S.M. (2005). Some results on fuzzy Banach spaces. *Journal of Applied Mathematics and Computing* 17 (1): 475–484. [cited on page(s) 229]

253 Sambuc, R. (1975). Fonctions Φ-floues, Application à l'aide au diagnostic en pathologie thyroidienne. PhD thesis. France: Université de Marseille. [cited on page(s) 22]

254 Samelson, H. (1990). *Notes on Lie Algebras*, 2e. New York: Springer-Verlag. [cited on page(s) 219]

255 Sezer, S. (2011). Vague rings and vague ideals. *Iranian Journal of Fuzzy Systems* 8 (1): 145–157. [cited on page(s) 222, 227]

256 Shackle, G.L.S. (2010). *Decision Order and Time in Human Affairs*, 2e. Cambridge: Cambridge University Press. [cited on page(s) 111]

257 Shafer, G. and Vovk, V. (2006). The sources of Kolmogorov's *Grundbegriffe*. *Statistical Science* 21 (1): 70–98. Freely available from https://projecteuclid.org/euclid.ss/1149600847. [cited on page(s) 111]

258 Shapiro, A.F. (2004). Fuzzy regression and the term structure of interest rates revisited. In: *Proceedings of the 14th International AFIR Colloquium 1*, 29–45. [cited on page(s) 151, 152]

259 Sharma, B.K., Syropoulos, A., and Tiwari, S.P. (2016). On fuzzy multiset regular grammars. *Annals of Fuzzy Mathematics and Informatics* 12 (5): 617–639. [cited on page(s) 213]

260 Shonkwiler, C. (2019). Stiefel manifolds and polygons. In: *Proceedings of Bridges 2019: Mathematics, Art, Music, Architecture, Education, Culture* (ed. S. Goldstine, D. McKenna, and K. Fenyvesi), 187–194. Phoenix, AZ: Tessellations Publishing. [cited on page(s) 285]

261 Siler, W. and Buckley, J.J. (2005). *Fuzzy Expert Systems and Fuzzy Reasoning*. New York: Wiley. [cited on page(s) 183]

262 Smarandache, F. (2006). Neutrosophic set—a generalization of the intuitionistic fuzzy set. In: *2006 IEEE International Conference on Granular Computing*, 38–42. [cited on page(s) 29]

263 Smithson, M. (1988). Possibility theory, fuzzy logic, and psychological explanation. In: *Fuzzy Sets in Psychology, Advances in Psychology*, vol. 56 (ed. T. Zétényi), 1–50. North-Holland. [cited on page(s) 122, 124]

264 Šostak, A. (1985). On a fuzzy topological structure. In: *Rendiconti del Circolo Matematico di Palermo Serie II*, Supplemento No. 11, 89–103. Available from http://dml.cz/dmlcz/701883. [cited on page(s) 258]

265 Šostak, A. (1999). Fuzzy categories related to algebra and topology. *Tatra Mountains Mathematical Publications* 16 (1): 159–185. Available from the web site of the Slovak Academy of Sciences. [cited on page(s) 103]

266 Šostak, A. (2002). Fuzzy functions and an extension of the category **L-TO** *P* of Chang-Goguen L-topological spaces. In: *Proceedings of the 9th Prague Topological Symposium* (ed. P. Simon). arXiv:math/0204165 [math.GN], 271–294. [cited on page(s) 103]

267 Stone, M.H. (1937). Applications of the theory of Boolean rings to general topology. *Transactions of the American Mathematical Society* 41 (3): 375–481. [cited on page(s) 316]

268 Stone, M.H. (1948). The generalized Weierstrass approximation theorem. *Mathematics Magazine* 21 (4): 167–184. [cited on page(s) 316]

269 Sugeno, M. and Sasaki, M. (1983). L-fuzzy category. *Fuzzy Sets and Systems* 11 (1): 43–64. [cited on page(s) 99]

270 Syau, Y.-R., Lee, E., and Jia, L. (2004). Convexity and upper semicontinuity of fuzzy sets. *Computers & Mathematics with Applications* 48 (1): 117–129. [cited on page(s) 242]

271 Syropoulos, A. (2006a). Fuzzifying P systems. *The Computer Journal* 49 (5): 619–628. [cited on page(s) 210, 212]

272 Syropoulos, A. (2006b). Yet another fuzzy model for linear logic. *International Journal of Uncertainty, Fuzziness and Knowledge-Based Systems* 14 (1): 131–135. [cited on page(s) 99, 263]

273 Syropoulos, A. (2012). On generalized fuzzy multisets and their use in computation. *Iranian Journal of Fuzzy Systems* 9 (2): 115–127. [cited on page(s) 210]

274 Syropoulos, A. (2014a). Fuzzy categories, arXiv:1410.1478v1 [cs.LO]. [cited on page(s) 104, 191]

275 Syropoulos, A. (2014b). *Theory of Fuzzy Computation, IFSR International Series on Systems Science and Engineering, No. 31*. New York: Springer-Verlag. [cited on page(s) 6, 187, 206, 209, 212]

276 Syropoulos, A. (2016). Fuzzy sets and fuzzy logic. In: *Encyclopedia of Computer Science and Technology* (ed. P.A. Laplante), 459–468. Boca Raton, FL: CRC Press. [cited on page(s) 107]

277 Syropoulos, A. (2017). On vague computers. In: *Emergent Computation: A Festschrift for Selim G. Akl* (ed. A. Adamatzky), 393–402. Cham: Springer International Publishing. [cited on page(s) 4]

278 Syropoulos, A. (2018). A (basis for a) philosophy of a theory of fuzzy computation. *Kairos. Journal of Philosophy & Science* 20 (1): 181–201. [cited on page(s) 209]

279 Syropoulos, A. and de Paiva, V. (2011). Fuzzy topological systems, arXiv:1107.2513v1 [cs.LO]. [cited on page(s) 265]

280 Szabo, R.J. (2003). Quantum field theory on noncommutative spaces. *Physics Reports* 378 (4): 207–299. [cited on page(s) 286]

281 Tahayori, H. and Antoni, G.D. (2007). A simple method for performing type-2 fuzzy set operations based on highest degree of intersection hyperplane. In: *NAFIPS 2007 - 2007 Annual Meeting of the North American Fuzzy Information Processing Society* (ed. M. Reformat and M.R. Berthold), 404–409. [cited on page(s) 23]

282 Taheri, S.M. (2016). Trends in fuzzy statistics. *Austrian Journal of Statistics* 32 (3): 239–257. [cited on page(s) 146]

283 Tanaka, H., Uejima, S., and Asai, K. (1982). Linear regression analysis with fuzzy model. *IEEE Transactions on Systems, Man, and Cybernetics* 12 (6): 903–907. [cited on page(s) 150]

284 Tarski, A. (1956). *Logic, Semantics, Metamathematics: Papers from 1923 to 1938*. Oxford: Oxford University Press. [cited on page(s) 339]

285 Thiele, H. (1995). On fuzzy quantifiers. In: *Fuzzy Logic and its Applications to Engineering, Information Sciences, and Intelligent Systems, Theory and Decision Library (Series D: System Theory, Knowledge Engineering and Problem Solving)*, vol. 16 (ed. Z. Bien and K. Min), 343–352. Dordrecht, The Netherlands: Springer. [cited on page(s) 177]

286 Timan, A.F. (1963). *Theory of Approximation of Functions of a Real Variable*. Oxford: Pergamon Press. [cited on page(s) 315]

287 Tremblay, J.-P. and Sorenson, P.G. (1985). *The Theory and Practice of Compiler Writing*. Singapore: McGraw-Hill. [cited on page(s) 192]

288 Tsaur, R.-C. (2014). Residual analysis using Fourier series transform in fuzzy time series model. *Iranian Journal of Fuzzy Systems* 11 (3): 43–54. [cited on page(s) 313]

289 Tu, L. and Wenxiang, G. (1994). Abelian fuzzy group and its properties. *Fuzzy Sets and Systems* 64 (3): 415–420. [cited on page(s) 224]

290 Vickers, S. (1989). *Topology via Logic, Cambridge Tracts in Theoretical Computer Science, No. 6*. Cambridge: Cambridge University Press. [cited on page(s) 11, 264, 266]

291 Walker, C.L. (2004). Categories of fuzzy sets. *Soft Computing* 8 (4): 299–304. [cited on page(s) 99]

292 Wang, H. (2005). Interval neutrosophic sets and logic: theory and applications in computing. PhD thesis. Georgia State University Atlanta, GA, USA. [cited on page(s) 29]

293 Wang, G.-j. and He, Y.-Y. (2000). Intuitionistic fuzzy sets and l-fuzzy sets. *Fuzzy Sets and Systems* 110 (2): 271–274. [cited on page(s) 28]

294 Wang, Z. and Klir, G.J. (2009). *Generalized Measure Theory*, IFSR International Series on Systems Science and Engineering, No. 25. New York: Springer Science+Business Media. [cited on page(s) 113]

295 Wardhaugh, R. and Fuller, J.M. (eds.) (2015). *An Introduction to Sociolinguistics*, 7e. Chichester, West Sussex: Wiley-Blackwell. [cited on page(s) 54]

296 Weber, S. (1983). A general concept of fuzzy connectives, negations and implications based on t-norms and t-conorms. *Fuzzy Sets and Systems* 11: 115–134. [cited on page(s) 24, 26]

297 Weisstein, E.W. Heine-Borel Theorem. From MathWorld–A Wolfram Web Resource. http://mathworld.wolfram.com/Heine-BorelTheorem.html. [cited on page(s) 243]

298 Wiedermann, J. (2004). Characterizing the super-Turing computing power and efficiency of classical fuzzy Turing machines. *Theoretical Computer Science* 317: 61–69. [cited on page(s) 206]

299 Wolf, A., Swift, J.B., Swinney, H.L., and Vastano, J.A. (1985). Determining Lyapunov exponents from a time series. *Physica D: Nonlinear Phenomena* 16 (3): 285–317. [cited on page(s) 323]

300 Wong, C. (1975). Fuzzy topology. In: *Fuzzy Sets and their Applications to Cognitive and Decision Processes* (ed. L.A. Zadeh, K.-S. Fu, K. Tanaka, and M. Shimura), 171–190. Academic Press. [cited on page(s) 247]

301 Wu, C. and Gong, Z. (2001). On Henstock integral of fuzzy-number-valued functions (I). *Fuzzy Sets and Systems* 120 (3): 523–532. [cited on page(s) 297]

302 Yager, R.R. (1986). A characterization of the extension principle. *Fuzzy Sets and Systems* 18 (3): 205–217. [cited on page(s) 32]

303 Yager, R.R. (2006). On the fuzzy cardinality of a fuzzy set. *International Journal of General Systems* 35 (2): 191–206. [cited on page(s) 18, 21]

304 Yuan, X. and Shen, Z. (2001). Notes on "Fuzzy plane geometry I, II". *Fuzzy Sets and Systems* 121 (3): 545–547. [cited on page(s) 269]

305 Zadeh, L.A. (1965). Fuzzy sets. *Information and Control* 8: 338–353. [cited on page(s) 6, 7, 14, 19, 32, 71]

306 Zadeh, L.A. (1968). Fuzzy algorithms. *Information and Control* 12: 94–102. [cited on page(s) 205]

307 Zadeh, L.A. (1971). Similarity relations and fuzzy orderings. *Information Sciences* 3 (2): 177–200. [cited on page(s) 71, 83, 87]

308 Zadeh, L.A. (1972). A fuzzy-set-theoretic interpretation of linguistic hedges. *Journal of Cybernetics* 2 (3): 4–34. [cited on page(s) 183]

309 Zadeh, L.A. (1975a). Calculus of fuzzy restrictions. In: *Fuzzy Sets and their Applications to Cognitive and Decision Processes* (ed. L.A. Zadeh, K.-S. Fu, K. Tanaka, and M. Shimura), 1–39. Academic Press. [cited on page(s) 111]

310 Zadeh, L.A. (1975b). Fuzzy logic and approximate reasoning. *Synthese* 30 (3): 407–428. [cited on page(s) 179]

311 Zadeh, L.A. (1975c). The concept of a linguistic variable and its application to approximate reasoning—I. *Information Sciences* 8 (3): 199–249. [cited on page(s) 19, 22, 32, 39, 54]

312 Zadeh, L.A. (1975d). The concept of a linguistic variable and its application to approximate reasoning—II. *Information Sciences* 8 (3): 301–357. [cited on page(s) 54]

313 Zadeh, L.A. (1975e). The concept of a linguistic variable and its application to approximate reasoning—III. *Information Sciences* 9 (1): 43–80. [cited on page(s) 54]

314 Zadeh, L.A. (1978). Fuzzy sets as a basis for a theory of possibility. *Fuzzy Sets and Systems* 1: 3–28. [cited on page(s) 111]

315 Zadeh, L.A. (1983). A computational approach to fuzzy quantifiers in natural languages. *Computers & Mathematics with Applications* 9 (1): 149–184. [cited on page(s) 177]

316 Zadeh, L.A. (1984). Fuzzy probabilities. *Information Processing & Management* 20 (3): 363–372. Special Issue Information Theory Applications to Problems of Information Science. [cited on page(s) 119]

317 Zadeh, L.A. (1995). Discussion: probability theory and fuzzy logic are complementary rather than competitive. *Technometrics* 37 (3): 271–276. [cited on page(s) 119, 120]

318 Zardecki, A. (1997). Fuzzy fractals, chaos, and noise. New Mexico. https://digital.library.unt.edu/ark:/67531/metadc674876/ (accessed 28 October 2019). University of North Texas Libraries, Digital Library, https://digital.library.unt.edu; crediting UNT Libraries Government Documents Department, May 1997. [cited on page(s) 325]

319 Zhang, J.-W. (1980). A unified treatment of fuzzy set theory and Boolean-valued set theory—fuzzy set structures and normal fuzzy set structures. *Journal of Mathematical Analysis and Applications* 76 (1): 297–301. [cited on page(s) 35]

Subject Index

a

A Modern Introduction to Fuzzy Mathematics, First Edition.
Apostolos Syropoulos and Theophanes Grammenos.

d

e

f

q

r

s

Author Index

A Modern Introduction to Fuzzy Mathematics, First Edition.
Apostolos Syropoulos and Theophanes Grammenos.

Printed and bound by CPI Group (UK) Ltd, Croydon, CR0 4YY